Seed Physiology
Volume I

Seed Physiology

Volume I
Development

Edited by

David R. Murray

*Biology Department
The University of Wollongong
New South Wales*

ACADEMIC PRESS
(Harcourt Brace Jovanovich, Publishers)
Sydney   Orlando   San Diego   Petaluma   New York
London   Toronto   Montreal   Tokyo
1984

ACADEMIC PRESS AUSTRALIA
Centrecourt, 25-27 Paul Street North
North Ryde, N.S.W. 2113

United States Edition published by
ACADEMIC PRESS INC.
Orlando, Florida 32887

United Kingdom Edition published by
ACADEMIC PRESS, INC. (LONDON) LTD.
24/28 Oval Road, London NW1 7DX

Copyright © 1984 by
ACADEMIC PRESS AUSTRALIA

All rights reserved. No part of this publication may be
reproduced or transmitted in any form or by any means,
electronic or mechanical, including photocopy, recording,
or any information storage and retrieval system, without
permission in writing from the publisher.

Printed in Australia

National Library of Australia Cataloguing-in-Publication Data

Seed physiology. Volume 1. Development.

Includes bibliographies and index.
ISBN 0 12 511901 1.

1. Seeds - Physiology. 2. Germination. I. Murray,
David R. (David Ronald), 1943- .

582'.0333

Library of Congress Catalog Card Number: 84-71959

Contents

List of Contributors ... ix

Preface ... xi

Contents of Volume 2 ... xiii

1. The Seed and Survival ... 1

DAVID R. MURRAY

I.	The Evolution of Seed-bearing Plants	2
II.	Identification of Wild Progenitors of Crop Species	13
III.	The Process of Domestication	28
IV.	The Importance of Conserving Genetic Diversity	34
	References	37

2. The Carbon and Nitrogen Nutrition of Fruit and Seed — Case Studies of Selected Grain Legumes ... 41

J. S. PATE

I.	Introduction	42
II.	Fruit Nutrition in Relation to Ontogenetic Profiles for Assimilation and Partitioning of Carbon and Nitrogen in the Whole Plant	42
III.	Translocatory Pathways for Photosynthetically Fixed Carbon in Fruiting Plants	46
IV.	Quantitative Studies of Transfer of Photosynthetically Fixed Carbon to Fruits	49

V.	The Photosynthetic Activity of Fruits	54
VI.	Identification of the Major Solutes That Supply Carbon and Nitrogen to Fruits in Xylem and Phloem	60
VII.	The Origin of Assimilates for Fruits and Seeds as Determined by Short-term Isotope Labelling Studies	64
VIII.	Partitioning of Carbon and Nitrogen in the Whole Plant and the Nutrition of Fruits and Seeds	70
IX.	The Water Economy of Fruits and Its Integration with Fruit Carbon and Nitrogen Nutrition	74
X.	The Conversion of Imported Solutes into Seed Reserves	79
	References	81

3. Accumulation of Seed Reserves of Nitrogen 83

DAVID R. MURRAY

I.	Albumins	84
II.	Lectins	86
III.	Seed Proteinase Inhibitors	98
IV.	Timing of Reserve Protein Synthesis	104
V.	Protein Accumulation in Protein Bodies	107
VI.	Processing of Seed Reserve Proteins	117
VII.	Conclusions	126
	References	127

4. Accumulation of Seed Reserves of Phosphorus and Other Minerals 139

J. N. A. LOTT

I.	Introduction	139
II.	Phosphorus Reserves in Seeds	140
III.	Distribution of Mineral Reserves within Seeds	147
IV.	Influence of External Soil Conditions on Mineral Levels in Seeds	157
V.	Calcium-rich Crystals	159
VI.	Procedures for Studying Minerals in Seeds	159
VII.	Future Research	162
	References	163

5. The Synthesis of Reserve Oligosaccharides and Polysaccharides in Seeds 167

N. K. MATHESON

I.	Introduction	167
II.	Oligosaccharides	168
III.	Polysaccharides	176
	References	202

6. Synthesis of Storage Lipids in Developing Seeds   209

C. R. SLACK AND J. A. BROWSE

I.	Introduction	209
II.	Occurrence and Composition of Storage Lipids	210
III.	Formation of Fatty Acids from Sucrose	214
IV.	Changes in Lipid Composition and Content during Seed Development	222
V.	Triacylglycerol Synthesis in Developing Seeds	223
VI.	Composition and Origin of Oil Bodies	236
VII.	Environmental Effects on the Fatty Acid Composition of Oil Seed Lipids	237
VIII.	Conclusions	239
	References	240

7. Toxic Compounds in Seeds   245

E. A. BELL

I.	Introduction	245
II.	The Alkaloids	246
III.	Cyanogenic Compounds	250
IV.	Non-Cyanogenic Nitriles	251
V.	Non-Protein Amino Acids	252
VI.	Amines	254
VII.	Saponins	257
VIII.	Glucosinolates	258
IX.	Miscellaneous Toxins	258
X.	Possible Roles of Toxic Compounds in Seeds	259
XI.	Seed Toxins and the Development of Food and Fodder Crops	261
	References	262

Plant Species Index   265

Subject Index   271

Contributors

Numbers in parentheses indicate the pages on which the authors' contributions begin.

E. A. BELL (245) Royal Botanic Gardens, Kew, Richmond Surrey, TW9 3AB, England.

J. A. BROWSE (209) Plant Physiology Division, Department of Scientific and Industrial Research, Palmerston North, New Zealand.

J. N. A. LOTT (139) Department of Biology, McMaster University, 1280 Main Street West, Hamilton, Ontario L8S 4K1, Canada.

N. K. MATHESON (167) Department of Agricultural Chemistry, The University of Sydney, New South Wales 2006, Australia.

DAVID R. MURRAY (1, 83) Biology Department, The University of Wollongong, New South Wales 2500, Australia.

J. S. PATE (41) Botany Department, The University of Western Australia, Nedlands, Western Australia 6009, Australia.

C. R. SLACK (209) Plant Physiology Division, Department of Scientific and Industrial Research, Palmerston North, New Zealand.

Preface

Seed development has been a much neglected area of seed biology until recently. Increasing interest in seed development now warrants a detailed synthesis of the subject to provide a basis for future research. This the authors have given. In this volume, comprehensive chapters describe and discuss the nutrition of the developing seed (J. S. Pate, Chapter 2), and the accumulation of the major categories of reserve material: protein (D. R. Murray, Chapter 3), mineral (J. N. A. Lott, Chapter 4), carbohydrate (N. K. Matheson, Chapter 5) and lipid (C. R. Slack and J. A. Browse, Chapter 6). Attention has been directed to the subcellular locations of these reserves and the factors regulating their synthesis. Areas requiring more detailed investigation are clearly identified.

We hope to promote an awareness that the events that follow imbibition and germination are not entirely divorced from the process of seed development. This underlying theme is emphasized in both Volume 1 of *Seed Physiology, Development*, and in Volume 2, *Germination and Reserve Mobilization*. A second theme is how seed characteristics have been modified in response to human selection. Methods for the identification of wild progenitors of cultivated species are illustrated in Chapter 1 of this volume. In Chapter 7, E. A. Bell classifies toxic compounds produced by seed tissues and gives examples of their removal from seeds by breeding and selection.

Both Volume 1 and 2 of *Seed Physiology* should become essential reading for teachers and researchers in plant physiology, agronomy and allied areas of plant science. Each author has thoroughly reviewed and brought together the existing literature, as well as contributing previously unpublished material. Insight and access to the literature are provided for intending research students as well as for those already actively engaged in research.

I am grateful to all who provided information and material for presentation, and to copyright owners who readily gave permission for the reproduction of items already published. Thanks are due also to the staff of Academic Press, to the staff of the University of Wollongong Library, to Ms Judy Ward for the final typing of my own chapters, and to Mr J. B. Murray for drawing Figure 6 of Chapter 3.

Contents of Volume 2

1. **Structural Aspects of Dormancy**
   V. N. TRAN AND A. K. CAVANAGH

2. **Metabolic Aspects of Dormancy**
   J. D. ROSS

3. **Early Events in Germination**
   E. W. SIMON

4. **Mobilization of Polysaccharide Reserves from Endosperm**
   ANNE E. ASHFORD AND FRANK GUBLER

5. **Mobilization of Nitrogen and Phosphorus from Endosperm**
   MICHAEL J. DALLING AND PREM L. BHALLA

6. **Mobilization of Oil and Wax Reserves**
   RICHARD N. TRELEASE AND DIANE C. DOMAN

7. **Axis-Cotyledon Relationships during Reserve Mobilization**
   DAVID R. MURRAY

CHAPTER 1

The Seed and Survival

DAVID R. MURRAY

I.	The Evolution of Seed-bearing Plants	2
	A. The Earliest Seeds	2
	B. The Origins of Angiosperms	3
	C. The Development of Endosperm	6
II.	Identification of Wild Progenitors of Crop Species	13
	A. Morphology and Geographic Distribution	14
	B. Cytogenetic Studies	14
	C. Comparative Electrophoretic Studies of Seed Proteins	18
	D. Genome Analysis	26
III.	The Process of Domestication	28
	A. Curtailment of Normal Seed Dispersal Mechanisms	28
	B. Changes in Seed Size	29
	C. Changes in the Seed Coats	33
IV.	The Importance of Conserving Genetic Diversity	34
	References	37

The present strong interest in all aspects of seed development, dormancy and germination arises partly from curiosity about the still mysterious process of seed germination, and partly from the knowledge that we will continue to rely heavily on seed crops as sources of dietary protein, carbohydrate and lipids in the foreseeable future (Section IV). It is therefore important that we determine the origins of our staple crop species, document their histories, and recognize the changes imposed by the process of domestication. In this way we can hope to introduce or reintroduce desirable characteristics from the gene pools available for each crop. Some of these modifications affect the reproductive structures or the seed itself, and these are discussed in Sections II and III of this chapter. The accelerated process of selection and evolution that followed cultivation may be compared with that which occurred by natural selection: recent advances in our knowledge of the origins of seed-bearing plants are discussed in Section I.

I. THE EVOLUTION OF SEED-BEARING PLANTS

The evolution of land plants has involved a progressive reduction in the size of the gametophyte generation. At the time of gamete production and fertilization, the female gametophyte of modern Angiosperms comprises usually 8 nuclei (sometimes 4 or 16, see p. 86 of Maheshwari, 1950). By comparison, the development of the megagametophyte of Gymnosperms is more substantial, but equally dependent on the parental sporophyte for nutrients. The male gametophyte (pollen grain) is greatly reduced in both groups.

As we currently use the term, a seed is a reproductive unit that contains an embryo sporophyte and food reserves, located either in the embryo itself or in some external storage tissue(s). These reserves are utilized rapidly when the embryo resumes growth, allowing it to develop the photosynthetic capacity necessary for autonomy. The first seed-bearing plants arose in the late Devonian about 350 million years ago, at least 200 million years before the earliest Angiosperms. In these early seeds, fertilization may have been delayed until after shedding, as still observed in *Ginkgo biloba*. The earliest remains of seeds containing an embryo date from about 270 Million Years Before Present (MYBP) (Raven *et al.*, 1981).

A. The Earliest Seeds

The oldest seed-bearing plant described to date comes from the Hampshire formation of Randolph County, West Virginia U.S.A. (Gillespie *et al.*, 1981). These fossils can be dated at about 350 MYBP, within the estimates 353–345 MYBP for Famennian strata given by Banks (1980).

This plant has not yet been named. The seeds were borne in cupules that terminated a bifurcating branch system (Fig. 1). They were 5 mm to 6 mm long, tapering from a mid-point diameter of 2 mm to 1 mm at the base. The integument was fused around the megasporangium only in the basal third of the seed, and divided distally into four or five separate lobes (Fig. 1). Clearly recognized features include the pollen chamber and the underlying nucellus and megaspore membrane. The apex of the pollen chamber was elongated into a definite tube (Fig. 1A).

1. The Seed and Survival 3

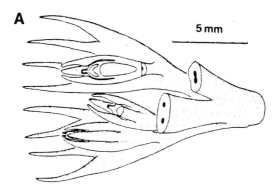

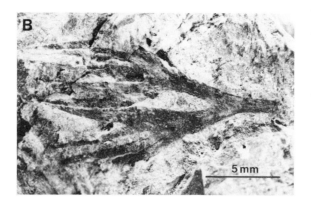

Fig. 1(A). Diagrammatic reconstruction of a cupule of 350 million-year-old seeds. Xylem is shown in black (from Gillespie *et al.*, 1981). Reprinted by permission from *Nature*, Vol. 293, No. 5832, pp. 462–464. Copyright © 1981 Macmillan Journals Limited. (B) Photograph of compressed ovulate cupule typical of that for which diagrammatic interpretation is given above (Fig. 1A) (kindly provided by G. W. Rothwell).

B. The Origins of Angiosperms

The seeds of Gymnosperms and Angiosperms have long been known to differ substantially in the development of integuments and surrounding fruit tissues. The possession of two integuments is thought to have arisen with the earliest Angiosperms (Doyle, 1978). There is no convincing evidence that Angiosperms existed before the Cretaceous period, that is, prior to 136 MYBP (Hickey and Doyle, 1977; Doyle, 1978). Tiffney (1983) concludes that the earliest seed remains that are clearly from Angiosperms are from the Albian, 105 MYBP (Table I). These belong to two species, *Caspiocarpus paniculiger* and *Ranunculaecarpus quinquiecarpellatus*, both from the U.S.S.R. Remains of seeds from the earlier Aptian strata (Table I) are not conclusively from Angiosperms. Leaf and pollen remains provide more substantial evidence for the development of Angiosperms before the Albian, but these are still within the Cretaceous (Hickey and Doyle, 1977).

Plant remains from the slightly younger Dakota formation (Cenomanian, Table I) have provided a clear association between leaves described originally as *Lyriophyllum* and reproductive organs described more recently as *Archaeanthus*. Carpel-like structures are arranged in a helical axis up to 12 cm long (Fig. 2). Each carpel may contain a row of seeds, each about 1.4 × 0.6 cm. The seed coats have been dissected, revealing a relatively simple cellular structure. Dilcher *et al.* (1976) have interpreted the structure of the fruit to be a conduplicate carpel with an adaxial suture which may also have served as the stigmatic surface. This suture is often found open, as in a modern follicle. *Archaeanthus* is believed to be an early member of the Magnoliidae.

Throughout most of the Cretaceous, Angiosperms were relatively scarce and the remains of most that have been described are of small-seeded types (Tiffney, 1977, 1980, 1983). A rapid increase in the numbers of Angiosperm species occurred from the late Cretaceous into the early Tertiary period. Modern genera first appear after 65 MYBP, with large-seeded forms dominant. Small-seeded types returned to prominence in the later Tertiary. An example of such a species is *Zanthoxylum rhabdospermum* (Rutaceae), preserved in the Brandon lignite from Vermont, U.S.A. (Tiffney, 1977, 1980). This formation is mid-Oligocene (about 30 MYBP). One or occasionally two seeds (Fig. 3) were borne per follicle, and were probably dispersed by birds. The elaborate seed coats, shown in section in Figure 4, must have contributed to dormancy in this species. Their presence has led to the preservation of measurable quantities of unsaturated fatty acids typical of reserve lipids (Chapter 6), usually oxidised and not detected in seeds of similar age with thinner coats (Niklas *et al.*, 1982).

We may never know the complete story of the rise of the Angiosperms and their coevolution with pollination and dispersal vectors in the Tertiary. Many opportunities for discovery have been lost through lack of interest (Tiffney, 1977). However, it is still true that a vast amount of material remains to be examined, and we can look forward to constant revision and refinement of ideas about primitive and advanced characteristics and the relationships among major lineages of Angiosperms.

Table I. Estimated ages of key stages of the Cretaceous period.

Period	Stage	Age (MYBP)
Upper Cretaceous	5 stages	65 to 90
	Cenomanian	90 to 98
Lower Cretaceous	Albian	98 to 108
	Aptian	108 to 115
	Neocomian	115 to 136
Jurassic		136 to 190
Triassic		190 to 225

Based on information given by Doyle (1978) and Tiffney (1977, 1983).

Fig. 2. Drawing of *Archaeanthus* (*Lyriophyllum*) (courtesy of D. L. Dilcher).

Fig. 3. Seed of *Zanthoxylum rhabdospermum* viewed from the apex. Note many features retained in modern species, including hilum, hilar crest, micropyle (at apex) and raphal pore, although no single modern species is exactly the same (Tiffney, 1980; scanning electron micrograph courtesy of Bruce H. Tiffney).

C. The Development of Endosperm

The megagametophyte storage tissue of Gymnosperm seeds was replaced by endosperm in Angiosperms. Endosperm is ultimately replaced in species where the embryo is adapted for storage. The origin of endosperm is a second fertilization event, occurring after the fusion of the egg cell with one of the two male sperm cells. Usually two polar nuclei from the female gametophyte combine with the nucleus from the second sperm cell. The resulting triple fusion nucleus is triploid, but if the number of participating polar nuclei deviates from the two (p. 426, Maheshwari, 1950) or if the species is polyploid, then the ploidy of the endosperm initial will vary accordingly.

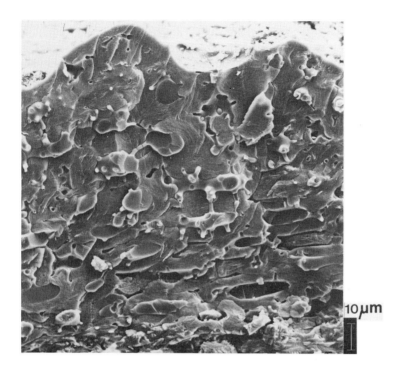

Fig. 4. Transverse section of the outer seed coat of *Zanthoxylum rhabdospermum*, showing the contour of the exterior surface (top) and about ten cell layers, the outermost consisting of isodiametric sclereids (scanning electron micrograph courtesy of Bruce H. Tiffney).

There is evidence that many types of endosperm cell undertake endoreduplication of the nuclear DNA. This occurs in dicotyledons (Kapoor, 1966) and the monocotyledons, in both the central endosperm regions (Ross and Duncan, 1950) and the aleurone layer (Keown et al., 1977). This phenomenon is discussed further in Section 3.IV. It would seem that the triploid endosperm cell is much scarcer than commonly imagined.

Although technically a second zygote, the endosperm is still functionally a megagametophyte, delayed in its full development until the trigger of successful fertilization. Strasburger (1900) first emphasized this economic aspect of endosperm development: reserves are not committed to a particular seed unless fertilization has been accomplished. The nutritive functions of female gametophyte and endosperm are identical. In tissue culture, it is difficult to encourage endosperm to develop further than a vegetative mass. Organ development is rare (Bhojwani and Johri, 1971).

In species such as castor bean (*Ricinus communis*), changes in the endosperm cells leading to mobilization of the major oil reserves as sucrose are initiated during germination independently of the presence of the embryo (Huang and Beevers, 1974). Nevertheless, certain species have evolved an endosperm with highly differentiated regions, such as the basal or placental cells adjacent to aleurone transfer cells (Rost, 1973), the site for hydrolysis of incoming sucrose in young developing maize (*Zea mays*) fruits (Shannon and Dougherty, 1972; Felker and Shannon, 1980). Most aleurone cells store oil and distinctive proteins (Table II), reserve materials that distinguish them from the central endosperm cells. These aleurone cells are capable of developing sensitivity to hormonal signals from the embryo that could regulate the production and release of hydrolytic enzymes into the interior of the endosperm on germination (Chapters 4, 5, Volume 2).

In describing different patterns of development of endosperm, the classical emphasis has been on whether cell wall formation keeps pace with the rate of nuclear division. Thus in free nuclear endosperm, the development of cell walls is deferred altogether for some time; in the helobial type, wall formation lags after the first division that separates chalazal and micropylar regions (Brink and Cooper, 1947; Chopra and Sachar, 1963). More functional alternatives might be as follows: (i) endosperm is formed and retained as a reserve tissue in the mature seed; (ii) endosperm is formed but substantially degraded before the embryo is mature; or (iii) endosperm nuclear divisions are terminated early.

Free nuclear, helobial and fully cellular types are all found in the first category. For example, castor bean (Singh, 1954), *Stellaria media* (Newcomb and Fowke, 1973) and wheat, *Triticum aestivum* (Mares *et al.*, 1975; Morrison and O'Brien, 1976) are all free nuclear. Endosperm formation occurs in several stages in *Haemanthus katherinae*. After helobial separation, free nuclear and cellular stages, endosperm cells lining the central vacuole are released into it, resuming nuclear division (Newcomb, 1978). Fully cellular endosperm is developed from the outset in carob (*Ceratonia siliqua*) and other legume seeds that store galactomannans (Section 5.III,C).

In category (ii) some important examples are shepherd's purse (*Capsella bursa-pastoris*) (Schultz and Jensen, 1974), cotton (*Gossypium hirsutum*) (Schultz and Jensen, 1977), sunflower (*Helianthus annuus*) (Newcomb, 1973), almond (*Prunus dulcis*) (Hawker and Buttrose, 1980) and celery (*Apium graveolens*) (Dwarte and Ashford, 1981). In almond seeds, a redistribution of reserves from perisperm (nucellus) to endosperm to embryo takes place over a lengthy developmental period (Fig. 5A). The

Table II. Amino acid composition of proteins from the aleurone layer compared to those from the balance of the endosperm from wheat grain (*Triticum aestivum* cv. Timgalen). Values as a percentage of total estimated.

Amino acid	Protein source	
	Aleurone	Endosperm
Aspartate[a]	9.3	5.7
Glutamate[a]	18.3	33.5
Threonine	3.8	3.1
Serine	5.0	6.1
Glycine	7.0	4.6
Alanine	6.5	3.9
Valine	5.5	4.1
Cysteine	n.d.	n.d.
Methionine	0.4	0.4
Isoleucine	3.2	3.5
Leucine	6.7	7.0
Tyrosine	2.8	3.1
Phenylalanine	4.2	4.4
Tryptophane	n.d.	n.d.
Histidine	4.3	2.3
Arginine	12.3	4.0
Lysine	5.9	3.1
Proline	4.6	11.5

[a] Aspartate and glutamate values include their respective amides.
Data of Fulcher *et al.* (1972).

relationships of these tissues to one another are shown in the remaining parts of Figure 5. The embryo displaces surrounding endosperm while the endosperm in turn displaces nucellar tissue. Rapid accumulation of oil and protein reserves in the embryo does not begin until after 16 weeks (Fig. 5B) as the endosperm starts to degenerate (Fig. 5A).

Consideration of category (iii) introduces the controversial subject of 'liquid endosperm'. It has generally been assumed that in legumes like pea (*Pisum sativum*) the embryo floats freely in a liquid free nuclear endosperm for much of its early development. However, the cytological and anatomical observations are unambiguous: the endosperm is short-lived and the embryo always develops firmly appressed to the chalazal end of the embryo sac (Fig. 6). Endosperm nuclei have been identified in a thin layer of cytoplasm that lines the embryo sac and surrounds the young embryo. These nuclei are associated with cell walls only in the immediate vicinity of the suspensor and the embryo proper (Marinos, 1970). The main function

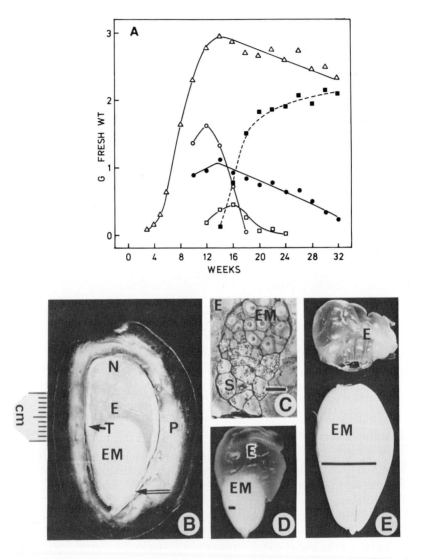

Fig. 5. Development of the almond kernel, *Prunus dulcis* cv. Chellaston. (A) Changes in the fresh weight of the kernel and its components: ■, embryo; □, endosperm; △, kernel; ○, pericarp; ●, testa. (B) Median longitudinal section of developing almond fruit at 16 weeks. E, endosperm; EM, embryo; N, nucellus; P, pericarp; T, testa. The forked arrow indicates a major vascular connection between the pericarp and testa. (C) Longitudinal section at 6 weeks showing the embryo with suspensor (S) surrounded by endosperm. Bar = 10 μm. (D) Embryo with attached endosperm at 14 weeks. Bar = 1 mm. (E) Embryo and detached endosperm at 18 weeks. Bar = 1 cm. (From Hawker and Buttrose, 1980).

1. The Seed and Survival

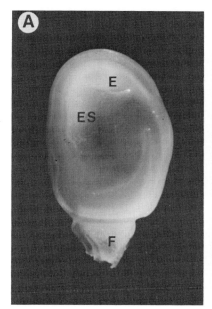 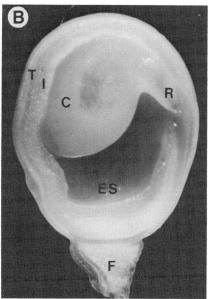

Fig. 6. Median longitudinal sections of developing pea seeds (*Pisum sativum* cv. Melbourne Market) at two early stages: 11 days after flowering (37 mg, left) and 16 days after flowering (130 mg, right). The embryo is always appressed to the wall of the sac at the chalazal end of the seed. C, cotyledon; E, embryo; ES, embryo sac; F, funiculus; I, inner seedcoat; R, radicle; T, testa (from Murray, 1980).

of these bridging cell walls is to hold the embryo in its characteristic position against the embryo sac. The embryo is first brought to this position by elongation of the suspensor (Cooper, 1938; Marinos, 1970). The orientation of the embryo in the sac determines its subsequent shape (Fig. 6), and I favour the view that orientation, not nutrition, is the prime function of both the suspensor and the vestigial endosperm.

When the endosperm nuclei cease division, some of them engage in endoreduplication (Kapoor, 1966). The reasons for their failing to continue to divide are not known, but this is possibly one effect of the potentially toxic ammonium ions secreted with nutrients by the surrounding seed coats (Murray, 1979, 1980). Labelling data (Table III) confirm that asparagine is the main source of the ammonium ions that accumulate in the embryo sac, as proposed earlier (Murray, 1979; Murray and Kennedy, 1980). Asparagine retained in the hull (carpel wall) is not extensively metabolized, whereas that delivered to the seeds is utilized within the seed coats. Surprisingly, much of the asparagine carbon is converted to alanine, glutamine and sucrose, compounds that are secreted.

Table III. Distribution of label in soluble metabolites of hull and seeds 24 h after feeding [^{14}C]asparagine to a pea[a] plant with a single 12-day old pod. Values as % of total for each tissue or compartment.[b]

Compound	Hull	Seed coats	Embryo sac
Sucrose	10.10	28.37	7.3
Malate	n.d.	4.09	n.d.
Alanine	1.64	6.58	15.1
Aspartate	0.61	1.44	0.4
Asparagine	81.52	12.35	5.6
Glutamate	0.49	7.60	1.2
Glutamine	2.23	21.87	52.4
Homoserine	0.49	4.34	5.3
Serine	0.83	2.04	1.3
Glycine	n.d.	0.31	n.d.
Threonine	0.37	2.07	2.1
Valine	0.43	7.00	8.8
Others	1.29	1.94	0.5

[a] cv. Melbourne Market, growth conditions as in Murray (1983).
[b] Embryos too small for analysis.
n.d., not detected.
Data of Murray and Cordova-Edwards (1984).

There are changes in the composition of embryo sac liquid during seed development (Table IV), but its total amino nitrogen content is maintained while the embryo is rapidly expanding and accumulating free amino nitrogen (Murray, 1983). This liquid therefore does represent 'an intermediate reservoir of nutrients made available continuously and asymmetrically to the expanding embryo' (Murray, 1980). Hedley and Ambrose (1980) put forward an alternative view, that 'the testa and embryo are sinks which both compete for the endosperm which may act as a common source'. This assessment disregards the known pathways of nutrient supply (Hardham, 1976).

Dore (1956) suggests that claims of liquid endosperm occurring in grasses should be treated cautiously. However, there is a very good example of authentic liquid endosperm in the coconut (*Cocos nucifera*). This species can be placed in category (i), since eventually some of the free endosperm nuclei lodge at the lining of the cavity and develop into layers of cellular endosperm (Cutter et al., 1955; Baptist, 1963).

The formation of vestigial endosperm, as in pea, can be seen as an arrested version of the free nuclear type, which is a modification of the fully

Table IV. Changes in free amino acid composition of embryo sac liquid from developing pea seeds.

Compound	Days after flowering	
	13	17
Alanine	+ + + +	+ + + +
Aspartate	tr	+
Asparagine	+	tr
Glutamate	+	+
Glutamine	+ + + +	+ + +
Homoserine	+ + + +	+ +
Serine	+ +	+ +
Glycine	+	tr
Threonine	+	+ +
Valine	+ +	+

tr = trace. Samples of embryo sac liquid were analysed by two-dimensional descending paper chromatography and the relative abundance of amino acids and amides indicated after treatment with ninhydrin.
From D. R. Murray (1983).

cellular condition. The formation and retention of fully cellular endosperm is the most primitive condition, a view supported by Swamy and Ganapathy (1959), who showed that in many dicotyledons the occurrence of free nuclear endosperm is positively correlated with open perforation of the end walls of early formed xylem vessels.

II. IDENTIFICATION OF WILD PROGENITORS OF CROP SPECIES

Angiosperms had been evolving for at least 100 million years before the earliest hominids arose between four and six million years ago (Johanson and Edey, 1981). It is remarkable that many Angiosperms should retain sufficient genetic flexibility to lose features that had been instrumental in their survival up to the time of domestication. In the transition from hunter-gatherer to agriculturist, our more immediate ancestors selected attributes of convenience, either inadvertently or deliberately. Uniformity of germination after planting, with shorter time to first appearance of seedlings, for example, are properties encouraged by cultivation (e.g., Lush

14   David R. Murray

et al., 1980). Such changes could be the result of altered seed coat structure and permeability (Section III.C and Chapter 1, Volume 2) or a metabolic shift eliminating the hormonal inhibition that would normally ensure a period of 'after-ripening' (Chapter 2, Volume 2). Comparison of wild anemones (*Anemone coronaria*) with de Caen forms cultivated for only 400 years demonstrates the removal of dormancy of the latter type (Fig. 7).

A. Morphology and Geographic Distribution

Extant wild plants closely resembling cultivars in many features of vegetative and floral morphology and annual life cycle have been discovered in areas corresponding to or near early agricultural sites. Modern geographic regions which host concentrations of genetic diversity in cultivated forms are no longer believed to necessarily coincide with the centre of origin for each crop. Erosion of the Vavilovian equation has resulted from the need to accommodate information on carbon dated plant remains from archaeological sites, the discovery of new wild forms, and their progressive elimination or confirmation as possible progenitors through increasingly skilled cytogenetic and biochemical assessment. There is no guarantee either that domestication has occurred where wild species are most abundant today. Sometimes it is not possible to pinpoint any single area as a centre of original agriculture (e.g., Africa, Harlan *et al.*, 1976). The possible origins of 86 major crops are outlined in Simmonds (1976).

B. Cytogenetic Studies

Karyotypes, and any irregularities in chromosomal behaviour during attempted crosses, are a useful source of information about the genotypes of particular species. Such studies allow the degree of relationship between wild forms and cultivars to be ranked. Some important examples follow.

*1. Pea (*Pisum sativum *L.)*
The carbonized remains of peas, lentils, vetches (*Vicia sativa*) and chick peas are found as contemporaries of the earliest domesticated cereals from the Mediterranean through to Asia (Zohary and Hopf, 1973). Two wild peas in this area are closely related to *P. sativum*: *P. elatius* is generally tall, erect, with large red-purple flowers, and grows in the more humid areas around the Mediterranean and *P. humile* is smaller, lacks pronounced apical dominance, and grows in drier regions at higher altitudes. Both are diploid (2n = 14). Experimentally, hybrids can be formed readily among

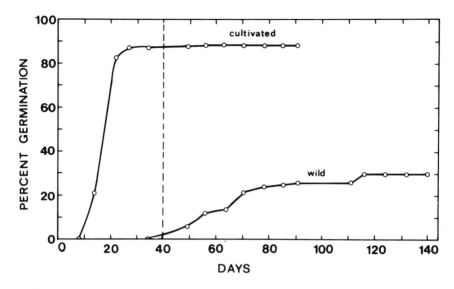

Fig. 7. Cumulative germination for seed of the cultivated de Caen anemone compared with the wild anemone, *Anemone coronaria* (from Horovitz et al., 1975).

P. elatius, *P. humile* and *P. sativum*. Usually the expected 7 bivalents occur during meiosis, but in some crosses 5 bivalents and one quadrivalent form. This indicates a single translocation and is typical of all *P. elatius* tested, and of *P. humile* from southern Israel (Zohary and Hopf, 1973; Ben Ze'ev and Zohary, 1973). By comparison, the chromosomes of *P. humile* from Turkey and north-eastern Israel are homologous with those of *P. sativum*. This evidence suggests that *P. sativum* is descended from northern forms of *P. humile* with some characteristics from *P. elatius*, as judged by apical dominance, increased height and climatic adaptability of many cultivars.

One wild pea (*P. fulvum*) can be eliminated as a progenitor of *P. sativum*. This species has a scrambling habit like *P. humile*, an orange flower and small, dark velvety seeds. In attempted crosses with *P. sativum*, *P. humile* or *P. elatius*, seeds set only when *P. fulvum* is the male parent. The hybrids are semi-sterile and set few seeds (Ben Ze'ev and Zohary, 1973). There are substantial differences in the karyotype of *P. fulvum*, but Marx (1977) considers all four should be designated a single species, either *P. sativum*, or *P. arvense*, which has priority.

2. Lentil (Lens culinaris)

Of only five species constituting the genus *Lens*, two are morphologically similar to the cultivated lentil (*Lens culinaris*). These are *L. orientalis* and *L. nigricans* (Zohary, 1972; Ladizinsky, 1979a). As for *Pisum*, all three possess diploid chromosome number 2n = 14. *Lens nigricans* from Turkey was shown by Ladizinsky (1979a) to differ from *L. culinaris* by three chromosomal translocations. By comparison, several lines of *L. orientalis* from Israel differed only by a single translocation. Despite the occurrence of chromosomal irregularities in meiosis, the fertility of hybrids formed by either *L. culinaris* × *L. orientalis* or *L. culinaris* × *L. nigricans* is reasonably good (Ladizinsky, 1979a, b). When the cross *L. orientalis* × *L. nigricans* was attempted, few viable seeds were formed and all seedlings were albino, unable to survive. Clearly *L. orientalis* is more closely related to the wild progenitor of *L. culinaris* than *L. nigricans*, but introgression from *L. nigricans* cannot be ruled out.

*3. Chick Pea (*Cicer arietinum *L.)*
The most recently discovered wild chick peas, *Cicer reticulatum* and *C. echinospermum* from Turkey, proved to be those most closely related to *C. arietinum* (Ladizinsky and Adler, 1975, 1976a, b). All have the same chromosome number, 2n = 16. Breeding experiments (Table II) show that *C. reticulatum* is the likely progenitor. The hybrids of crosses of *C. arientinum* with *C. reticulatum* (with one exception) breed as successfully as the parental species. In contrast, the cross *C. echinospermum* × *C. arientinum* is typified by low seed yield, low pollen fertility, and chromosomal irregularities during meiosis (Table V). A reciprocal translocation has occurred between two of the chromosomes of *C. echinospermum*, thus further isolating it from *C. reticulatum*. One line of *C. arietinum*, the exception referred to above (58F in Table II), has independently undergone the same translocation and has, in addition, a paracentric inversion (Ladizinsky and Adler, 1976a).

The members of a second group, *C. judaicum*, *C. pinnatifidum* and *C. bijugum* could form hybrids among themselves, but not with any of *C. arietinum*, *C. reticulatum*, or *C. echinospermum*, nor with the geographically isolated climber, *C. cuneatum* (Ladizinsky and Adler, 1976b).

*4. Bread Wheat (*Triticum aestivum*)*
Wheat shows a progression from diploid (2n = 14) to tetraploid to hexaploid forms. Cultivated diploid wheat (einkorn) shares the AA genome with the wild diploid, *T. boeoticum*. Although named *T. monococcum*, many have argued that its differences do not warrant distinction from *T. boeoticum* at species level (Zohary, 1969; Harlan *et al.*, 1973).

Cultivated emmer so prevalent in Egypt (*T. dicoccum*) and the later *T. durum* are tetraploids sharing the genome AA BB with the wild tetraploid

Table V. Fertility of parental lines and hybrids used in assessing relationships of chick pea (*Cicer arietinum*) to the wild species (*C. echinospermum* and *C. reticulatum*).

Species and source	Irregularities in meiosis	Pollen fertility (%)	Seed set (%)
Parents			
C. arietinum: IX Ethiopia	no	91.0	55.5
58F Ethiopia	no	92.2	68.9
77 Greece	no	98.2	49.5
127 India	no	99.7	55.3
187 Israel	no	37.0	31.0
C. echinospermum Turkey	no	98.2	40.8
C. reticulatum Turkey	no	97.9	43.0
F1 Hybrids			
C.a. 127 × *C.a.* 187	no	85.9	76.8
C.r. × *C.e.*	yes	20.0	nil
C.e. × *C.a.* 58F	yes	28.0	nil
C.e. × *C.a.* 77	yes	25.0	1.9
C.a. IX × *C.r.*	no	80.9	72.2
C.a. 58F × *C.r.*	yes	11.1	nil
C.a. 77 × *C.r.*	no	84.3	56.8
C.a. 127 × *C.r.*	no	96.0	51.4
C.r. × *C.a.* 187	no	67.5	46.3

Data of Ladizinsky and Adler (1976a).

T. dicoccoides. This is the only wild tetraploid with this genome. Its parents are believed to be *T. boeoticum* (AA donor) and another diploid wheat, *T. urartu*, donor of a genome converted from AA to BB (for a review of this subject see Johnson and Dhaliwal, 1978).

The hexaploid bread wheat, *T. aestivum*, appeared 1000 to 2000 years later with the cultivated einkorn and emmer wheats. Its genomic constitution is AA BB DD, and there is no corresponding wild hexaploid. The donor of the D genome is the wild diploid 'goat-face' grass, *Aegilops squarrosa* (Zohary, 1969), also named *T. tauschii* (Feldman and Sears, 1981). The tetraploid parent must have been cultivated emmer, rather than the wild *T. dicoccoides*, since if the cross *A. squarrosa* × *T. dicoccoides* is made, the hybrids are weak dwarf plants (Zohary, 1969). The gene controlling the dwarf habit is present in most lines of *T. dicoccoides*, but generally absent from cultivated tetraploids of genome AA BB.

C. Comparative Electrophoretic Studies of Seed Proteins

The proteins accumulated by seeds as reserves of nitrogen and sulphur (Chapter 3) are products of genes shared by species with common ancestries. The closer the phylogenetic relationship, the more similar are the seed proteins — a conclusion reached long ago by Osborne (1909, 1924). Although it will be some time yet before we know the amino acid sequence of every reserve protein, much faster comparative electrophoretic studies can be used to support distinctions or alliances between plant accessions.

In their studies of *Sinapis* and *Brassica*, Vaughan and Denford (1968) found that the greater number and spread of the albumin bands compared to the globulins gave a better correlation with the conventional taxonomic treatment. The maximum amount of discriminatory information can be obtained from the albumin fraction when size and charge relationships of protein bands are assessed by varying the average pore size of the gels (Murray *et al.*, 1978; Murray and Porter, 1980). Proteins belonging to all four of Osborne's main solubility categories have been studied in attempts to establish progenitor–cultivar relationships. Such information complements that obtained by other methods and occasionally can be used to predict which species would repay more intensive cytogenetic investigation, as with chick pea (Ladizinsky and Adler, 1975).

1. Pea (Pisum sativum)

Four groups have been recognized within the genus *Pisum* on the basis of inherited variation in the electrophoretic pattern of albumins soluble at pH 4.6 (Przybylska *et al.*, 1973, 1977). The main group includes *Pisum sativum*, *P. elatius* and *P. humile*, which all display patterns I and II (Table VI). Pattern I is most common. Pattern II is characterized by the lack of a prominent band of mobility 0.50 (Przybylska *et al.*, 1977). One of the type II lines, Jordan Valley ecotype W 1293, has been shown by Guldager (1978) to lack two of the three major albumins, A1 and A2, as well as the pea lectin. Occasionally, patterns intermediate between I and II occur in this group.

The other three groups comprised a single line of *P. cinereum* (W 1490; pattern III), *P. abyssinicum* (pattern IV) and *P. fulvum* (pattern V). Most of the protein bands used to distinguish these groups were subsequently found to belong to fraction S2 from gel filtration on Sephadex G-100 (Table VI; Jakubek and Przybylska, 1979).

The occurrence of bands distinguishing pattern III from pattern I is governed by two alleles at a single locus in chromosome 6 (Blixt *et al.*, 1980). The constituent polypeptides of proteins from fraction S2 are of

Table VI. Distribution of albumin protein bands belonging to the S2 fraction from *Pisum*.

Band mobility	Pattern				
	I	II	III	IV	V
0.26				+	
0.34				+	+
0.42	+	+	+		
0.50	+				+
0.56			+		
0.64			+		

Data of Jakubek and Przybylska (1979).

23 000 daltons (Jakubek and Przybylska, 1979, 1982) and are discussed further in Section 3.I.

This assessment links *P. sativum* with its progenitors, *P. humile* and *P. elatius* (Section II.B) and clearly distinguishes *P. fulvum*, in accordance with its cytogenetic isolation (Ben Ze'ev and Zohary, 1973). Furthermore, isoelectric focusing of legumin preparations from more than 100 *Pisum* lines has shown that *P. fulvum* can always be distinguished from the others by the absence of small (basic) legumin subunits of isoelectric point, pI 9.5-9.7 (Przybylska *et al.*, 1981).

*2. Lentil (*Lens culinaris*)*
Ladizinsky (1979d) compared buffer-extractable seed proteins from *Lens culinaris* (15 lines), *L. orientalis* (6 lines), *L. nigricans* (1 line) and *L. ervoides* (4 lines). Electrophoresis was performed in 10% polyacrylamide gels towards the cathode (running buffer pH 3.7). The characteristic pattern of 11 protein bands obtained for all 15 lines of *L. culinaris* was shown also by *L. orientalis* and *L. nigricans*. However, only 4 bands in this profile had counterparts of identical mobility in the 12-banded profile typical of *L. ervoides*. This result is in keeping with the failure of all attempts to cross *L. ervoides* with *L. culinaris* (Ladizinsky, 1979d) and supports the proposed close relationship of *L. culinaris* to *L. orientalis* and *L. nigricans* (Section II.B).

*3. Chick Pea (*Cicer arietinum*)*
Ladizinsky and Adler (1975) used a similar electrophoretic system (running buffer pH 4.3, 15% polyacrylamide gels) to compare 88 accessions of *Cicer arietinum* with six wild species (Table VII). These fractions correspond to albumins, having been prepared from water extracts at 0°C. Despite wide

Table VII. Distribution of protein bands following electrophoresis of seed extracts from representatives of *Cicer*.

Band mobility (R_f)	arietinum	judaicum	pinnatifidum	bijugum	cuneatum	echinospermum	reticulatum
0.37	+	+		+	+	+	+
				+			
				+	+		
0.41	+	+		+		+	+
0.50	+					+	+
			+	+	+	+	
			+	+			
					+		
0.61	+	+		+	+	+	+
			+				
			+				
				+			
0.74	+	+		+		+	+
			+				
					+		
0.92	+	+				+	+
1.00		+					
Total no.:	7	9	7	9	7	8	6

Data of Ladizinsky and Adler (1975).

variation in morphology, the profiles obtained for *C. arietinum* were identical in all but two instances, where the band of R_f 0.61 was divided into two distinct bands. All seven bands of the typical *C. arietinum* profile were represented by bands of identical mobility in two or more wild profiles. Ladizinsky and Adler (1975) judged the profile of *C. reticulatum* to most closely resemble that of *C. arietinum*. Both the albumin and globulin fractions from *C. arietinum*, *C. reticulatum* and *C. echinospermum* have been compared using different systems, including gels set at a range of acrylamide contents to obtain plots of mobility in relation to pore size (Vairinhos and Murray, 1983). Pore-gradient electrophoresis under alkaline conditions shows that on the basis of size (Manwell, 1977), the albumin profiles of the three cannot be distinguished (Fig. 8A). It follows that the differences observed in uniform pore gel systems (Ladizinsky and Adler, 1975; Vairinhos and Murray, 1983) result mainly from differences

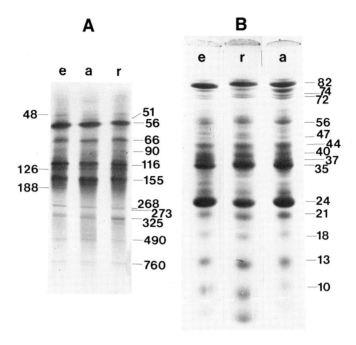

Fig. 8. (A) Gradient polyacrylamide ('Gradipore', 2.5%-27%) gel electrophoresis at pH 8.2 of water-soluble protein fractions obtained from the embryos of *Cicer echinospermum* (e), *C. arietinum* (a) and *C. reticulatum* (r). (B) The polypeptide composition of the water-soluble protein fractions from chick pea embryos (e, a, r as above). Disc gels of 10% polyacrylamide were used, with SDS and 2-mercaptoethanol both present. Estimated MW × 10^{-3} daltons (from Vairinhos and Murray, 1983).

in net charge (i.e., amino acid composition) between proteins of almost identical sizes. The amino acid composition of the albumin fraction shows minor differences according to source: that from *C. arietinum* has lower contents of lysine and methionine, but slightly higher contents of alanine, valine and phenylalanine (Table VIII). The close relationship of these three species is revealed also in the subunit composition of the albumin fractions on sodium dodecyl sulphate (SDS) gels (Fig. 8B).

The globulin profiles following pore-gradient electrophoresis were indistinguishable (Vairinhos and Murray, 1982, 1983). Differences were revealed on dissociation of globulin subunits with SDS, either in the absence or presence of 2-mercaptoethanol (Fig. 9). In summary (Table IX), both wild species are distinguished from *C. arietinum* by possession of polypeptide I. However, *C. reticulatum* and *C. arietinum* are alike in their contents of polypeptides II and III and their lack of polypeptide IV.

Table VIII. Amino acid composition of the albumin fraction from embryos of *Cicer arietinum*, *C. reticulatum* and *C. echinospermum*. Values as % of total estimated (mole basis).

Amino acid	Cicer arietinum	Cicer reticulatum	Cicer echinospermum
Aspartic[a]	13.65	13.37	13.23
Glutamic[b]	13.93	13.93	13.58
Threonine	5.46	5.35	5.40
Serine	7.37	7.42	7.21
Glycine	8.74	8.60	8.61
Alanine	7.54	7.06	6.79
Valine	4.03	3.70	3.76
Cysteine	2.51	3.13	3.45
Methionine	2.00	2.37	2.30
Isoleucine	3.45	3.43	3.53
Leucine	7.37	7.55	7.42
Tyrosine	2.54	2.31	2.45
Phenylalanine	3.33	3.04	3.14
Histidine	1.97	1.97	2.03
Arginine	4.10	3.96	4.35
Lysine	6.83	7.61	7.70
Proline	4.97	5.01	4.91
Unknowns (2)	0.22	0.18	0.14

[a] Includes asparagine.
[b] Includes glutamine.
Data of Murray and Roxburgh (1984).

Table IX. The distribution of disulphide-linked polypeptides in globulin fractions from embryos of *Cicer echinospermum*, *C. reticulatum* and *C. arietinum*.

Associated polypeptides (MW × 10^{-3} daltons)		C.e.	C.r.	C.a.
I.	46 + 21 = 67	+ + +	+ + +	absent
II.	41 + 21 = 62	+ +	+ + +	+ + +
III.	39 + 21 = 60	+	+ + +	+ + +
IV.	36 + 21 = 57	+ + +	+	+
V.	22 + 22 = 45	?	?	+ +
VI.	21 + 21 = 43	trace	trace	+ +
VII.	31 + 14 = 45	+	+	+

Data of Vairinhos and Murray (1982, 1983).

1. The Seed and Survival   23

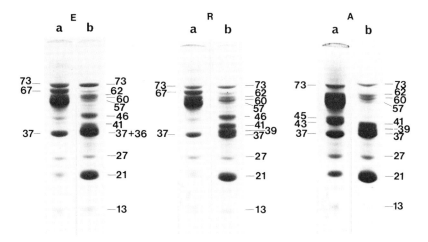

Fig. 9. The polypeptide composition of the globulin fractions from embryos of *Cicer echinospermum* (E), *C. reticulatum* (R) and *C. arietinum* (A). Disc gels of 10% polyacrylamide run with SDS in the absence (a) or presence (b) of 2-mercaptoethanol to reduce disulphide bridges. Estimated MW × 10^{-3} daltons (from Vairinhos and Murray, 1982, 1983).

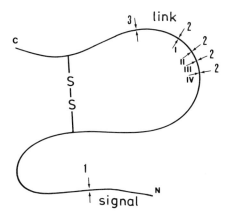

Fig. 10. Variable link model for legumin subunits in *Cicer*: disulphide-linked pairs of large and small polypeptides are formed by removal of similar signal sequences (1), formation of disulphide bonds, then excision of linking polypeptides of variable length, according to the positions of susceptible peptide bonds (2, 3) in different precursor polypeptides (after Vairinhos and Murray, 1982, 1983).

Polypeptides I to IV are components of legumin resembling those occurring in members of the tribe Vicieae (Vairinhos and Murray, 1982). The model proposed to account for the multiplicity of large subunits (Fig. 10) was put forward assuming that there should not be large differences in the signal sequences removed cotranslationally (Section 3.VI.A).

The subunit differences found for *Cicer* legumin indicate multiple gene copies, and confirm that *C. reticulatum* rather than *C. echinospermum* is the likely progenitor of *C. arietinum*, in accordance with the cytogenetic studies (Section II.B).

*4. Bread Wheat (*Triticum aestivum*)*
In a comprehensive study of tetraploid and hexaploid wheats, Johnson *et al.* (1967) established that there were indeed genome specific components in the electrophoretic profiles of albumins and gliadins. Tetraploid and protein fractions were compared using 15% acrylamide gels and a running buffer of pH 4.3 (towards the cathode). The highly conserved albumin profile from hexaploids (AA BB DD) comprised 12 bands, of which 9 were held in common with *Triticum dicoccum* and other AA BB tetraploids. Extension of this study to *Aegilops squarrosa*, the proposed source of the D genome (Section II.B), allowed Johnson (1972a) to nominate *T. dicoccum* accession 497 and *A. squarrosa* accession 962 as the most likely progenitors of the original hexaploid bread wheat. Both Johnson (1972a, b) and Nishikawa (1973), who studied the distributions of α-amylase isoenzymes, have provided data supporting a monophyletic origin of *T. aestivum*.

Altogether, six diploid members of *Aegilops* were characterized by a prominent albumin band of higher mobility than the fastest band from all the tetraploid wheats (Johnson, 1972b). This comparison included *A. speltoides*, believed by many to be the original donor of the B genome. Johnson (1972b) concluded that none of the known *Aegilops* species could possibly be the donor of the B genome, which is now believed to have diverged from the A genome after the original tetraploid had formed (Johnson and Dhaliwal, 1978).

Nevertheless, all three genomes of *T. aestivum* specify distinct proteins in several categories. Body *et al.* (1969) attributed four glutenins to the D genome. Waines (1973) allocated albumins not only to genome, but also to chromosome within each genome. In the same way, Wrigley and Shepherd (1973) allocated gliadins, after separating them first by isoelectric focusing, then by uniform pore gel electrophoresis (Fig. 11). These allocations were possible because of the development of nullisomic/tetrasomic stocks, in which a missing chromosome pair from one genome was compensated for by the doubling of corresponding chromosomes from another.

1. The Seed and Survival   25

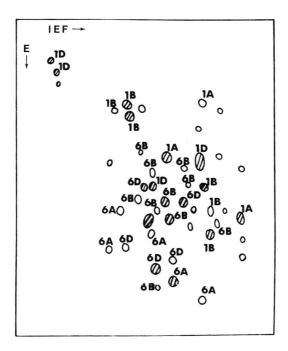

Fig. 11. Allocation of gliadins to controlling chromosome (number and genome, A, B or D) in endosperm of *Triticum aestivum* cv. Chinese Spring. The gliadins have been separated by isoelectric focusing (IEF) in the first direction and gel electrophoresis (E) in the second (from Wrigley and Shepherd, 1973).

The use of isoelectric focusing in the first dimension followed by pore-gradient gel electrophoresis in the second has since been advocated for routinely monitoring the variability of the gliadin fraction among the numerous cultivars of *T. aestivum* (du Cros and Wrigley, 1979). To gain the best spread and resolution of the gliadins, the albumins are run off the gel completely.

The embryo, too, contains proteins specified by each nuclear genome. Rice (1976) separated three distinct forms of wheat germ agglutinin from *T. aestivum* by ion exchange chromatography. The tetraploid *T. turgidum* (AA BB) yielded only two forms, types I and III. Thus the D genome specifies type II agglutinin, whereas the A and B genomes apparently specify one each of the other two forms.

*5. Broad Bean (*Vicia faba *L.)*
A number of studies have been inconclusive. For example, it has not been possible to identify the wild progenitor of the broad bean (*Vicia faba*). The

electrophoretic profile of proteins obtained from *V. narbonensis* is much simpler than than of *V. faba* (Ladizinsky, 1975; Abdalla and Günzel, 1979). Profiles from *V. galilaea* and *V. hyaeniscyamus* are very similar to one another, but different from those of *V. narbonensis* and *V. faba* (Ladizinsky, 1975). Situations like this repay the application of a wider range of electrophoretic procedures, but it is possible that the wild progenitor is in this case already extinct.

Ladizinsky and Hymowitz (1979) review other instances of comparative electrophoretic studies of seed proteins.

D. Genome Analysis

Double-stranded DNA can be cleaved by restriction endonucleases at sites determined by base sequence. Separation of the resulting fragments by gel electrophoresis provides a fairly direct parameter of genome construction. Using several endonucleases of varied specificities, altered base sequences in related genomes from different taxa can be detected as differences in the fragment patterns, and cleavage sites can be mapped (e.g., Gordon et al., 1981). These techniques have been applied to the problem of determining the relationships between maize (*Zea mays*, 2n = 20) and the annual diploid teosintes (formerly *Euchlaena mexicana*, also 2n = 20). These are now widely believed to include the progenitor of maize (Galinat, 1975; Wilkes, 1977). The major differences between the two-rowed female spike of teosinte and the multi-rowed cob of maize are illustrated in Figure 12.

Kempton (1938) thought that teosinte was 'the most maize-like relative' when he wrote that 'No more useless grasses from the standpoint of human consumption could be devised than the American relatives of maize'. Maize itself escapes Kempton's assessment, as there is evidence that substantial population increases in the Amazonian basin were supported by a diet comprising at least 80% maize (van der Merwe et al., 1981).

Beadle (1939, 1978) countered the structural difficulties (Fig. 12) that Kempton had emphasized by pointing out that when subjected to heat, the kernels of teosinte pop in the same way as modern popcorn cultivars. Beadle (1939) suggested further that this method of using teosinte probably led to its cultivation and eventual domestication.

The four main Mexican races of teosinte are Chalco, Central Plateau, Nobogame and Balsas. All except Balsas, the least like maize, constitute *Zea mays* subsp. *mexicana* (Doebley and Iltis, 1980; Iltis and Doebley, 1980). In the same systematic treatment, Balsas is distinguished as *Z. mays* subsp. *parviglumis* var. *parviglumis*; the northern Guatemalan variety Huehuetenango as *Z. mays* subsp. *parviglumis* var. *huehuetenangensis*, and

1. *The Seed and Survival* 27

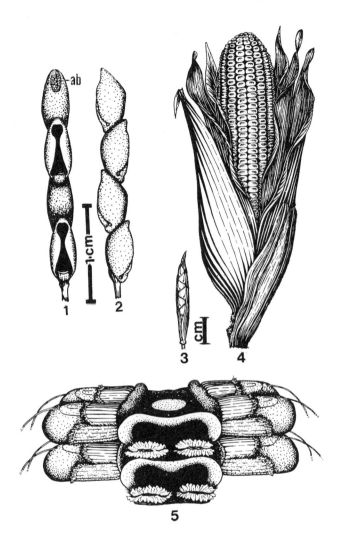

Fig. 12. The female spikes of teosinte and maize. 1, front view of four alternate fruit cases from a spike of teosinte (race Nobogame). The single spikelet that is normally held erect within each cupule has been removed to reveal each cavity. An abscission layer (ab) develops between adjacent fruit cases. 2, side view of 1. 3, the teosinte spike is enclosed in a single husk. 4, the multiple husk system of maize protects a many-rowed spike (ear) of paired spikelets each bearing an exposed kernel. 5, section of an eight-rowed maize ear. Each pair of spikelets is borne outside its associated cupule; condensation of the cupules has resulted in loss of interspace within a row. Note that the spikelet pairs have been removed from the front in order to reveal the cupules. (Reproduced from Galinat, Walton C. 1975. III. The evolutionary emergence of maize. Bull. Torrey Bot. Club. 102: 313-324).

Guatemala teosinte as *Zea luxurians*. A recently discovered diploid perennial teosinte, *Zea diploperennis* (Iltis *et al.*, 1979) is also interfertile with maize, and is grouped in Section Luxuriantes with *Zea luxurians*, although the latter fails to hydridize with maize (Wilkes, 1977).

In comparisons of maize with wild teosintes, maternally inherited genomes of the chloroplast and mitochondrion have been investigated (Timothy *et al.*, 1979). All of the above races were included in the genome fragment analyses.

Chloroplast DNA patterns with three endonucleases consistently grouped Guatemala and perennial teosintes, distinguishing them from all the other teosintes and maize B37. With one endonuclease, *Eco*RI, maize was grouped with Balsas and Huehuetenango teosintes, distinguished from the other Mexican teosintes. Mitochondrial DNA patterns with four endonucleases all distinguished perennial and Guatemala teosintes from one another, and from all the other teosintes. The *Bam*HI pattern, the easiest to interpret, linked maize and Nobogame, the pattern differing by only a single band (one fewer) when compared with Chalco, Central Plateau, Balsas and Huehuetenango.

This novel method of assessment illustrates the independent evolution of chloroplast and mitochondrial genomes. It confirms the close links between maize and the Mexican teosintes, the relative distance of the Guatemala teosinte, and the systematic treatment of *Z. diploperennis*.

III. THE PROCESS OF DOMESTICATION

Once the progenitors of crop species have been identified by some or all of the criteria discussed in Section II, then speculation about the timing and sequence of subsequent changes in morphology can be more soundly based. The gross changes in seed arrangement, dispersal mechanism, seed coat appearance and properties, seed size and composition often distort the phylogenetic relationship between wild and domesticated forms so that their relationships have initially been lost in traditional taxonomic treatments.

Many of the characteristics that now distinguish cultivars from their wild progenitors are determined by the double recessive condition in one or a few genes that are Mendelian in every sense, a single dominant allele governing the wild type.

A. Curtailment of Normal Seed Dispersal Mechanisms

In the wild forms of wheat and many other cereals, the rachis becomes brittle and disarticulates at maturity, scattering seed that will bury itself

(Zohary, 1969) and thereby provide the next generation. In cultivated forms, the ear is less brittle. This difference is under the control of one or few genes, and is critical, since it aids collection of ripe seeds. Harlan et al. (1973) and others have described how such a characteristic could become established quickly once seeds were being planted. Alleles that govern less brittle ears automatically favour collection and increased representation in the cultivated population. The grains of the early domesticated wheats (*Triticum monococcum*, *T. dicoccum*) still retained investing glumes, but the alleles for complete liberation of the grain on threshing have been incorporated in later cultivated tetraploids and hence in *T. aestivum* (Dunstone and Evans, 1974). In maize, Galinat (1975) considers that the suppression of lateral spikes together with condensation of rachis segments could have allowed a sudden transition to a fruit that fails to disarticulate (Fig. 12).

A feature of legumes comparable to the non-shattering rachis of wild cereals is the indehiscent pod. A single dose of the dominant allele *Dpo* (pea, Waines, 1975) or *Pi* (lentil, Ladizinsky, 1979b) is sufficient for the dry pod to dehisce along both seams at maturity, thereby releasing the enclosed seeds. In these species, pod indehiscence results from the double recessive condition at this one locus. Pod indehiscence in cowpea (*Vigna unguiculata*), similarly controlled, is expressed as a reduction in the thickness of the spirally thickened layer of the endocarp, but there are no differences in the valves of dehiscent and indehiscent pods (Lush and Evans, 1981).

Pod indehiscence would become established in a cultivated population very quickly, and it is difficult to imagine that lentil could be collected in the quantities discovered in village sites without it. By comparison, wild chick peas differ from wild peas and lentils by retaining seeds in the mature pods for a relatively long time before shedding (Ladizinsky, 1979c). Indehiscence is controlled by several genes rather than one. If pod indehiscence were already established in wild chick peas before their deliberate cultivation, this might explain their spread from a small centre of origin in Turkey (Section II.B) to predominance among the grain legumes of early Old World agriculture.

B. Changes in Seed Size

Enlargement of seeds is frequently but not always a consequence of domestication (Ladizinsky, 1979c). The sunflower, from the U.S.A., provides a good example (Heiser, 1976). Deliberate cultivation of wild sunflowers was probably practised by the Indians before a simple but drastic modification to flowering occurred. This modification eliminated the succession of flower heads borne on branches and maintained a solitary capitulum at the apex of the main stem. This change may have predisposed an increase in the size of each achene (Fig. 13). There is no doubt that the Indians subsequently developed cultivars with seeds as large or larger than those of many modern cultivars (Fig. 14). Heiser (1976) indicates that the earliest cultivated sunflower remains in the U.S.A. (1500 BC) predate the arrival of Mexican agriculture, a penetration that relegated sunflower to the margins of the maize fields.

Seed enlargement that followed domestication of chick pea, pea and cowpea (Table X) has involved an increase in the size of the embryo and a decrease in the proportion of total seed material committed to the mature seed coats (Section III.C). With larger seeds the yield is greater for the same effort in harvesting, but perhaps more importantly, the larger the seed, the faster the embryo establishes itself after germination (e.g., Lush and Wein, 1980; Table XI).

The enlargement of the wheat grain following domestication has involved an increase in the size of cells that comprise the bulk of the endosperm in the transition from diploid to tetraploid forms, but

Table X. Distribution of seed mass between embryo and seed coats in wild and cultivated legumes.

Identity	Seed mass (mg)	Embryo (%)	Seed coats (%)
Pisum humile W 936	67.8	73.7	26.3
Pisum humile 717[a]	113.7	71.0	29.0
Pisum elatius 723[a]	174.5	86.4	13.6
Pisum sativum cv. Dunn	209.7	91.7	8.3
Pisum sativum cv. Telephone	403.5	90.8	9.2
Cicer echinospermum[b]	151.6	67.2	32.8
Cicer reticulatum[b]	131.4	76.9	23.1
Cicer arietinum	414.2	95.7	4.3
Vigna unguiculata			
subsp. *mensensis*[c]	18	74	26
subsp. *dekindtiana*[c]	32	85	15
subsp. *cylindrica*[c]	60	87	13
subsp. *unguiculata* (smooth)[c]	128	89	11
subsp. *unguiculata* (rough)[c]	116	96	4
subsp. *sesquipedalis*[c]	141	87	13

[a] Reference numbers of Ben Ze'ev and Zohary (1973). Seeds of wild peas were obtained from Dr G. J. Berry and grown in Wollongong N.S.W. in 1980 to provide those dissected.
[b] Seeds of wild chick peas were provided by Dr G. Ladizinsky. Data of Vairinhos and Murray (1983).
[c] Data of Lush and Evans (1980).

1. The Seed and Survival   31

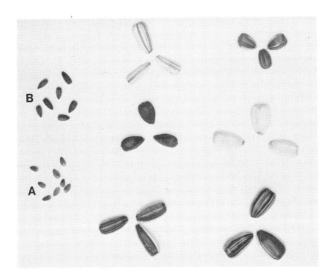

Fig. 13. Achenes of wild sunflower (A), weed sunflower (B) and a selection of modern cultivars. All are considered to belong to a single species, *Helianthus annuus*. (Reproduced with permission from *The Sunflower*, by Charles B. Heiser Jr. Copyright 1976 by the University of Oklahoma Press).

Fig. 14. Sunflower achenes from the Cramer Village site in Ohio, U.S.A. Actual dimensions of this sample of 299 achenes were: mean length, 10.3 mm (range 8.2–12.7 mm); mean width, 5.3 mm (range 3.5–7.6 mm). Since the achenes have been charred by fire, they are approximately 10%–15% smaller than originally. (Reproduced with permission from *The Sunflower*, by Charles B. Heiser Jr. Copyright 1976 by the University of Oklahoma Press).

Table XI. The sizes of seedlings of wild and cultivated accessions of cowpea (*Vigna unguiculata*) of differing seed weights ten days after sowing.

Subspecies	Seed mass (mg)	Plant mass (mg)	Primary leaf area (cm^2)
mensensis[a]			
mean	14.2	33.3	5.1
lowest	8.1	22.2	3.6
highest	20.9	39.1	6.4
dekindtiana[b]			
mean	30.3	88.1	11.6
lowest	23.8	77.4	9.7
highest	38.2	107.4	14.2
unguiculata[c]			
mean	152.1	347.0	37.6
lowest	65.9	197.3	24.5
highest	285.1	549.6	59.8

[a] 4 accessions.
[b] 3 accessions.
[c] 8 accessions.
Data of W. M. Lush and H. C. Wien (1980): 'The importance of seed size in early growth of wild and domesticated cowpeas', *J. Agric. Sci.* **94**, 177–182 (Cambridge University Press).

subsequently there has been an increase in the number of endosperm cells, especially of those in central regions that are enriched in starch but relatively deficient in protein (Dunstone and Evans, 1974; Parker, 1982). Qualitative changes among wheat endosperm protein fractions have been considered (Section II.D).

Among these changes, one is directly attributed to domestication and was clearly inadvertent. The inhibitory activity acting specifically against the proteinase of grain beetle larvae (*Tribolium castaneum*) has increased about 20-fold, comparing wild tetraploid wheat (*Triticum dicoccoides*) with cultivated tetraploids (Applebaum and Konijn, 1966). This higher level of inhibitor must have been transmitted to *T. aestivum* from its cultivated tetraploid ancestor, as the activities in *Aegilops* are generally as low or lower than those of *T. dicoccoides*. This elevation can be seen as a response with survival value once wheat grain was stored for long periods and thereby provided a nutritious habitat for *Tribolium* larvae.

C. Changes in the Seed Coats

Reduction in the thickness of the seed coats (Fig. 15) is one factor contributing to the lower proportion of final seed weight made up by the seed coats compared to the embryo (Table X). Such changes involve differences in cell number and degree of thickening and also in the surface structure and texture. The elaborate surface of the testa of *Cicer reticulatum* is very much flattened in many cultivars of *C. arietinum* (Fig. 16). The gritty or papillose texture of wild pea seed coats (Marbach and Mayer, 1974) is a feature lost following domestication. The texture of the seed coat can be diagnostic even in carbonized remains (Zohary and Hopf, 1973). However, surface texture seems to be unrelated to permeability. The roughest seed coats among cowpea cultivars are also the thinnest, and freely permeable to water (Lush et al., 1980).

Metabolic changes affecting pigment synthesis can be major determinants of permeability. Comparison of wild peas with *P. sativum* (Marbach and Mayer, 1974, 1975) reveals that catechol oxidase activity normally increases in the seed coats of *P. elatius*, acting on potential *o*-diphenol substrates as they are released from intracellular compartments in the last stages of desiccation. Such seeds are impermeable to water. If developing seeds of *P. elatius* are allowed to mature in an oxygen-free atmosphere, catechol oxidase-mediated pigmentation is inhibited and the seed coats are permeable to water. Both the increase in particulate catechol oxidase activity and the preliminary accumulation of *o*-diphenols are curtailed in the seed coats of *P. sativum* (Table XII).

Table XII. Changes in catechol oxidase activity and *o*-diphenol content of the seed coats during development of wild and cultivated pea seeds.

| | Seed coat | | Catechol | |
	fresh wt (mg)	dry wt (mg)	oxidase activity[a]	*o*-Diphenol content[b]
P. elatius	64	17	2.5	11.2
	72	21	1.5	8.2
	54	21	15.0	12.4
	42	21	50.5	20.3
	25	22	12.0	33.0
P. sativum	99	23	0	0.2
(cv. Alaska)	79	20	0.3	0.4
	18	16	1.5	0.5

[a] as μL O_2 consumed per min per g fresh weight.
[b] mg per g fresh weight.
Data of Marbach and Mayer (1975).

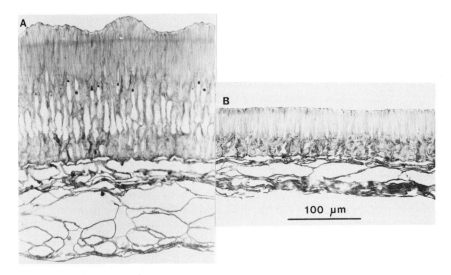

Fig. 15. Radial transverse sections of the seed coats from mature seeds of *Pisum elatius* (A) and *P. sativum* cv. Greenfeast (B). Note the major difference in the thickness of the external macrosclereid layer, and the scalloped surface of *P. elatius* (from Lush and Evans, 1980).

Changes leading to the loss of impermeability from the seed coats of cultivated legumes have contributed greatly to their more uniform and predictable germination response following sowing, but it is unlikely that such changes were necessary for cultivation. In some species, cultivars with coloured seed coats are still more numerous than those without. Continued selection against coat pigmentation may be worthwhile. White-coated cultivars of *Phaseolus vulgaris* have been shown to lack the heat-resistant tannin-derived factors extractable from coloured seed coats (Elias *et al.*, 1979). These factors retain trypsin-inhibitory activity even after cooking and so coloured seed coats may erode the nutritive value of the seeds of those cultivars that still possess them.

IV. THE IMPORTANCE OF CONSERVING GENETIC DIVERSITY

As Evans (1975) has pointed out, our reliance on seed crops to fulfil the major proportion of our food requirements is not going to diminish. This is clear from Table XIII. In particular, protein production is still a limiting factor in many countries. Advocates of single-cell protein sources concede that such sources are not to be seen as 'substitutes for long-established

1. The Seed and Survival 35

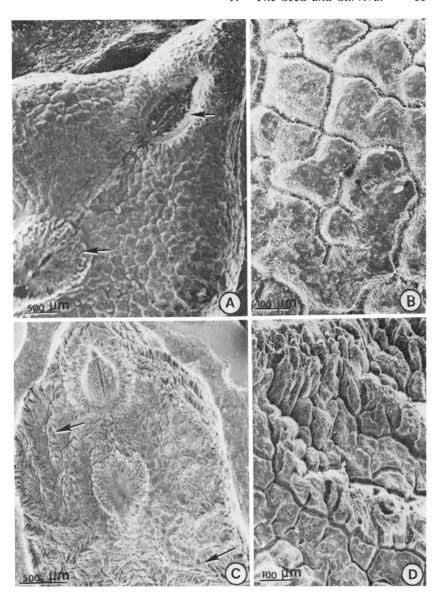

Fig. 16. Topography of the testa in *Cicer arietinum* (A, B) and *C. reticulatum* (C, D). (A) Seed of *C. arietinum*, with arrows indicating hilum (upper) and lens (lower). (B) *C. arietinum*: enlarged view of testa surface plates. (C) Seed of *C. reticulatum*, with arrows indicating prominent ridges. (D) *C. reticulatum*: enlarged view of testa surface, showing plates elevated to varying degrees. (From Lersten and Gunn, 1981).

agricultural products' (Mauron, 1980/81). The world's fisheries are now suffering the inevitable penalty for over-fishing and neglect of breeding grounds. Improvements in the quality and yield of protein-rich seed crops will be welcome in the coming centuries.

Population pressures and decreasing areas of arable land highlight the importance of at least maintaining existing yields and avoiding losses through inadequate seed storage conditions. In this context, the genetic uniformity of the most popular cultivars of many species is risky. The prevalence of type T cytoplasm in American cultivars of maize led to the disastrous loss of this crop in 1970 (Leaver, 1980). Maize is the second or third largest individual crop component of world food production (Table XIII).

It should be emphasized that it is not the crop monoculture itself that is undesirable, as evidenced by the capacity of wild cereals to form natural

Table XIII. FAO estimates of world annual production for 1979 to 1982 (metric tons × 10^{-6}).

	1979	1980	1981	1982
Cereals: Wheat	428.7	446.1	453.7	481.0
Rice	376.8	399.1	411.7	411.9
Maize	418.9	395.9	450.6	455.4
Barley	158.0	159.3	154.6	160.3
Others	171.6	164.5	180.7	186.5
Pulses: Total	40.5	40.4	42.4	44.7
Nuts	3.53	3.56	4.07	3.79
Oil crops (oil equivalent)	51.3	49.8	53.5	57.0
Root crops: Potatoes	288.4	230.3	258.0	254.9
Others	295.4	299.4	292.5	301.5
Vegetables and melons	350.1	349.7	357.2	367.7
Fruits: Grapes	69.0	66.7	61.4	70.6
Citrus	50.9	56.0	55.7	53.8
Bananas	37.6	40.1	40.6	41.2
Apples	36.4	33.8	33.2	39.4
Others	89.4	93.0	93.3	96.7
Sugar (raw)	88.9	84.0	92.6	101.4
Cocoa beans	1.66	1.62	1.64	1.59
Coffee beans (green)	5.05	4.80	6.03	4.93
Tea	1.82	1.87	1.86	1.93
Fibres: Cotton lint	13.94	13.92	15.32	14.70
Jute and similar	4.47	4.04	4.08	3.95
Others	2.43	2.49	2.39	2.47
Meat	137.9	141.1	143.4	144.6
Milk	465.5	470.6	473.0	482.7
Hen eggs	27.4	28.3	29.2	29.8

Reproduced with permission from FAO Production Yearbook 1982: Volume 36 (1983).

stands. Rather, it is the genetic constitution of any crop that will help or hinder it under stress. Adverse circumstances strike suddenly, and it is important for plant breeders to have immediately available lines with resistance to climatic extremes and as complete a range of pathogens and pests as possible. It is obviously easier to preserve the diversity represented by numerous existing cultivars and wild forms than it would be to generate new viable mutations by artificial means each time disaster threatens. For this reason, governments must be encouraged to support the collection and maintenance of such diversity by publicly funded institutions that have already begun this task. Heiser (1976) has argued that we should not rely on subsistence farmers to maintain low yielding primitive cultivars on our behalf. Nor should we have to rely on so-called private enterprise, with restrictive variety rights legislation. The diversity engendered under these schemes will largely concern cosmetic characteristics that facilitate discrimination rather than substantial improvements in the unseen reserve materials of the seed.

Plant breeding institutions must have unhindered access to plant genotypes already in existence, and should be encouraged to undertake conservation and domestication of many plants of great potential from all parts of the world.

REFERENCES

Abdalla, M. M. F., and Günzel, G. (1979). *Z. Pflanzenzüchtg.* **83**, 148–154.
Applebaum, S. W., and Konijn, A. M. (1966). *J. Insect Physiol.* **12**, 665–669.
Banks, H. P. (1980). In 'Biostratigraphy of Fossil Plants' (D. L. Dilcher and T. N. Taylor, eds.), pp. 1–24. Dowden, Hutchinson, and Ross, Stroudsberg, Pennsylvania.
Baptist, N. G. (1963). *J. Exp. Bot.* **14**, 29–41.
Beadle, G. W. (1939). *J. Heredity* **30**, 245–247.
Beadle, G. W. (1978). In 'Maize Breeding and Genetics' (D. B. Walden, ed.), pp. 113–128. Wiley, New York.
Ben-Ze'ev, N., and Zohary, D. (1973). *Isr. J. Bot.* **22**, 73–91.
Bhojwani, S. S., and Johri, B. M. (1971). *New Phytol.* **70**, 761–766.
Blixt, S., Przybylska, J., and Zimniak-Przybylska, Z. (1980). *Genetica Polonica* **21**, 153–161.
Boyd, W. J. R., Lee, J. W., and Wrigley, C. W. (1969). *Experientia* **25**, 317–319.
Brink, R. A., and Cooper, D. C. (1947). *Bot. Rev.* **13**, 423–541.
Chopra, R. N., and Sachar, R. C. (1963). In 'Recent Advances in the Embryology of Angiosperms' (P. Maheshwari, ed.), pp. 135–170. Catholic Press, Ranchi, India.
Cooper, D. C. (1936). *Bot. Gaz.* **98**, 169–177.
Cooper, D. C. (1938). *Bot. Gaz.* **100**, 123–132.
Cutter, V. M., Wilson, K. S., and Freeman, B. (1955). *Am. J. Bot.* **42**, 109–115.
Dilcher, D. L., Crepet, W. L., Beeker, C. D., and Reynolds, H. C. (1976). *Science* **191**, 854–856.
Doebley, J. F., and Iltis, H. H. (1980). *Am. J. Bot.* **67**, 982–993.

Dore, W. G. (1956). *Bull. Torrey Bot. Club* **83**, 335-337.
Doyle, J. A. (1978). *Annu. Rev. Ecol. Syst.* **9**, 365-392.
du Cros, D. L., and Wrigley, C. W. (1979). *J. Sci. Food Agric.* **30**, 785-794.
Dunstone, R. L., and Evans, L. T. (1974). *Aust. J. Plant Physiol.* **1**, 157-165.
Dwarte, D., and Ashford, A. E. (1982). *Bot. Gaz.* **143**, 164-175.
Elias, L. G., de Fernandez, D. G., and Bressani, R. (1979). *J. Food Sci.* **44**, 524-527.
Evans, L. T. (1975). *In* 'Crop Physiology' (L. T. Evans, ed.), pp. 1-22. Cambridge University Press, Cambridge.
Feldman, M., and Sears, E. R. (1981). *Sci. Am.* **244**, 98-109.
Felker, F. C., and Shannon, J. C. (1980). *Plant Physiol.* **65**, 864-870.
Fulcher, R. G., O'Brien, T. P., and Simmonds, D. H. (1972). *Aust. J. Biol. Sci.* **25**, 487-497.
Galinat, W. C. (1975). *Bull. Torrey Bot. Club* **102**, 313-324.
Gillespie, W. H., Rothwell, G. W., and Scheckler, S. E. (1981). *Nature* **293**, 462-464.
Gordon, K. H. J., Crouse, E. J., Bohnert, H. J., and Herrmann, R. G. (1981). *Theor. Appl. Genet.* **59**, 281-296.
Guldager, P. (1978). *Theor. Appl. Genet.* **53**, 241-250.
Hardham, A. R. (1976). *Aust. J. Bot.* **24**, 711-721.
Harlan, J. R., de Wet, J. M. J., and Price, E. G. (1973). *Evolution* **27**, 311-325.
Harlan, J. R., de Wet, J. M. J., and Stemler, A. B. L. (1976). *In* 'Origins of African Plant Domestication' (J. R. Harlan, J. M. J. de Wet and A. B. L. Stemler, eds.), pp. 3-19. Mouton, The Hague.
Hawker, J. S., and Buttrose, M. S. (1980). *Ann. Bot.* **46**, 313-321.
Hedley, C. L., and Ambrose, M. J. (1980). *Ann. Bot.* **46**, 89-105.
Heiser, C. B. Jr. (1976). 'The Sunflower'. University of Oklahoma Press, Oklahoma.
Hickey, L. J., and Doyle, J. A. (1977). *Bot. Rev.* **43**, 3-104.
Horovitz, A., Bullowa, S., and Negbi, M. (1975). *Euphytica* **24**, 213-220.
Huang, A. H. C., and Beevers, H. (1974). *Plant Physiol.* **54**, 277-279.
Iltis, H. H., and Doebley, J. F. (1980). *Am. J. Bot.* **67**, 994-1004.
Iltis, H. H., Doebley, J. F., Guzman, M. R., and Pazy, B. (1979). *Science* **203**, 186-188.
Jakubek, M., and Przybylska, J. (1979). *Genetica Polonica* **20**, 369-380.
Jakubek, M., and Przybylska, J. (1982). *Pisum Newsl.* **14**, 26-28.
Johanson, D. C., and Edey, M. A. (1981). 'Lucy — The Beginnings of Humankind'. Book Club Associates, London.
Johnson, B. L. (1972a). *Am. J. Bot.* **59**, 952-960.
Johnson, B. L. (1972b). *Proc. Nat. Acad. Sci. USA* **69**, 1398-1402.
Johnson, B. L., and Dhaliwal, H. S. (1978). *Am. J. Bot.* **65**, 907-918.
Johnson, B. L., Barnhart, D., and Hall, O. (1967). *Am. J. Bot.* **54**, 1089-1098.
Kapoor, B. M. (1966). *Genetica* **37**, 557-568.
Kempton, J. H. (1938). *Smithsonian Rep.* **1937**, 385-408.
Keown, A. C., Taiz, L., and Jones, R. L. (1977). *Am. J. Bot.* **64**, 1248-1253.
Ladizinsky, G. (1975). *Euphytica* **24**, 785-788.
Ladizinsky, G. (1979a). *Euphytica* **28**, 179-187.
Ladizinsky, G. (1979b). *J. Heredity* **70**, 135-137.
Ladizinsky, G. (1979c). *Econ. Bot.* **33**, 284-289.
Ladizinsky, G. (1979d). *Bot. Gaz.* **140**, 449-451.
Ladizinsky G., and Adler, A. (1975). *Isr. J. Bot.* **24**, 183-189.
Ladizinsky, G., and Adler, A. (1976a). *Theor. Appl. Genet.* **48**, 197-203.
Ladizinsky, G., and Adler, A. (1976b). *Euphytica* **25**, 211-217.
Ladizinsky, G., and Hymowitz, T. (1979). *Theor. Appl. Genet.* **54**, 145-151.

Leaver, C. J. (1980). *TIBS* **5**, 248-252.
Lersten, N. R., and Gunn, C. R. (1981). *System Bot.* **6**, 223-230.
Lush, W. M., and Evans, L. T. (1980). *Field Crops Res.* **3**, 267-286.
Lush, W. M., and Evans, L. T. (1981). *Euphytica* **30**, 579-587.
Lush, W. M., and Wien, H. C. (1980). *J. Agric. Sci.* **94**, 177-182.
Lush, W. M., Evans, L. T., and Wien, H. C. (1980). *Field Crops Res.* **3**, 173-187.
Maheshwari, P. (1950). 'An Introduction to the Embryology of Angiosperms'. McGraw-Hill, New York.
Manwell, C. (1977). *Biochem. J.* **165**, 487-495.
Marbach, I., and Mayer, A. M. (1974). *Plant Physiol.* **54**, 817-820.
Marbach, I., and Mayer, A. M. (1975). *Plant Physiol.* **56**, 93-96.
Mares, D. J., Norstog, K., and Stone, B. A. (1975). *Aust. J. Bot.* **23**, 311-326.
Marinos, N. G. (1970). *Protoplasma* **70**, 261-279.
Marx, G. A. (1977). *In* 'The Physiology of the Garden Pea' (J. F. Sutcliffe and J. S. Pate, eds.), pp. 21-43. Academic Press, London.
Mauron, J. (1980/81). *Nestlé Research News*, pp. 70-79.
Morrison, I. N., and O'Brien, T. P. (1976). *Planta* **130**, 57-67.
Murray, D. R. (1979). *Plant Physiol.* **64**, 763-769.
Murray, D. R. (1980). *Ann. Bot.* **45**, 273-281.
Murray, D. R. (1983). *New Phytol.* **93**, 33-41.
Murray, D. R., and Cordova-Edwards, M. (1984). *New Phytol.* **97**, 253-260.
Murray, D. R., and Kennedy, I. R. (1980). *Plant Physiol.* **66**, 782-786.
Murray, D. R., and Porter, I. J. (1980). *Plant System. Evol.* **134**, 207-214.
Murray, D. R., and Roxburgh, C. McC. (1984). *J. Sci. Food Agric.* **35**.
Murray, D. R., Ashcroft, W. J., Seppelt, R. D., and Lennox, F. G. (1978). *Aust. J. Bot.* **26**, 755-771.
Newcomb, W. (1973). *Can. J. Bot.* **51**, 879-890.
Newcomb, W. (1978). *Can. J. Bot.* **56**, 483-501.
Newcomb, W., and Fowke, L. C. (1973). *Bot. Gaz.* **134**, 236-241.
Niklas, K. J., Tiffney, B. H., and Leopold, A. C. (1982). *Nature* **296**, 63-64.
Nishikawa, K. (1973). *In* 'Proceedings 4th International Wheat Genetics Symposium' (E. R. Sears and L. M. S. Sears, eds.), pp. 851-855. University of Missouri, Columbia.
Osborne, T. B. (1909, 1924). 'The Vegetable Proteins'. 1st and 2nd editions. Longmans Green, New York.
Parker, M. L. (1982). *Plant Cell Environ.* **5**, 37-43.
Przybylska, J., Mikolajczyk, J., and Zimniak-Przybylska, Z. (1973). *Genetica Polonica* **14**, 383-387.
Przybylska, J., Blixt, S., Hurich, J., and Zimniak-Przybylska, Z. (1977). *Genetica Polonica* **18**, 27-38.
Przybylska, J., Hurich, J., and Blixt, S. (1981). Pisum *Newsl.* **13**, 44-45.
Raven, P. H., Evert, R. F., and Curtis, H. (1981). 'Biology of Plants'. 3rd edition. Worth Publishers, New York.
Rice, R. H. (1976). *Biochim. Biophys. Acta* **444**, 175-180.
Ross, J. G., and Duncan, R. E. (1950). *J. Heredity* **41**, 259-268.
Rost, T. L. (1973). *Bot. Gaz.* **134**, 32-39.
Schulz, P., and Jensen, W. A. (1974). *Protoplasma* **80**, 183-205.
Schulz, P., and Jensen, W. A. (1977). *Am. J. Bot.* **64**, 384-394.
Shannon, J. C., and Dougherty, C. T. (1972). *Plant Physiol.* **49**, 203-206.
Simmonds, N. W. (1976). 'Evolution of Crop Plants'. Longman, London.

Singh, R. P. (1954). *Phytomorphol.* **4**, 118-123.
Strasburger, E. (1900). *Bot. Ztg.* **II 58**, 293-316.
Swamy, B. G. L., and Ganapathy, P. M. (1959). *Bot. Gaz.* **119**, 47-50.
Tiffney, B. H. (1977). *J. Seed Technol.* **2**, 54-71.
Tiffney, B. H. (1980). *J. Arnold Arboretum* **61**, 1-40.
Tiffney, B. H. (1983). *Ann. Missouri Bot. Gard.*
Timothy, D. H., Levings, C. S. III, Pring, D. R., Conde, M. F., and Kermicle, J. L. (1979). *Proc. Nat. Acad. Sci. USA* **76**, 4220-4224.
Vairinhos, F., and Murray, D. R. (1982). *Z. Pflanzenphysiol.* **107**, 25-32.
Vairinhos, F., and Murray, D. R. (1983). *Plant System Evol.* **142**, 11-22.
Van der Merwe, N. J., Roosevelt, A. C., and Vogel, J. C. (1981). *Nature* **292**, 536-538.
Vaughan, J. G., and Denford, K. F. (1968). *J. Exp. Bot.* **19**, 724-732.
Waines, J. G. (1973). *In* 'Proceedings 4th International Wheat Genetics Symposium' (E. R. Sears and L. M. S. Sears, eds.), pp. 873-877. University of Missouri, Columbia.
Waines, J. G. (1975). *Bull. Torrey Bot. Club* **102**, 385-395.
Wilkes, H. G. (1977). *Econ. Bot.* **31**, 254-293.
Wrigley, C. W., and Shepherd, K. W. (1973). *Ann. N.Y. Acad. Sci.* **209**, 154-162.
Zohary, D. (1969). *In* 'The Domestication and Exploitation of Plants and Animals' (P. J. Ucko and G. W. Dimbleby, eds.), pp. 47-66. Duckworth, London.
Zohary, D. (1972). *Econ. Bot.* **26**, 326-332.
Zohary, D., and Hopf, M. (1973). *Science* **182**, 887-894.

CHAPTER 2

The Carbon and Nitrogen Nutrition of Fruit and Seed — Case Studies of Selected Grain Legumes

J. S. PATE

I.	Introduction	42
II.	Fruit Nutrition in Relation to Ontogenetic Profiles for Assimilation and Partitioning of Carbon and Nitrogen in the Whole Plant	42
III.	Translocatory Pathways for Photosynthetically Fixed Carbon in Fruiting Plants	46
IV.	Quantitative Studies of Transfer of Photosynthetically Fixed Carbon to Fruits	49
V.	The Photosynthetic Activity of Fruits	54
VI.	Identification of the Major Solutes That Supply Carbon and Nitrogen to Fruits in Xylem and Phloem	60
VII.	The Origin of Assimilates for Fruits and Seeds as Determined by Short-term Isotope Labelling Studies	64
	A. Feeding of Foliar Organs	64
	B. Feeding of Roots or Nodulated Roots	65
	C. Feeding Labelled Compounds to Cut Shoots through Xylem	66
	D. Leaf Flap Feeding	68
VIII.	Partitioning of Carbon and Nitrogen in the Whole Plant and the Nutrition of Fruits and Seeds	70
IX.	The Water Economy of Fruits and Its Integration with Fruit Carbon and Nitrogen Nutrition	74
X.	The Conversion of Imported Solutes into Seed Reserves	79
	References	81

I. INTRODUCTION

This chapter describes and defines quantitatively the physiological and biochemical processes concerned with the supply of carbon and nitrogen to the fruit and its seeds. I will discuss the function of vegetative parts of the plant in synthesis and further metabolism of the range of carbon and nitrogen-containing materials that are supplied to fruits through xylem and phloem, especially the siting of these processes, their timing in relation to the growth cycle of the plant, and their quantitative interrelationships with individual plant parts. The functioning of the fruit and its parts will also be assessed, concentrating on the economy of the whole fruit in terms of carbon, nitrogen and water, and on the respective roles of pod wall and seed in processing assimilates derived from the parent plant and in converting the carbon and nitrogen of these materials into seed reserves.

By concentrating on a few species, I hope to achieve continuity between the accounts of different facets of fruit nutrition presented in the various sections of the chapter, and to provide an element of authenticity which would not be possible were the chapter to extend to species with which the author was not familiar.

II. FRUIT NUTRITION IN RELATION TO ONTOGENETIC PROFILES FOR ASSIMILATION AND PARTITIONING OF CARBON AND NITROGEN IN THE WHOLE PLANT

Before describing in detail how assimilates are supplied to fruits and how fruits and seeds metabolize incoming solutes, it is important to have in perspective the overall profiles for partitioning of net photosynthate and fixed nitrogen (N) within the legume species under consideration. Substantial differences can exist between species in these respects, as shown in data for percentage allocations of net photosynthate during growth of cowpea (*Vigna unguiculata* cv. Caloona) and white lupin (*Lupinus albus* cv. Neutra) (Fig. 1).

Before flowering of either species, the principal items of expenditure of net photosynthate are respiration of root and nodules, and investment of carbon (C) into dry matter of new leaf tissue and roots. After flowering, the partitioning patterns of the species diverge. The lupin cultivar shows a continued high allocation of photosynthate to roots and nodules, so that the increasing proportional demands of reproductive organs are met largely at the expense of dry matter gain by vegetative parts of the shoot. The cowpea cultivar, by contrast, exhibits an abrupt decline in proportional allocation of photosynthate to below-ground parts, progressive abscission of leaves,

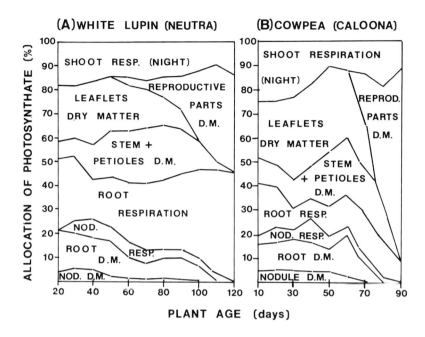

Fig. 1. Allocation of the carbon of net photosynthate during growth of two species of grain legumes (data from Herridge and Pate, 1977; Pate and Herridge, 1978).

and an increasing monopolization of photosynthate by developing fruits.

As roots of white lupin still command more than 40% of the plant's net photosynthate well into late fruiting, nitrogen fixation is able to continue until the end of the growth cycle (Pate and Herridge, 1978), whereas symbiotic activity of cowpea has virtually ceased by the end of flowering (Herridge and Pate, 1977). Therefore almost 90% of the N accumulated over the life cycle of white lupin, and about 75% of the nitrogen finally accumulated in its seed is derived from symbiotic nitrogen-fixing activity after flowering (Pate et al., 1980c). By contrast, fruits of cowpea (cv. Vita 3) gain 69% of their N requirement from atmospheric nitrogen (N_2) fixed before anthesis — most of this through mobilization to the fruit of N from the protein reserves of leaflets (Peoples et al., 1983).

When comparisons of this nature are extended to other legumes (Table I) it becomes apparent that the relative proportions of the plant's total photosynthesis and N_2 fixation which are accomplished during different phases of growth vary widely between different legumes. However, in each of the cited cases (Table I) proportionally more N_2 is fixed by a species

Table I. Ontogenetic relationships between nitrogen fixation and net photosynthesis[a] in various grain legumes.

		Proportion of total CO_2[a] and N_2 assimilated over growth cycle (%)		
		Vegetative growth to fruiting	Flowering and early fruiting	Seed filling
Cowpea (*Vigna unguiculata*)				
cv. Caloona[b]	CO_2	37	44	19
	N_2	50	42	8
cv. Vita 3[c]	CO_2	51	31	18
	N_2	56	40	4
White lupin (*Lupinus albus*)				
cv. Neutra[d]	CO_2	6	59	35
	N_2	11	62	27
cv. Ultra[e]	CO_2	9	58	33
	N_2	18	59	23
Mung bean (*Vigna radiata*)[f]	CO_2	13	43	44
	N_2	16	45	39
Field pea (*Pisum sativum* var. *arvense*)[f]	CO_2	13	58	39
	N_2	24	60	16

[a] As net photoperiod gain of CO_2 by shoot from atmosphere.
[b] Data from Herridge and Pate (1977).
[c] Data from Pate *et al.* (1983a); Peoples *et al.* (1983).
[d] Data from Pate and Herridge (1978).
[e] Data from Pate *et al.* (1980c).
[f] Data from Pate and Minchin (1980).

relative to carbon dioxide (CO_2) during vegetative growth, whereas the reverse of this applies during fruiting, as proportional allocation of assimilates to nodules becomes much reduced (Pate and Minchin, 1980).

Agronomists frequently express yield effectiveness of grain crops in terms of 'harvest index', defining this as the fraction of the total weight of above-ground dry matter of the crop at final harvest which is recoverable as harvested seed. We have extended this concept in our studies of pot-grown grain legumes by making separate measurements of harvest indices for C and N, taking care to include all shed leaves and underground parts as well as vegetative shoot, fruits and seeds when assessing the total dry matter yield of the plant over its growth cycle. The resulting information (Table II) can then be viewed against other indices of plant efficiency in usage of net photosynthate; for instance, estimates of the amounts of

2. Seed Nutrition 45

Table II. Functional economies for carbon and nitrogen in parent and improved cultivars of two species of grain legumes depending on N_2 as sole source of nitrogen.

Item	Cowpea (*Vigna unguiculata*)		White lupin (*Lupinus albus*)	
	Parent cultivar[a]	Improved cultivar[b]	Parent cultivar[a]	Improved cultivar[b]
Translocate utilized by nodulated root over whole growth cycle (mg CH_2O/mg N fixed)	25.7	24.1	55.0	48.1
Net photosynthate generated per unit protein synthesized in vegetative parts by end of flowering (g CH_2O/g protein)	17.2	13.4	24.7	27.1
Percentage of C of plant net photosynthate lost in plant respiration in 10-day period before flowering	28	29	45	46
Harvest index for carbon	0.29	0.35	0.27	0.46
Harvest index for nitrogen	0.34	0.62	0.61	0.84
Net photosynthate produced during growth cycle per unit yield of seed dry matter (g CH_2O/g seed DM)	6.6	5.1	9.9	5.3
Net photosynthate produced during growth per unit yield of seed protein (g CH_2O/g protein)	32.5	18.2	31.0	16.8

[a] used as green manure crop.
[b] selected for high seed yield and determinate habit.
Data from Herridge and Pate (1977); Pate and Herridge (1978); Pate *et al.* (1980c, 1983); Peoples *et al.*, (1983).

photosynthate consumed per unit nitrogen fixed, per unit protein synthesized in vegetative parts before flowering, or per unit dry matter or protein in seeds.

This approach proves particularly instructive when cultivars selected for high grain-yielding capacity are compared with the low-yielding parental stocks or wild types from which these improved cultivars have been selected. The data of Table II, for example, illustrate how selection for high seed yield and a more determinate habit in cowpea and lupin has caused noticeable improvements in harvest indices for both C and N species (items

4 and 5, Table II), although the elevation of the index is proportionately greater for N than for C in lupin, but less so for N than C in cowpea. Together with these changes in harvest index, the improved cultivars of both species use much lower amounts of plant net photosynthate in producing unit weight of seed dry matter or seed protein than their unselected counterparts (items 6 and 7, Table II). Conversely, the improved and parental types of a species do not differ significantly in the mean amounts of translocate consumed by their nodulated roots. They also show closely similar proportional losses of net photosynthate in respiration, and do not differ greatly from each other in amounts of net photosynthate consumed per unit of protein produced in vegetative parts up to the end of flowering. Nevertheless, values for each of these first three quantities (items 1-3, Table II) turn out to be much higher in white lupin than in cowpea, a difference which undoubtedly reflects the large investments of C by lupin in the growth and maintenance of an extensive tap root and in the secondary thickening of its stout, fibrous stem (Pate *et al.*, in press).

III. TRANSLOCATORY PATHWAYS FOR PHOTOSYNTHETICALLY FIXED CARBON IN FRUITING PLANTS

Experiments in which $^{14}CO_2$ or [^{14}C]urea is fed to different foliar organs and the distribution of the resulting labelled assimilates traced among different organs of the plant, represent a widely used technique for identifying a plant's main sources of photosynthate. If performed sequentially on the same batch of plants at different stages of development, such experiments also allow one to follow how translocation patterns change as flowering and fruiting proceed. The resulting profiles for assimilate translocation are usually described as percentages of the exported label ending up in different sink organs, thus enabling comparisons to be made of the relative abilities, say, of young leaves, root and stem, and of fruits of different age and position to draw upon the current photosynthate resource of a particular foliar surface.

When studied in the above manner all three species; field pea (*Pisum sativum* var. *arvense*), white lupin and cowpea, show a fully 'integrated' pattern of assimilate distribution in early vegetative growth, in which all dependent plant parts express a capacity to draw freely and non-preferentially from a 'common' pool of net photosynthate provided by the mature leaves. By late vegetative growth, however, both pea (Flinn and Pate, 1970) and lupin (Pate *et al.*, 1980a; Pate and Farrington, 1981) already show a marked stratification of assimilate distribution in which photosynthate from a specific source leaf moves preferentially to nearby sink organs, especially those with direct vascular connections from the source leaf. Lower leaves thus generally favour roots, upper leaves and the shoot apex, whereas, among leaves, mature leaves tend to direct assimilates mainly to younger leaves of the same orthostichy. In pea, for example, with a simple alternate leaf arrangement, each leaf nourishes preferentially those leaves situated two and four nodes vertically above it, although these same leaves still supply some assimilate to root and shoot apex and to young leaves on the other side of the stem. In lupin, with a phyllotaxis approximately 5/13 the situation is far more complex, particularly since the mid and lateral traces serving a leaf each interconnect differently with the vasculature of the rest of the shoot (O'Neill, 1961).

From flowering onwards each of the three legumes develops a highly distinctive pattern of assimilate distribution. In the terminally flowering lupins, white (*Lupinus albus*) and narrow-leaved (*L. angustifolius*), the top leaves immediately below each inflorescence are the main contributors to that inflorescence, each leaf distributing photosynthate preferentially to flowers or young fruits of the same or interconnected orthostichies within the adjacent inflorescence (Fig. 2; Pate *et al.*, 1974; Farrington and Pate, 1981). Later, however, once secondary thickening of the inflorescence stalk has taken place, peripheral exchange of assimilates through secondary phloem becomes increasingly possible, so that the most actively growing fruit at any one time can command a major share of photosynthate from nearly every nurse leaf, regardless of its orthostichy. In certain indeterminate lupin cultivars (e.g., *L. angustifolius* cv. Unicrop) lateral shoots grow out above the primary inflorescence and thereby cause premature senescence of the leaves they shade on the main stem. Fruits of the primary inflorescence then draw upon the photosynthate of leaves of these lateral branches (Farrington and Pate, 1981). These laterals carry the additional responsibility of supplying assimilates to roots and to their own terminal (secondary) inflorescences.

The distribution of post-anthesis assimilate in cowpea, an axillary flowering species, shows a fully integrated pattern of nutrition in which all currently active sources share fairly equally in supplying all sink organs. There is no evidence that blossom leaves are specialized to supply their subtended fruits (Pate *et al.*, 1983).

Our third example, field pea, resembles cowpea in showing sequentially upward ripening of its axillary fruits. Unlike cowpea, each set of leaflets is deeply committed (up to 90%) to its subtended fruit and remains as the major source of assimilates for that fruit until almost the end of seed filling. The stipules of field pea leaves, by contrast, make greater photosynthetic contributions to the rest of the plant than to the adjacent fruit (Flinn and Pate, 1970).

48 J. S. Pate

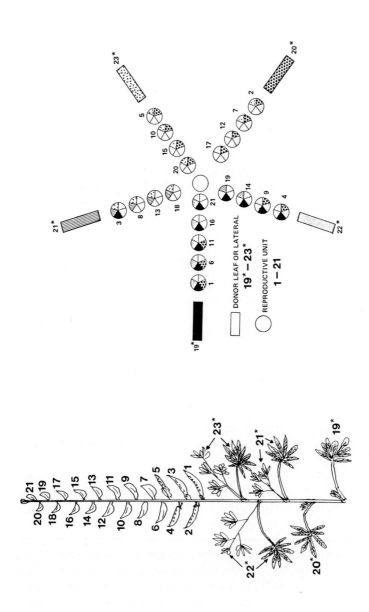

Fig. 2. Specificities of assimilate flow from upper main stem leaves and leaves of upper lateral shoots to flowers and fruits of the main stem inflorescence of narrow-leafed lupin (*Lupinus angustifolius*). The relative sizes of the contributions of ^{14}C assimilates from each foliar surface to specific reproductive units are shown sectorially according to the phyllotaxis of donor and receptor organs (data from Pate and Farrington, 1981).

IV. QUANTITATIVE STUDIES OF TRANSFER OF PHOTOSYNTHETICALLY FIXED CARBON TO FRUITS

Having assessed the degree of commitment of each photosynthetic organ in supplying assimilates to specific fruits, and determined the exportable surplus of C generated by each of these source organs, it becomes possible to estimate the net flows of C which take place between specific sources and fruits during particular stages of reproductive development. The data of Table III illustrate the main principles of this approach, using as an example the performance of the uppermost main stem leaf of narrow-leafed lupin in supplying C to the adjacent terminal inflorescence. This leaf generated a fairly constant surplus of C from two to four weeks after emergence of the inflorescence and donated a steady proportion (14%-20%) of this surplus to the inflorescence. In so doing it supplied an amount of photosynthate equivalent to from one-fifth to one-third of the total C acquired by the inflorescence. In the fifth week of inflorescence growth, however, this leaf became shaded by the developing laterals (Fig. 2), and then graduated into an insignificant source of assimilates. At this point, the first-formed fruits on the inflorescence started to grow rapidly, and, as mentioned earlier, drew upon expanded leaves on lateral shoots as their main sources of assimilates (Farrington and Pate, 1981; Pate and Farrington, 1981).

As a second example, detailed flow profiles for C have been constructed for the three consecutive ten-day periods after flowering of the Vita 3 cultivar of cowpea (Fig. 3). Because of early senescence of its subtending leaf, each filling fruit relies on sources other than its blossom leaf for most of its supplies of assimilates. For example, from 10-20 days after anthesis, when the first formed fruit (F4, Fig. 3) was growing most rapidly, it commanded 44% of the total exportable surplus of all source leaves but acquired only 4% of its assimilate supply from its subtending leaf (L4). Similarly fruit 5 (F5) received 30% of the plant's net photosynthate 20-30 days after anthesis, yet its subtending leaf (L5) contributed only one-fifth of the fruit's requirement of C.

Studies of this kind enable one to assess the overall importance of each foliar organ in supplying C during the whole growth period of the plant. In Vita 3 cowpea, we found that of the 6733 mg C translocated from leaves during the 30 day post-flowering period, 32% originated from the top leaf (L6), 28% from leaf 5, 16% from the earliest-senescing leaf (L4) and the remaining 24% from the lowest three trifoliate leaves (L1-L3) not subtending fruits (Fig. 3). Despite their small total contribution of photosynthate over the whole growth cycle, these lower vegetative leaves clearly

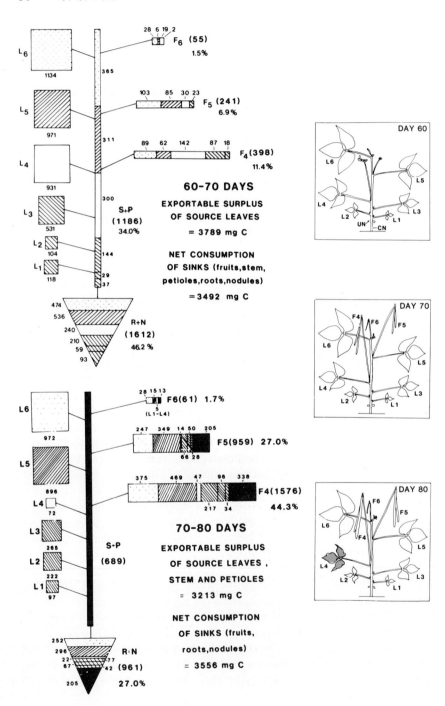

2. Seed Nutrition 51

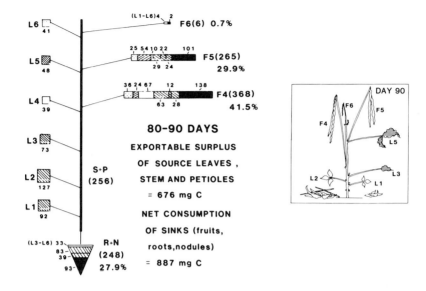

Fig. 3. Post-anthesis carbon flow to fruits of cowpea (*Vigna unguiculata*, subsp. *unguiculata* Vita 3). The top row of figures denotes the morphological condition of plants at the four times of harvest, the lower row the net contributions of carbon from individual leaves (L1-L6) and stem plus petioles (S + P) to nodulated root (R + N) and individual fruits (F4-F6) during the intervals between these harvests. Donor organs are hatch-coded so that their contributions to sinks organs can be readily recognized. Amounts (mg) of C donated and received are given and percentages of total exported C received by each fruit are indicated (data from Pate *et al.*, 1983).

helped support root function and provided assimilates to fruits during the final stages of fruiting. Information of this kind is obviously useful to programmes aimed at improving leaf canopy characteristics of seed crop species.

Our third example is field pea, for which day by day budgets have been compiled for C transfer from pod and subtending leaf to ripening seeds (Flinn and Pate, 1970). As shown in Figure 4, pod and blossom leaf supply about two-thirds of the C required by the seeds borne at that node. Leaflets, stipules and pod supply C to the seeds in approximate proportions of 3:1:2 respectively. In late development the seeds acquire the bulk of their C from sources outside the blossom node; at this time the fruit is possibly able to draw upon C mobilized from a range of senescing tissues of the shoot and root.

Table III. Budget for translocation of photosynthetically fixed carbon from the uppermost main stem leaf to the adjacent main stem inflorescence of Lupinus angustifolius cv. Unicrop.

Item	Days from emergence of main inflorescence					
	14-17	17-21	21-24	24-28	28-31	31-35
1. Carbon fixed in leaf as net daily photosynthesis (mg C/leaf)	47.9	51.8	40.4	40.1	3.5	-5.7
2. Carbon incorporated into leaf dry matter (mg C/leaf)	2.9	2.3	0.7	0.7	-2.4	-1.2
3. Net carbon available for export from leaf [(1) - (2)] (mg C/leaf)	45.0	49.5	39.7	39.4	5.9	0
4. Proportion of carbon exported from leaf translocated to inflorescence (%)	15.6	19.7	20.1	15.5	15.0	13.6
5. Estimated carbon transfer from leaf into inflorescence [(3) × (4)] (mg C/inflorescence)	7.0	9.8	7.9	6.1	0.9	0
6. Carbon consumption by inflorescence (gain of DM + respiration) (mg C/inflorescence)	26.5	48.4	23.3	21.7	23.9	-0.4
7. Proportion of carbon intake of inflorescence supplied by leaf [(5) ÷ (6) × 100] (%)	26.4	20.1	34.2	28.1	3.8	0

Data from Pate and Farrington (1981).

It appears to apply generally to all seed crops that C assimilated before flowering is proportionately much less readily available to developing fruits than is N assimilated before flowering. This principle is well displayed in studies on field pea in which the fate of the ^{14}C of photosynthetically assimilated $^{14}CO_2$ and of the ^{15}N of root-applied $^{15}NO_3^-$ was followed after feeding these materials during early vegetative growth or in mid fruiting (Pate and Flinn, 1973). Of the early-fed ^{14}C only 2% was finally recovered

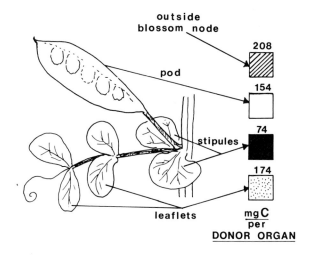

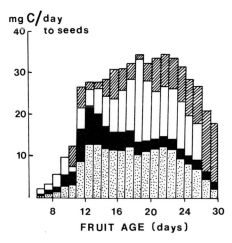

Fig. 4. Carbon nutrition of seeds borne at the lowest reproductive node of field pea (*Pisum arvense* cv. Black-eyed Susan). Contributions (mg C) from leaflets, stipules, pod and sources outside the blossom node to the seeds are shown for the whole maturation period (values to left) and on a daily basis (histogram on right) (data from Flinn and Pate, 1970).

in seeds, mainly because of respiratory losses from the plant before fruiting, but also because any ^{14}C remaining in the structural framework of vegetative parts of the plant was not readily mobilized to seeds during fruiting. By contrast, 76% of the ^{14}C fed during fruiting was available to seeds because the relevant photosynthate was mainly translocated directly to the seeds. For the $^{15}NO_3^-$ feedings the comparable percentage transfers to seeds were 51% for early-fed ^{15}N and 74% for late-fed ^{15}N, with evidence that much of the early-fed nitrogen was cycled through successive foliar organs before being finally donated to the seeds.

The situation in lupin resembles closely that of field pea, as demonstrated in the data for economy of C (Pate and Herridge, 1978) and N (Pate et al., 1981) over the growth cycle of white lupin. Studies on the narrow-leafed lupin (Pate et al., 1980a) showed that labelled carbon from $^{14}CO_2$ fed before flowering or during early flowering eventually contributed mainly to the C of protein reserves of the seed, whereas the ^{14}C of photosynthate formed during mid or late fruiting was recovered mainly in non-protein components, especially in wall reserves and oil (Fig. 5). Reflecting this effect, fruit phloem sap of plants fed early with $^{14}CO_2$ showed heavy labelling of amino compounds (60% of total ^{14}C recovered) whereas the corresponding phloem sap of plants fed $^{14}CO_2$ in late fruiting showed the bulk (78%) of the ^{14}C in non-amino compounds, principally sucrose (Pate et al., 1980a).

V. THE PHOTOSYNTHETIC ACTIVITY OF FRUITS

The green colour of aerially-borne legume fruits results from the presence of chloroplast-containing cells in the outer mesocarp of the fruit, and, also in some cases, in the inner epidermis of the pod wall that lines the gas cavity of the fruit (Pate and Kuo, 1981; Fig. 6). Developing seeds may also be noticeably green, through possessing chloroplasts in their developing embryos (Atkins and Flinn, 1978). In certain cases (e.g., pea and cowpea) the embryo of the seed becomes less green as it matures because plastids act as foci for storage of starch (Flinn and Pate, 1968), but where starch reserves are absent, or present at only low level, as in white lupin (Pate et al., 1977), embryos remain dark green until late in seed development (Atkins and Flinn, 1978).

Distribution of CO_2-assimilating enzymes tends to follow closely the distribution of chlorophyll within the fruit. Mesocarp and inner epidermis of pods of garden pea, for example, were found to contain 70% and 25% respectively of the chlorophyll of the pod, and roughly commensurate proportions of the fruit's contents of ribulose-1,5-bisphosphate carboxylase (RuBPCase, EC 4.1.1.39) and phosphoenolpyruvate carboxylase (PEPCase, EC 4.1.1.31). Also, as the pod ripened, chlorophyll was lost earlier from the mesocarp than from the inner epidermis and losses of CO_2-fixing enzymes reflected this effect (Atkins et al., 1977). The PEPCase

2. Seed Nutrition   55

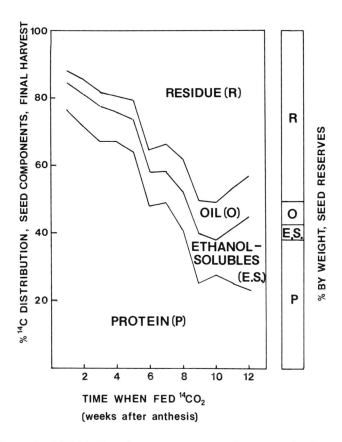

Fig. 5. Proportional ^{14}C-labelling of reserve components of mature seeds of narrow-leafed lupin (*Lupinus angustifolius*) after feeding $^{14}CO_2$ to whole plants at different times after anthesis. Note the accentuated labelling of protein from early-fixed ^{14}C. The bar diagram on the right indicates the proportions by weight of the various reserves of the seeds. The term 'residue', R, refers mainly to the fibre and wall reserves of the seed. Starch reserves were insignificant (data from Pate *et al.*, 1980a).

levels and CO_2-fixing capacity of seeds of the starch-storing species cowpea and pea are much less than in the dark green seeds of the non starch-storing white lupin (Atkins *et al.*, 1977; Atkins and Flinn, 1978).

There is ample evidence that well-illuminated legume fruits are capable of effecting significant net fixation of CO_2, and thus contributing positively to their overall carbon balance. Indeed a net uptake of CO_2 from the atmosphere may occur during the day in early fruit development of pea (Fig. 7A; Flinn and Pate, 1970), and lupin (Pate *et al.*, 1977, 1978). Alternatively, losses of CO_2 may be sufficiently lower in the day than at

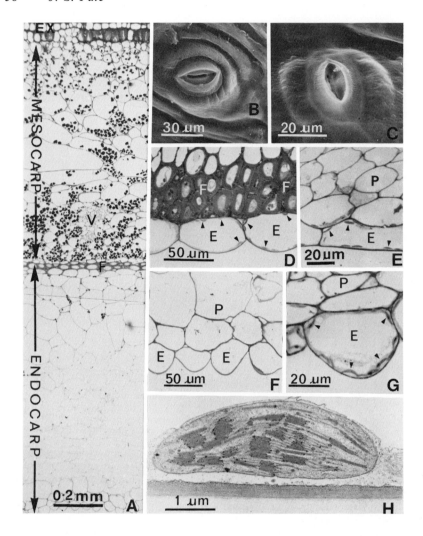

Fig. 6. Anatomical features of the pod walls of various grain legumes. A — transverse section of pod of *Phaseolus vulgaris* showing three major layers of pod (EX = exocarp). Note abundance of chloroplasts in outer mesocarp and starch grains in inner mesocarp, positioning of vascular tissue (V) in relation to fibre layers (F) of the outer endocarp. B, C — scanning electron micrographs of stomata of soybean (*Glycine max*) and lentil (*Lens culinaris*) respectively, showing epidermal ridges partially obstructing the stomatal openings. D-G — details of inner endocarp of lupin (*Lupinus albus*) (D), chick pea (*Cicer arietinum*) (E), bean (*Phaseolus vulgaris*) (F) and field pea (*Pisum sativum* var. *arvense*) (G); P, parenchyma; F, fibre layers; E, inner epidermis. Note presence of chloroplasts (arrows) in inner epidermis of all species except bean. H — electron micrograph of chloroplast of inner epidermis of lupin (from Pate and Kuo, 1981).

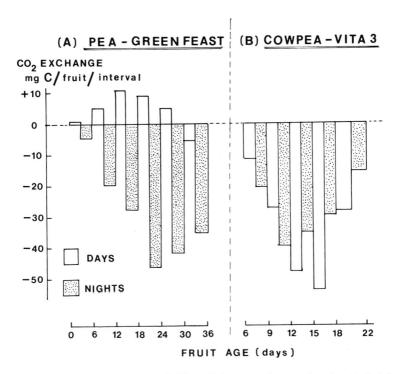

Fig. 7. Net day and night exchanges of CO_2 with the external atmosphere by attached fruits of two legumes over different intervals of fruit development. Note daytime gains of CO_2 in pea (*Pisum sativum*) fruit up to 30 days, but not at any stage in the development of the cowpea fruit (*Vigna unguiculata*, subsp. *unguiculata*) (data from Flinn et al., 1977, pea; and M. B. Peoples, J. S. Pate and C. A. Atkins, unpublished data, cowpea).

night to suggest significant photoassimilation of CO_2 from the pod gas space (cowpea, Fig. 7B). Late in development, when seeds represent the major respiratory output of the fruit, net losses of CO_2 from whole fruits can occur in the day as well as the night (30–36 days pea fruit, Fig. 7A), and, because of higher day temperatures, may even be higher in the day than at night (cowpea 12–22 days, Fig. 7B).

The functioning of the photosynthetic inner epidermis of the fruit (pea) in reassimilation of respired C can be demonstrated by injecting $^{14}CO_2$ into the pod gas cavity, whereupon the inner pod epidermis, whole pod and seeds become sequentially labelled with ^{14}C, most of this attached to sugar (Atkins and Flinn, 1978; Pate, 1979). By contrast, neither pod nor seeds become labelled if $^{14}CO_2$ is injected in darkness. Similar functioning of the outer mesocarp of the pod is shown by feeding $^{14}CO_2$ to the atmosphere surrounding the fruit or by applying drops of [^{14}C]urea solution locally to the pod wall. When such experiments are performed on cowpea, ^{14}C-labelled sugars can then be recovered in phloem exudates from the dorsal strands of the fruit close to the vascular connections to the seeds, a result which again points definitively to the role of pod walls in nourishing seeds with photoassimilated C (Pate *et al.*, 1984a).

Despite its obvious photosynthetic capabilities, the pod possesses a heavily cutinized outer epidermis and shows much lower densities of stomata than are found on adjacent leaf surfaces (Fig. 6; Atkins *et al.*, 1977; Pate and Kuo, 1981). Rates of exchange of CO_2 and transpiration are accordingly low in fruits in comparison with leaves, and, as a consequence of poor ventilation, high concentrations of CO_2 may build up in the fruit gas space, for example, 0.05%–2.5% in garden pea (Flinn *et al.*, 1977), 0.3%–2.2% in cowpea (M. B. Peoples, unpublished data). Gas space concentrations are especially high at night or if a pod is darkened during the day, since under these circumstances CO_2-assimilating systems of the fruit are no longer effective.

The CO_2-rich environment of the gaseous phase of the fruit may be particularly relevant to the functioning of PEPCase in pod and seed tissues, and may also have a noticeable effect on fruit C economy by depressing respiration rates of seeds. This is suggested from the observation that freshly detached seeds respire much less when exposed to high levels of CO_2 than when in air of ambient CO_2 level (pea, Atkins *et al.*, 1977; Flinn *et al.*, 1977; and white lupin, Atkins and Flinn, 1978).

Based on studies of CO_2 exchanges of pods and seeds and information on gains of dry matter C of fruit parts during different stages of development, a quantitative inventory of C flow between pod and seeds and their respective atmospheres has been compiled for the maturation period of a pea fruit (Flinn *et al.*, 1977) and a cowpea fruit (Peoples, Pate and Atkins, unpublished data). Each budget (Fig. 8) depicts the fate of 100 units by weight of C entering the fruit from the parent plant, indicating how C entering or lost from dry matter and the daytime (striped lines) and night time (black lines) exchanges of C (as CO_2) by pod and seeds relate to the total economy of C in the developing seeds.

The budget for the garden pea fruit (cv. Greenfeast) (Fig. 8A) showed a net loss by the fruit over its life of 13.1 units of C as CO_2 to the atmosphere, while photosynthesis of the pod wall conserved 16.2 units of C, made up of 7.1 units C as pod daytime respiration, 7.1 units of seed daytime respiration and the 2.1 units of C absorbed from the surrounding atmosphere. By the time of its maturity, the fruit had transferred the equivalent of 69% of its intake of translocated carbon to the seeds, with 18 units of C left behind in its pod walls and the remaining 13% lost as respired CO_2.

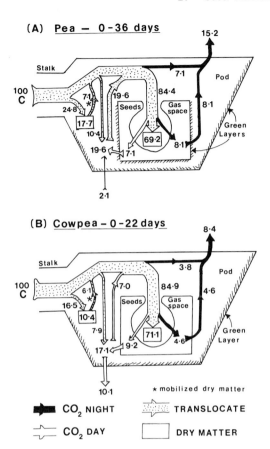

Fig. 8. Carbon economy for the complete maturation period of fruits of two species of legumes. Items of the budgets of each fruit are expressed relative to a net intake of 100 units by weight of C through the fruit stalk. Contributions of mobilized dry matter (asterisks) from pod to seed during late fruit development are indicated, and details given for the day and night exchange of CO_2 by pod and seed (data from Flinn et al., 1977; pea, *Pisum sativum*; and M. B. Peoples, J. S. Pate and C. A. Atkins; cowpea, *Vigna unguiculata* subsp. *unguiculata*).

Considerable benefits flow from the ability of the fully illuminated fruit to conserve respired CO_2 from pod and seed during the day. In the budget of Figure 8A, for example, an illuminated fruit would have required 16% less photosynthate from the parent plant when laying down a unit amount of seed dry matter than would a fruit developing in continuous darkness, when these conservation processes would not have operated effectively (Flinn et al., 1977).

Using similar types of information, a C flow profile has recently been assembled for cowpea (M. B. Peoples, J. S. Pate and C. A. Atkins, unpublished data). Net losses of C from fruit to the atmosphere turned out to be somewhat higher in the Vita 3 cultivar (18%) than in pea (13%), possibly due to the absence of a photosynthetic inner epidermis to the pod of cowpea. Counteracting this, however, the cowpea fruit made a lesser investment of C in its pod structure than did pea (Fig. 8B), so that it was still able to effect as great a net transfer of C to its seed dry matter (71%) as in pea (69%).

The situation in white lupin is complicated by detached seeds exhibiting a net uptake of CO_2 when exposed to conditions simulating those experienced within the fruit, that is, a light flux equivalent to that transmitted by the pod wall and a CO_2 concentration equal to that of the fruit gas space (Atkins and Flinn, 1978). Seeds in intact attached pods are therefore likely to be capable of reassimilating much of their own respired CO_2, as borne out by their ability to fix $^{14}CO_2$ injected into the fruit in either light or darkness (Atkins and Flinn, 1978).

VI. IDENTIFICATION OF THE MAJOR SOLUTES THAT SUPPLY CARBON AND NITROGEN TO FRUITS IN XYLEM AND PHLOEM

The fruit and its seeds are connected to the rest of the plant by conventional vascular systems that comprise both xylem and phloem, and most, if not all, of the solutes and water supplied to the fruit presumably enter through these channels. An indication of the main solutes likely to be supplied to fruits of the three species via the xylem has been obtained by analysing bleeding sap from root xylem or of tracheal (xylem) sap extracted by mild vacuum from upper stems or fruit stalks (Pate, 1976). All three legumes are also amenable to techniques that induce spontaneous phloem exudation from fruit stalks or other parts of attached fruits. Lupins are most profusive phloem bleeders (Pate et al., 1974, 1980a; Farrington and Pate, 1981), yielding phloem sap from shallow cuts in the vasculature of practically any organ of the shoot: inflorescence stalks, stalks of individual flowers or fruits, tip and mid regions of fruits, and stems and petioles of leaflets. Translocation streams can thus be monitored for solutes at a variety of points in the pathways between the major sources and sinks for solutes (Pate et al., 1979b). Garden pea is harder to work with, however, and exudes only from the phloem of its fruit stalks (Lewis and Pate, 1973), whereas cowpea usually fails to bleed from phloem of cut vascular tissue except, occasionally from pedicels or peduncles bearing actively growing fruits. However, phloem bleeding can be regularly induced in cowpea by puncturing the fruit stalk or the dorsal suture of the cowpea fruit with a fine needle cooled in liquid nitrogen. This technique has been termed 'cryopuncturing' (Pate et al., 1984a).

Sucrose represents 85% or more of the translocated C in phloem exudates of all three species. Monosaccharides, by contrast, are present in only trace amounts (usually less than 5% of total C in phloem sap) despite their high concentration in tissues adjacent to the site of collection of the sap. The remaining C of the phoem exudate is composed mainly of nitrogenous solutes and organic acids. As with phloem exudates of other non-leguminous species, the pH is relatively high (pH 6.6-8.2), the potassium ion (K^+), is the major mineral cation, and the ratio of magnesium to calcium ions ($Mg^{2+}:Ca^{2+}$) is high. By contrast, xylem sap samples contain little or no sugar, have nitrogenous solutes as their major organic constituents, are acidic (pH 5.5 ± 0.5) and have narrower ratios of $K^+:Ca^{2+}$ and $Mg^{2+}:Ca^{2+}$ than in corresponding phloem sap (Pate, 1976, 1980). Because of its high sugar content, phloem sap has a high ratio of C:N — usually from 10 to 40 in fruits (Pate et al., 1977, 1984a) or even higher (60-115) in the case of translocate from mature leaves (Pate and Layzell, 1981; Pate and Atkins, 1983). Xylem sap samples, by contrast, typically show low C:N ratios (usually within the range 1.0-3.0), close to the C:N ratio of their major nitrogenous solutes (Pate, 1980).

The nitrogenous solutes supplied to legume fruits in xylem and phloem are particularly interesting in view of their obvious significance to protein synthesis in seeds. Also of importance are any differences which may exist in composition of these transport streams when plants rely on symbiotically fixed N_2 rather than nitrate from the soil. Table IV presents relevant data for the three legumes — the information for nitrate nutrition relating to plants supplied with a level of nitrate-nitrogen (10 mM) capable of eliciting a rate of growth roughly commensurate with that of effectively nodulated plants not supplied with nitrate.

Considering nitrate nutrition and the xylem export from roots of cowpea, a species with weak nitrate-reducing systems in its root (Atkins et al., 1980b), one finds a much greater spillover of free nitrate to the shoot, than in pea or lupin, species where the nitrate reductase system of the root can be induced readily by supplying the roots with nitrate (Wallace and Pate, 1965; Atkins et al., 1979). The major organic N fraction of xylem in all three species consists of the amides asparagine and glutamine, the remainder mostly a mixture of various other amino acids. .In cowpea (Atkins et al., 1980b; Pate et al., 1980b) and pea (Pate and Wallace, 1964; Wallace and Pate, 1965), trace amounts of ureides (allantoic acid and allantoin) are also present in the xylem sap of plants supplied with nitrate.

Symbiotic plants of lupin (Table IV) carry roughly similar proportions

Table IV. Composition of nitrogenous fractions of xylem (root bleeding) and phloem (fruit) sap of cowpea (*Vigna unguiculata* cv. Vita 3), lupin (*Lupinus albus* cv. Ultra) and pea (*Pisum sativum* var. *arvense* cv. Black-eyed Susan) utilizing N_2 or NO_3^- as sole source of nitrogen.

Nutrition type	Sap collected	Plant species	Nitrogenous compounds (% N basis)			
			Amide	Amino acids	Ureides	NO_3
Symbiotic (N_2-fed)	Xylem	Pea	67	23	10	—
		Lupin	85	15	0	—
		Cowpea	13	10	77	—
	Phloem	Pea	58	40	2	—
		Lupin	72	28	0	—
		Cowpea	46	43	11	—
NO_3-fed (non-nodulated)	Xylem	Pea	41	22	3	34
		Lupin	71	15	0	14
		Cowpea	11	9	2	78
	Phloem	Pea	62	38	Tr	0
		Lupin	68	32	0	0
		Cowpea	42	48	10	0.1

Data from Atkins *et al.* (1979, 1980b); Pate *et al.* (1965, 1975, 1977, 1979b, 1980b, 1983b); Peoples *et al.* (1983).

of the major classes of organic nitrogenous constituents as do comparable nitrate-fed plants. Symbiotically effective pea and cowpea, however, show significantly greater proportions of ureide in their xylem than in nitrate-fed counterparts, and this also applies to a range of other ureide-forming species (Pate *et al.*, 1980b). Differences in composition of xylem sap with varying N supply are large in cowpea in which 78% of the transported N in nitrate-dependent (non-nodulated) plants is present as unreduced nitrate compared with 77% of N in the xylem transported as ureide when plants are without nitrate and are fixing N_2 (Atkins *et al.*, 1980b).

Corresponding data for phloem exudates (Table IV) provide evidence of higher ratios of amino compounds to ureides in phloem than in xylem (pea and cowpea) and only trace (cowpea) or non-detectable levels (pea and lupin) of free nitrate in phloem, despite the abundance of nitrate in xylem sap and in tissues of the fruits from which the phloem sap was sampled. Thus, despite the substantial differences that can exist in composition of xylem sap in the two forms of N nutrition, the phloem stream that delivers N to the fruits of a nitrate-fed plant resembles closely that of the fruits of corresponding symbiotic plants (Table IV).

Differences in the amino acid composition of phloem and xylem exudates of these species require special comment. Using cowpea as an example (Table V; Pate *et al.*, 1984a), in nitrate-fed plants asparagine

Table V. Percentage composition of ninhydrin-positive amino compounds of fruit cryopuncture (phloem) sap and root (xylem) bleeding sap of symbiotically dependent and NO$_3$-fed, non-nodulated cowpea (*Vigna unguiculata* subsp. *unguiculata*).

Amino compound	Fruit phloem sap[a] symbiotic	Fruit phloem sap[a] NO$_3$-fed	Root xylem sap[b] symbiotic	Root xylem sap[b] NO$_3$-fed
	(% total ninhydrin positive N)			
Amides				
Glutamine	29.9	25.2	37.4	15.8
Asparagine	22.5	21.8	22.5	40.9
(Total amide)	(52.4)	(47.0)	(59.9)	(56.7)
Amino acids				
Arginine	1.8	2.7	17.4	10.9
Histidine	9.1	9.4	1.7	4.9
Aspartic acid	0.4	0.6	5.1	7.5
Serine	5.9	4.5	0.4	1.0
Valine	5.3	4.2	2.6	1.8
Lysine	0.1	3.8	0.3	5.1
Threonine	4.9	3.1	1.1	1.3
γ Aminobutyrate	0.9	0.1	4.0	1.8
Alanine	3.7	2.0	0.4	0.3
Isoleucine	3.4	3.4	2.3	2.2
Leucine	3.4	3.3	2.1	2.2
Phenylalanine	3.1	1.9	0.5	0.6
Tyrosine	1.7	0.9	0.3	0.6
(Other compounds[c])	(3.9)	(13.1)	(1.9)	(3.1)

[a] Bulked sample for first two hours of bleeding from 20 fruits of age ranging from 12 to 15 days after anthesis.
[b] Bulked sample from same plants as used for phloem sap samples.
[c] Includes glutamic acid, proline, glycine, cystine, methionine.
Data from Pate *et al.* (1984a).

comprises a much greater proportion of the total amino N of xylem sap than does glutamine in the xylem sap of symbiotic plants, however, and in the phloem exudates of fruit of both nitrate-fed and symbiotic plants, asparagine is at a lower level than glutamine. Regardless of the form of nutrition, arginine, aspartic acid and γ-aminobutyric acid comprise greater proportions of the amino N of xylem sap than of phloem exudates, whereas the reverse is true of histidine, serine, threonine, valine, alanine and phenylalanine. A similar pattern can be found in white lupin and pea; root xylem bleeding sap of both of these species consistently shows higher proportions of total sap N as aspartate, γ-aminobutyric acid and arginine and wider ratios for asparagine:glutamine than in corresponding fruit phloem sap (Pate *et al.*, 1965, 1979b; Lewis and Pate, 1973).

VII. THE ORIGIN OF ASSIMILATES FOR FRUITS AND SEEDS AS DETERMINED BY SHORT-TERM ISOTOPE LABELLING STUDIES

These studies aim to trace the immediate fate of specific labelled substrates as they and their metabolic products move from vegetative parts of the plant to fruit and seed.

Some of the experimental strategies employed in short-term labelling experiments are shown in Figure 9. These include (i) application of $^{14}CO_2$ or [^{14}C]urea to photosynthetic organs, such as leaves, pods, and stipules (e.g., Flinn and Pate, 1970; Atkins *et al.*, 1975), and the subsequent collection and assay of labelled assimilates in other parts of the plant, especially in the phloem exudates of fruits; (ii) feeding of $^{15}NO_3$ or $^{15}N_2$ to below-ground organs and assay of ^{15}N in shoot parts and in xylem and phloem exudates (e.g. Pate *et al.*, 1975); (iii) application to cut shoots through the transpiration stream of $^{15}NO_3$ or ^{14}C-labelled or ^{15}N-labelled ureides or amino compounds and study of the subsequent distribution of the label among shoot organs by autoradiography, or by assays for labelled solutes in petiole and fruit phloem exudates (Atkins *et al.*, 1975, 1980a), or in aphids feeding on the plant (Atkins *et al.*, 1982); and (iv) leaf flap feeding to the xylem of leaves of compounds typical of xylem transport and recovery of label in fruits and seeds and fruit phloem sap of the fed plants (Pate *et al.*, 1984a). Information gained from each of these four classes of feeding study is discussed below. Unless otherwise stated the conclusions drawn refer to all three species, lupin, pea and cowpea.

A. Feeding of Foliar Organs

These experiments highlight the role of sucrose in translocation of photoassimilated C from nurse leaf to fruit or pod, this sugar usually carrying over 90% of the ^{14}C label recovered in fruit phloem sap of pea (Lewis and Pate, 1973), lupin (Pate *et al.*, 1974) and cowpea (Pate *et al.*, 1984a). The remainder of the ^{14}C is associated mainly with organic acids and amino compounds, especially serine, glycine, alanine, and aspartic acid (pea and lupin). Studies of lupin fruits in which time courses of phloem labelling have been examined after $^{14}CO_2$ has been fed to leaves at different times during the day (Sharkey and Pate, 1976) reveal a close integration of the daily sugar:starch cycle of the leaf with the loading of sucrose into the phloem. Thus the previous day's pools of free sugars and starch contribute to phloem loading the following night.

2. Seed Nutrition 65

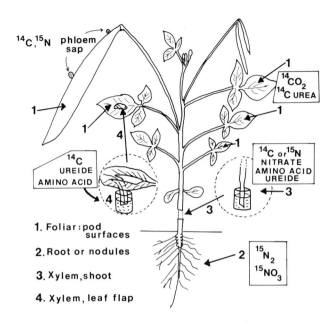

Fig. 9. Techniques for application of ^{14}C- or ^{15}N-labelled substrates used in studies of the origin of solutes delivered to legume fruits in phloem. Phloem sap is collected from fruits as indicated within a specific time interval (2-24 h) after feeding a labelled substrate through one of the sites marked 1-4 (see Pate et al., 1984a for details of feeding techniques).

B. Feeding of Roots or Nodulated Roots

The feeding of $^{15}NO_3^-$ to intact plants followed by collection and analysis of fruit phloem exudates for ^{15}N confirms how effectively vegetative organs of the plant reduce nitrate and translocate products of its reduction, but not nitrate itself, in phloem to fruits. For example, within 24 hours of feeding $^{15}NO_3^-$ (96 atom % excess) to roots of cowpea, phloem exudates of fruit showed labelling of N at 6 atom % excess, with certain compounds (asparagine 10.2 atom % excess ^{15}N and glutamine 7.6 atom % excesss ^{15}N) together carrying over 70% of the ^{15}N of the exudate. Low specific activity of ureides (0.4 atom % excess) suggested that these compounds represented an insignificant catchment for currently reduced N (Pate et al., 1984a). The complementary experiment, feeding $^{15}N_2$ to nodulated cowpea, produced the opposite result, with root xylem sap showing higher labelling of ureides than amino compounds, consistent with the relative importance of ureides in transporting fixed nitrogen in this species (Herridge et al., 1978).

Labelling of fruit phloem exudates of $^{15}N_2$-fed plants has yet to be studied, but in view of the high ureolytic activity of shoots (Atkins et al., 1982), it is to be expected that much of the labelled nitrogen initially attached to ureides in the xylem would transfer to amino compounds before being translocated to the fruits.

C. Feeding Labelled Compounds to Cut Shoots through Xylem

In these studies the labelled substrate is supplied to the cut base of the shoot in a solution of dilute unlabelled xylem sap, so that its uptake and metabolism takes place against the background of other solute transformations normally occurring as a shoot transpires xylem fluids from the root (Pate et al., 1975; Pate, 1976). Most of these studies have been undertaken on lupin (e.g., Sharkey and Pate, 1975; Pate, 1976, 1980), and their main outcome has been to demonstrate great diversity in patterns of primary uptake of different xylem solutes by different species.

The use of autoradiography of whole shoots to study overall patterns of uptake, and studies of intensities and distributions of ^{14}C in amino compounds and other solutes in shoot organs and fruit phloem sap to follow subsequent metabolic transfers of label (McNeil et al., 1979), showed that some amino acids (e.g., arginine) are so effectively abstracted by vascular tissue of stem, petiole and major leaf veins, that only a small amount of the compound or its labelled derivatives move into the leaf mesophyll. With compounds of this kind (arginine and serine), labelling of fruit phloem sap was generally weak, and any label that was present in the sap was mainly no longer associated with the fed compound. In sharp contrast to this, a second set of amino compounds, including the anionic dicarboxylic amino acids, aspartic acid and glutamic acid, were absorbed only weakly by stems and therefore tended to accumulate largely at the termini of transpiration; that is, in the non-vascular tissues of leaves. There the compounds were broken down, and the resulting metabolic products loaded into the phloem stream to the fruits. A third set of compounds, such as valine, asparagine, glutamine and threonine, were distributed more uniformly between stem, petiole and leaf. Unlike either of the previous two groups of compounds, this set of compounds was subsequently transferred in quantity direct to the phloem.

These different patterns of primary distribution and metabolism of xylem-borne amino compounds in shoots of lupin are interpreted to represent collectively a versatile system for the allocation and effective utilization of the diverse forms of N normally supplied by roots via the xylem (McNeil et al., 1979; Pate et al., 1981). Those compounds that are not readily transferred as such to phloem ensure that at least some of the N donated by the root is regularly available to mature organs of the shoot, whereas those unmetabolized compounds that pass readily from xylem to phloem may be regarded as having an equally vital role in providing essential sources of N to the shoot apex and developing fruits.

A detailed study of utilization of xylem-fed asparagine by fruiting shoots of lupin (Atkins et al., 1975) demonstrated that virtually all (92%-93%) of the ^{15}N and ^{14}C recovered from phloem sap of fruit tips of plants fed [^{14}C]asparagine labelled also with ^{15}N in the amide group was still associated with this compound. Once asparagine was unloaded from phloem it was immediately metabolized, apparently by the active asparaginase [EC 3.5.1.1] of the seeds (Atkins et al., 1975). Consistent with the testa of the young seed implementing much of the seed's asparagine catabolism (Section 1.I.C), the 'endospermic' fluid of the seed became extensively labelled in compounds other than asparagine; ^{15}N appeared in ammonium, alanine, glutamine and other amino acids, ^{14}C appeared largely in non-amino compounds. However, when the double-labelled asparagine was fed to older shoots with near mature fruits, the ^{15}N and ^{14}C of the amide became widely distributed across a range of amino acids of cotyledon protein. At this stage the embryo was the major site of asparaginase activity (Atkins et al., 1975).

Labelling studies of fruiting peas using transpiration-fed $^{15}NO_3^-$, ^{15}N (amide label)-glutamine or [^{15}N]glutamic acid (Lewis and Pate, 1973) demonstrated the integrated activity of vegetative organs of the legume shoot in processing root-derived N. The ^{15}N of each substrate was shown to be donated to a wide range of amino compounds in the soluble and insoluble fractions of the shoot and ultimately in the protein of seeds. Alanine, glutamic acid, homoserine and γ-aminobutyric acid were by far the most significant catchments for ^{15}N in the soluble fraction of leaves, and the amino acids serine, glycine, alanine, threonine, and glutamyl-nitrogen and aspartyl-nitrogen residues achieved especially high specific activity in leaf protein. By 24 hours after feeding of all substrates, ^{15}N was distributed with almost uniform specific activity among the amino acids of seed protein, indicating how effectively the whole fruiting shoot had utilized the different nitrogen sources.

In view of the absence of nitrate in fruit phloem sap of pea and the heavy reliance of seeds on phloem in delivering N (Lewis and Pate, 1973), most nitrate metabolism must be accomplished before this source of nitrogen reaches the fruit. This is true also for lupin: Pate et al. (1975) showed that feeding of $^{15}NO_3$ to cut shoots through the xylem led to high ^{15}N-enrichment of asparagine and glutamine (amino-N and amide-N of both compounds) and of other amino acids in fruit phloem sap, with no evidence of labelled or unlabelled nitrate in the phloem sap.

Experiments feeding labelled ureide ([2-^{14}C] allantoin) to cut shoots through xylem have been restricted to cowpea because of the significance of ureides in this species. All vegetative and reproductive parts of cowpea showed significant ureolytic activity throughout growth (Atkins et al., 1982), with evidence that the resulting $^{14}CO_2$ was reassimilated photosynthetically and subsequently appeared as ^{14}C-sugars and other non-nitrogenous organic solutes in fruit phloem sap (Pate et al., 1983b), or in the bodies of aphids feeding on fruits (Atkins et al., 1982). However, a significant proportion of the ^{14}C of phloem sap or aphids was still located in allantoin, indicating that the compound was indeed mobile in the phloem, and had probably been subjected to xylem to phloem transfer within the vasculature of the stem or the minor veins of leaves. Fruits of cowpea display significant activity of allantoinase [EC 3.5.2.5] as well as of asparaginase, indicating the capability of using either ureide-N or amide-N as sources of amino groupings for synthesis of protein amino acids (M. B. Peoples, D. R. Murray, J. S. Pate and C. A. Atkins, unpublished data). As shown earlier (Tables IV and V), asparagine and ureides represent major fractions of the nitrogen supplied to the fruit in both xylem and phloem.

D. Leaf Flap Feeding

In this form of xylem feeding the radiosubstrate is supplied directly to the cut mid vein of an attached leaf (see inset 4 to Fig. 9), with the opportunity for phloem-borne labelled assimilates to filter back to the parent plant via the peripheral vein network of the fed leaf. Recent experiments (Pate et al., 1984a) on cowpea have demonstrated that distribution of ^{14}C in phloem exudates following selective feeding of a leaf by the 'flap' technique as described above, may be different from that obtained when the same ^{14}C-labelled substrate is supplied to the whole cut shoot through the transpiration stream (Table VI). Thus, when [^{14}C]asparagine was fed to the shoot through the xylem, 84% of the ^{14}C in fruit phloem sap was still attached to asparagine — evidence of bulk transfer of this compound, as described earlier for lupin. However, following feeding of the labelled amide through the leaf flap, it was shown that fruit phloem sap carried only 54% of its ^{14}C label as asparagine. A high labelling of sugars and organic acids and other amino acids in the sap suggested that the amide had been partly broken down in the mesophyll of the fed leaf and that products of this metabolism had then been loaded onto the phloem. The comparison using ^{14}C-allantoin (Table VI) provided an essentially similar result, with 47% of the ^{14}C still as allantoin in fruit phloem sap following xylem feeding of the whole shoot, compared with only 21% when fed directly to the xylem of a single leaf.

Table VI. Distribution of radiocarbon among solutes of fruit[a] phloem sap after application of [^{14}C]asparagine or [^{14}C]allantoin through the transpiration stream[b] or to the cut mid vein xylem of blossom leaflets[c] of cowpea (*Vigna unguiculata* subsp. *unguiculata*)

[^{14}C]Substrate and mode of application	Percentage distribution of ^{14}C among labelled solutes							
	Asp	Thr/Ser	Asn	Gln	Ala	Val	Ureides	Sugars + organic acids
[^{14}C]Asn, xylem fed	0.4	—	83.9	4.6	—	—	—	11.1
[^{14}C]Asn, leaf flap fed	3.1	5.3	54.3	21	6.3	—	—	28.9
[^{14}C]Allantoin, xylem fed	—	—	—	—	—	—	46.5	53.5
[^{14}C]Allantoin, leaf flap fed	—	—	7.3	9.1	5.5	6.8	21.3	50.9

[a] Fruits used for study 12 to 15 days after anthesis.
[b] Applied as 5 µCi of radiosubstrate to cut shoot through xylem. Six hour collection period of fruit cryopuncture phloem sap (see shoot xylem feeding, Fig. 9).
[c] Applied as 3 µCi of radiosubstrate through the xylem of the mid vein of the attached mid leaflet of the leaf subtending the bleeding fruit (see leaf flap feeding, Fig. 9).
[d] Data from Pate *et al.* (1984a).

VIII. PARTITIONING OF CARBON AND NITROGEN IN THE WHOLE PLANT AND THE NUTRITION OF FRUITS AND SEEDS

By exploiting the phloem-bleeding capacities of white lupin, we have been able to construct a series of empirically based models of organ and whole plant functioning, depicting intake, utilization and exchange of net photosynthate and fixed N by the plant during different stages of development (Pate et al., 1979a, c; Atkins et al., 1980b; Layzell et al., 1981; Pate and Layzell, 1981).

Each model uses data for net increments or losses of C and N in dry matter by plant parts, respiratory losses of C as CO_2 from plant organs, and values for C:N weight ratios in xylem and phloem streams serving different regions of the plant. For example, data on C:N ratios of the phloem sap collected from the petiole of a leaf and of the tracheal (xylem) sap supplied to that leaf when combined with data on photosynthetic inputs and changes in C and N content of the leaf allow calculation of how much N and C is being translocated from that leaf to the rest of the plant and what proportion of this translocated N cycles through the leaf from xylem to phloem (e.g., Pate and Atkins, 1983). Similarly, by examining data on C:N ratio of stem-base phloem sap alongside data on the consumption of C by roots and nodules in growth and respiration, the amounts of N supplied to the root in the phloem can be determined (e.g., Pate et al., 1979a). Also, by comparing the C:N requirements of apical regions of the shoot, such as vegetative shoot, apices, young leaves, inflorescences and fruits, with C:N ratios of upper-stem tracheal (xylem) sap and stem-top phloem sap, an assessment can be made of the proportional involvement of xylem-delivered and phloem-delivered C and N in the nourishment of these organs (Pate et al., 1981; Pate and Layzell, 1981).

The data utilized in one study to model C and N partitioning in the 10-day period following initiation of the main stem inflorescence of white lupin are summarized in Figure 10. The information presented for intake and consumption of C relative to N (Fig. 10A) shows that at this stage of development net photosynthate equivalent to 30 units by weight of C are generated by the leaflets for each unit of N taken up by the roots, and that the C:N ratios for consumption of these assimilates by plant parts vary greatly. The ratios vary from 137:1 for the older segments of the stem (SP 1, Fig. 10A) engaged in secondary thickening of tissues, to 13:1 in the protein-rich tissues of young leaves or developing inflorescences. The complementary information on C:N ratios of transport fluids is provided in Figure 10B. It can be seen that phloem sap intercepted as it enters the inflorescence has a C:N ratio significantly lower (20:1) than that leaving the

2. *Seed Nutrition* 71

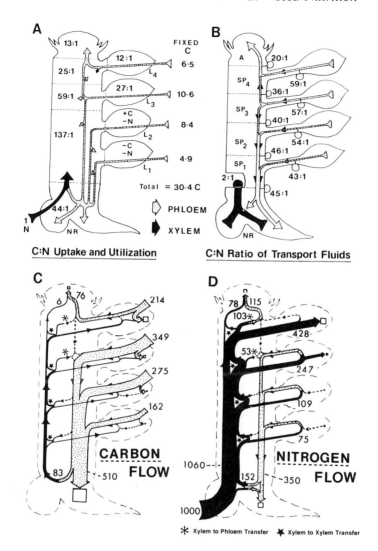

Fig. 10. Partitioning of C of net photosynthate and N fixed by nodules in early post flowering development (51-58 days) of lupin (*Lupinus albus*). (A) Inputs and ratios of consumption by weight of C and N by different plant parts; (B) C:N weight ratios of phloem and xylem fluids collected at different locations in the plant's circulatory system; (C) Flow profile for C; (D) Flow profile for N. Flow lines in C and D are drawn proportional to amounts of C or N transported and the numerical values for flow along different pathways are given relative to a net input by the plant of 1000 units by weight of C or N. A — apical organs of shoot (including top lateral shoots and terminal inflorescence); SP1-SP4 — successive strata of stem + petiole segments; L1-L4 — strata of leaflets; NR — nodulated root (data from Pate and Layzell, 1981).

leaflets that feed the same inflorescence (ratios 59:1, 57:1). This is interpreted as evidence that substantial amounts of N are added from the xylem to the phloem stream in the upper region of the stem, and that by this process of xylem to phloem exchange, the shoot apex is provided with significantly more N than it would receive were it to rely solely on translocate derived from upper leaves.

The final models for flow of C and N (Figs 10C and D respectively) suggest that another type of vascular exchange occurs, namely transfer from one xylem strand to another within the vasculature of the stem. This is depicted by the looped pathways marked by asterisks (Figs 10C and D). This pathway involves the progressive removal of N solutes from the xylem streams of vascular traces that supply leaves and the subsequent transfer of an equivalent amount of N back into stem traces that supply higher regions of the shoots. Organs at the top of the shoot thus receive N in the xylem at significantly higher concentration than is present initially in the xylem stream that leaves the root. The rate of acquisition of N per unit of water transpired is therefore significantly higher than that of the lower leaves whose supply of N in the xylem was reduced by the xylem to xylem transfer system of the stem.

Transfer cells of xylem parenchyma line the departing leaf traces at stem nodes (Gunning *et al.*, 1970; Gunning and Pate, 1974; Kuo *et al.*, 1980) and have been suggested as the probable uptake sites for N in the xylem to xylem exchange system of the stem. Experimental support of this came from demonstrations of the intensity with which such cells became labelled when ^{14}C-labelled amino compounds were fed through the xylem (Pate and O'Brien, 1968; Pate *et al.*, 1970). Direct demonstration of a progressive increase in concentration of N in xylem from base to top of the shoot was obtained by measuring N levels in xylem (tracheal) sap extracted from lower, mid and top segments of a shoot. The tracheal sap from the upper stem segments was then found to be from 2 to 4 times more concentrated in nitrogenous solutes than that collected at the base of the shoot (Pate *et al.*, 1980c; Pate and Layzell, 1981).

The information presented in the models of C and N flow shown in Figure 10 enables a model to be constructed of the various sources of C and N for growing structures of the plant such as root, developing inflorescence and vegetative shoot apex. The relevant model (Fig. 11), still fully consistent with all experimental data, aids in quantifying the role of xylem to phloem and xylem to xylem transfer in supplying the apical shoot regions with N and of integrating these activities with the complementary function of the top strata of leaves in providing the inflorescence with a bulk source of C. The model also deals with C and N supply to the nodulated root, the roots being shown to be supplied with sufficient C and a slight oversupply of N by the lower strata of leaves. The surplus N that arrives in the root is viewed

2. Seed Nutrition   73

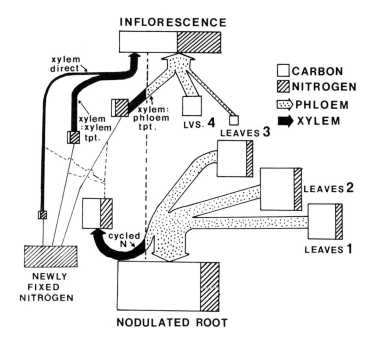

Fig. 11. Source agencies within the shoot supplying C and N to the apical inflorescence and nodulated root during early post flowering development (51-58 days after sowing) of white lupin (*Lupinus albus*). Relative amounts of C and N supplied from each source are indicated using flow lines of varying thickness and rectangles (sinks) or squares (sources) of varying area, all drawn proportional to the relative amounts of C or N transferred. In the area scale used 10 units by weight of C are equivalent to 1 unit by weight of N. Model fully consistent with data of Figure 10 (adapted from Pate and Layzell, 1981).

as cycling back to the shoot via the xylem and of thus becoming available along with newly fixed N for further differential partitioning within the shoot system.

According to this model (Fig. 11) four distinct sources of N are available to the lupin shoot apex or inflorescence. These are (i) xylem to xylem transfer within the stem, (ii) direct translocation from leaves, (iii) xylem to phloem transfer in upper stem, and (iv) direct xylem transfer from root. The last item is defined as the amount of N which the apical organ would have been expected to acquire were it to have attracted xylem fluid through transpiration at a concentration of N equal to that leaving the root.

In the models of C and N partitioning each mature organ of the lupin plant is envisaged as contributing in a unique fashion to the total nutrition of growing parts of shoot and root. It thus becomes possible to predict more adequately how partitioning for C and N might be regulated as assimilatory inputs of C and N change during the course of plant development, and how the partitioning processes might respond to short-term, environmentally induced changes in rates of assimilation of either C or N.

IX. THE WATER ECONOMY OF FRUITS AND ITS INTEGRATION WITH FRUIT CARBON AND NITROGEN NUTRITION

As a result of the low density of their stomata, and the occurrence, in some species, of epidermal plates or ridges that partly obstruct the stomatal openings (e.g., Fig. 6B,C), legume fruits typically show low rates of transpiration relative to their surface areas. Moreover the amplitude of diurnal fluctuations in their water loss is usually low: about three or four times higher loss in the day than at night in lupin compared with, for example, a 10–20 fold difference in whole plant transpiration (Hocking et al., 1978). In pea fruits (Flinn et al., 1977) transpiration losses are maximal halfway through development, fall slightly in mid pod fill, and then rise again as fruits and seeds dry out. Cowpea fruits show similar but less pronounced trends, whereas lupin fruits exhibit a mid-growth peak in transpiration when rates are several times higher than in early development or at subsequent maturity (Pate et al., 1977, 1978; Pate and Hocking, 1978).

The economy of water use by whole plants is often referred to in terms of transpiration ratio, namely the total amount (mL) of water transpired per unit weight (g) of dry matter gain by the whole plant. During their respective growth cycles, the legumes studied show values for this ratio of from 300 mL/g to 800 mL/g (e.g., Pate et al., 1980c). Extending the concept of transpiration ratio to the water economy of a fruit, one finds values of 28 mL/g for pea (Flinn et al., 1977), 23 mL/g for white lupin (Pate et al., 1977), 20 mL/g for soybean (Glycine max) (Layzell and LaRue, 1982), and the surprisingly low value of only 8 mL/g for cowpea (M. B. Peoples, J. S. Pate and C. A. Atkins, unpublished data). These low values come about because the fruit depends largely on its parent plant for photosynthate and the transpirational penalty in forming this photosynthate is paid in vegetative plant parts, not the fruits. Nevertheless, fruits of cowpea must clearly have an unusually efficient economy of water usage.

A second major component of fruit water balance is the amount of water held in the tissues of the fruit. This amount increases rapidly in early fruit development as the fleshy pod and then the young seeds increase in mass and water content. Fruit water content then declines equally rapidly in late development as the pod and then seeds dehydrate. A third minor component concerns metabolic exchanges of water in the respiration and

2. Seed Nutrition 75

condensation reactions within the fruit, but as these items at all times represent less than 3% of the net water budget of the whole fruit (Layzell and LaRue, 1982), they are discounted in subsequent discussion.

By combining data for changes in tissue water content and transpirational loss, the net amount of water imported by the fruit from the parent plant can be computed for different stages of development. Such data provide a basis for relating inputs of C and N to those for water, and ultimately for constructing models of proportional respective intakes of C, N and water through xylem and phloem of the fruit stalk (e.g., Pate *et al.*, 1977, 1978).

As shown earlier when discussing partitioning of C and N in whole plants, three basic classes of information are required for constructing models of exchange of C and N between a specific plant organ and the rest of the plant. When considering a fruit these would be the amounts of C and N incorporated into the dry matter of the fruit, the net atmospheric exchanges of carbon in respiration and photosynthesis, and the C:N weight ratios of the solutes delivered to a fruit in xylem and phloem. Model construction is based on the assumption that there is unidirectional mass flow into the fruit in xylem and phloem. As long as the concentrations of C and N in transport fluids are known, the validity of the model can be tested by examining whether the mixture of xylem and phloem streams estimated by the model to meet precisely the recorded C and N intake of the fruit does indeed correspond in water content with the estimated water consumption of the fruit.

Owing to absence of information on phloem sap composition for most of fruit development of pea, model building in relation to a fruit's economies of C, N and water has proved possible only for lupin (data from Pate *et al.*, 1977) and cowpea (unpublished data, M. B. Peoples, J. S. Pate and C. A. Atkins). To highlight the differences and similarities between these last two species itemized budgets have been constructed for two stages of fruit ripening, the first stage spanning early pod growth and early seed growth — up to the late liquid endosperm stage (0–6 weeks lupin, 0–11 days cowpea), the second covering the remainder of fruit development, and dominated by rapid growth and laying down of reserves in embryos and mobilization to seeds of reserves of C and N laid down in pods during the first half of fruit development (6–12 weeks lupin, 11–22 days cowpea).

During the first half of fruit development the two fruits performed differently. Lupin (0–6 weeks after anthesis) showed inputs through the fruit stalk of 416 mg C and 28.4 mg N, with a net loss of C to the atmosphere as CO_2 of only 6 mg (1.5% of that entering). Water input was 22.8 mL, 7.8 mL of this into tissues, the remaining 15.0 mL transpired. The

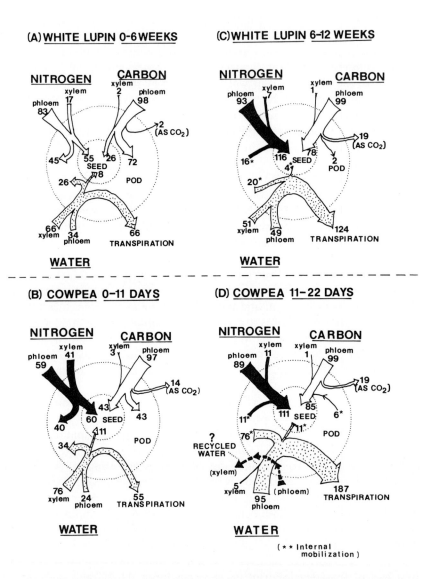

Fig. 12. Proportional intakes of C, N and water through xylem and phloem in early (A and B) and late (C and D) development of two legume fruits, white lupin (*Lupinus albus*) and cowpea (*Vigna unguiculata* subsp. *unguiculata*). All components of a fruit's budget are expressed relative to a net intake of 100 units of C, N or water through xylem and phloem. Values for internal mobilization of C, N or water (asterisks) are given for the second halves of fruit development. See text for discussion of the hypothetical component marked 'recycled water' in the 11–22 day budget for cowpea. Ratios of absolute amounts by weight of C, N and water consumed by fruits are given in the text for each stage of fruit development (data of Pate *et al.*, 1977, for lupin; Peoples *et al.*, 1984a, for cowpea).

ratio of consumption of water:C:N in the young lupin fruit was therefore 802:15:1. In cowpea (0–11 days after anthesis) 569 mg C and 37.4 mg N entered the fruit as assimilates from the parent plant and 78 mg (14%) of the entering C was lost in respiration. Of 20.4 mL water entering during this stage 9.2 mL was incorporated into tissues and the remaining 11.2 mL lost in transpiration, giving an overall weight ratio for consumption of water:C:N of 545:15:1. In effect, therefore, lupin showed a poorer efficiency in water usage, but a considerably better performance in conservation of respired CO_2 than cowpea.

During the second half of fruit development, further differences emerge among the species. The lupin fruit (6–12 weeks after anthesis) received from the parent plant 468 mg C (19% of this lost as respired CO_2), 44.1 mg N and 20.5 mL water, a ratio of inputs of water:C:N of 465:9.1:1. The cowpea fruit (11–22 days after anthesis) acquired from the plant 1104 mg C and 84.7 mg N for a net intake of only 9.9 mL water, with 21% of the C intake subsequently lost as CO_2 and tissue water losses contributing 9.6 mL of the total 18.5 mL lost in fruit transpiration. The ratio for intake of water:C:N was 117:13:1. Thus, by this stage of development the better economy in water usage by cowpea had become even more accentuated, although the earlier superiority of lupin over cowpea in refixation of respired CO_2 was now lost, with fruits of both species losing about one-fifth of their imported C as net losses of CO_2 to the atmosphere.

By incorporating information on the C:N ratios of xylem and phloem streams serving the fruit with the above-mentioned data on C, N and water consumption by the fruit, it becomes possible to develop empirically based models that depict the proportional involvements of phloem and xylem in supplying a fruit with C, N and water (Fig. 12).

During the first half of development, fruits of lupin (Fig. 12A) and cowpea (Fig. 12B) showed phloem intake as the dominant source of C (98% and 97% of intake respectively). Proportionately more N (83%) and water (34%) entered the fruit of lupin by the phloem than in cowpea (59% of N and 24% of water).

In the second half of fruit development, lupin (Fig. 12C) showed lesser intake of C, N and water through xylem than in its first stage of growth, and a substantial mobilization of earlier-established reserves of N, but not of C, from pod walls to fruit. As a result the seeds acquired 78% of the fruit's net intake of C and all of the intake of N plus an additional 16 units of N mobilized from the pod walls. Whereas the young lupin fruit (Fig. 12A) transpired an amount of water equivalent to its net intake through xylem, transpiration greatly exceeded total water intake in the second half of its maturation as a result of drying out of pod and seeds. Xylem intake through the fruit stalk was equivalent to only one-fifth of the water loss of the fruit.

The corresponding model for the second half of cowpea fruit development (Fig. 12D) proved to be particularly interesting. The mean C:N ratios for intake through phloem and xylem at this stage were 14.5:1 and 1.1:1, so that with a ratio of intake through the fruit stalk of 13.0 C:1 N, proportional intake through phloem was 99% for C and 89% for N. The xylem contributed the remaining 1% of the C and 11% of the N. However, with a concentration of 117.6 mg C/mL in phloem sap, 9.4 mL of water would have been predicted to have entered the fruit in a mass flow stream bringing in the recorded fruit's intake of 1104 mg C. Intake of water through xylem would then have been only 0.3 mL. With a recorded N intake of 9.3 mg through xylem, the mean concentration of N in the xylem stream to the fruit would have been 30 mg N/mL — an absurdly high value, more than 30 times the level of N recorded for root xylem sap. This apparently anomalous situation might be resolved if it were assumed that water entering via the phloem cycles back to the parent plant via the xylem, that is, along the lines indicated in the hypothetical component marked 'recycled water' in Figure 12D. It is conceivable, for example, that excess water entering the pod dorsal strands through the phloem might exit to the parent plant through the xylem of these same strands and, indeed, since the dorsal vasculature is directly involved in supplying the seeds it is likely to show a positive balance for water in view of its high intake of translocate. By contrast, the corresponding ventral strand, serving mainly the non-growing transpiring walls of the pod, would be likely to attract proportionately much less water through the phloem and would therefore probably have to effect a considerable net intake of water through the xylem to meet local requirements in transpiration.

Evidence for or against such a bipartite system for water exchanges within the fruit is inconclusive. Experiments in which cut, fruiting shoots were fed through the transpiration stream with the xylem-mobile dye, acid fuchsin, or the apoplast-marker substance [^{3}H]inulin, showed that during the second half of cowpea fruit development the xylem strands of the ventral half, but not of the dorsal half, of a fruit consistently recorded a positive intake of the marker materials in the night and early morning, but not the afternoon. In early fruit development, when the fruit transpired more water in relation to its intake of C and N, both dorsal and ventral vasculature of the fruit recorded intake of the tracers for most or all of the day (J. S. Pate, A. J. M. Van Bel, M. B. Peoples, C. A. Atkins, unpublished data). Also, in nitrate-fed plants, fruits 12 days old or older show free nitrate and associated nitrate reductase actively restricted to their ventral halves, whereas young fruit show a more uniform distribution of nitrate and reductase activity (M. B. Peoples, C. A. Atkins, J. S. Pate, unpublished data). With nitrate mobile in xylem, but not in phloem, the data for nitrate and nitrate reductase distribution are consistent with the

2. Seed Nutrition 79

concept of a transport system in which only the ventral strands of older fruits are involved in xylem intake of water.

This model obviously requires further testing, particularly its anatomical aspects, since its successful operation would require that xylem of dorsal and ventral strands remain physically separate, or that exchange between their respective conducting elements at least be impeded. Furthermore, the water balance of the seeds needs to be understood much better than at present, particularly in relation to how seeds rid themselves of the considerable excess of phloem-derived water which they must acquire when converting a mass stream of phloem solutes to dry matter of their cotyledon reserves. In the model of Figure 12D, seeds utilized 766 mg C for a net increase of only 2.7 mL of water. Thus, if fed with phloem sap at 109 mg C/mL, they would have generated an excess of 5.9 mL water. If this were returned to the parent plant via the dorsal xylem connection with the fruit stalk in the manner suggested above. An equivalent amount of water, and therefore more N, would then be able to enter the lower parts of the fruit through the xylem.

X. THE CONVERSION OF IMPORTED SOLUTES INTO SEED RESERVES

As already indicated, analyses of phloem and xylem exudates and estimates of proportional intakes of C and N through these channels provide the basis for determining the absolute amounts of the different major classes of organic solutes imported by a fruit during different stages of its growth. By comparing the total weight of these imported solutes with the total amounts of dry matter and reserve materials such as oil, protein and carbohydrate laid down in the seeds of the fruit, a measure can readily be obtained of the overall effectiveness with which the fruit has converted translocate to these various seed materials. Such conversion indices might well be regarded as the ultimate yardstick of efficiency of fruit ripening (Pate, 1979; Pate et al., 1984a).

Table VII sets out relevant information for the fruits of the three species studied. As shown earlier (Fig. 12) lupin makes a large, and essentially non-retrievable investment of C in the thick fibrous pod, and this is clearly reflected in the lower percentage conversion of translocate to seed dry matter in this species (50%) than in either pea (77%) or cowpea (69%). Species differences become even more accentuated when conversion efficiencies are expressed as synthesis of the major reserves (oil + protein + perchlorate-soluble carbohydrate): cowpea (61% conversion) has almost

Table VII. Indices of conversion of translocated organic solutes into seed products by fruits of three species of grain legumes

	Species		
Item	Garden pea (Pisum sativum cv. Greenfeast)	Cowpea (Vigna unguiculata cv. Vita 3)	White lupin (Lupinus albus cv. Neutra)
1. Solutes entering fruits in xylem and phloem (mg/fruit)			
Sucrose + organic acids	2404	3437	1756
Amides + amino acids	363	660	384
Ureides	30	72	—
Total	2797	4169	2140
2. Reserve materials laid down in seeds (mg/fruit)			
Dry matter	1930	3220	1070
Protein	387	719	412
Oil	19	30	132
Perchloric acid-soluble carbohydrate	857	1803	110
3. Conversion efficiency by weight of imported solutes to seed components (%)			
To dry matter	69	77	50
To protein	14	17	19
To protein + oil + $HClO_4$- soluble carbohydrate	45	61	31

Data from Pate (1979); Pate et al. (1984).

twice the efficiency of lupin (31%), with pea (45%) intermediate between the two. These last differences relate especially to the high content of starch (perchlorate-soluble) in pea and cowpea as opposed to the virtual absence of starch and the low-levels of other forms of extractable carbohydrate in lupin. When expressed in terms of conversion by weight of translocate to unit weight of seed protein, however, the superiority of pea and cowpea is no longer evident, with lupin showing a slightly higher conversion index (19%) of translocate to protein than either cowpea (17%) or pea (14%). It would be especially interesting to extend comparisons of this kind to different cultivars of a species, and to see whether fruit efficiency would be improved by selecting, for example, for pod characteristics leading to a more effective channeling of C and N into seeds.

ACKNOWLEDGEMENTS

The author is greatly indebted to the many persons who have contributed over the years to our programmes of work on fruit nutrition at The University of Western Australia, especially to Craig Atkins, Pat Sharkey, Mark Peoples, Peter Hocking, John Kuo, Peter Farrington, Alastair Flinn, David Layzell, David McNeil and David Herridge. The assistance is also acknowledged of funding agencies supporting the work, namely the Australian Research Grants Scheme, the Wheat Industry Council (Australia) and the United Nations Development Project.

REFERENCES

Atkins, C. A., and Flinn, A. M. (1978). *Plant Physiol.* **62**, 486-490.
Atkins, C. A., Pate, J. S., and Sharkey, P. J. (1975). *Plant Physiol.* **56**, 807-812.
Atkins, C. A., Kuo, J., Pate, J. S., Flinn, A. M., and Steele, T. W. (1977). *Plant Physiol.* **60**, 779-786.
Atkins, C. A., Pate, J. S., and Layzell, D. B. (1979). *Plant Physiol.* **64**, 1078-1082.
Atkins, C. A., Pate, J. S., and McNeil, D. M. (1980a). *J. Exp. Bot.* **31**, 1509-1520.
Atkins, C. A., Pate, J. S., Griffiths, G. J., and White, S. T. (1980b). *Plant Physiol.* **66**, 978-983.
Atkins, C. A., Pate, J. S., Ritchie, A., and Peoples, M. B. (1982). *Plant Physiol.* **70**, 476-482.
Farrington, P., and Pate, J. S. (1981). *Aust. J. Plant Physiol.* **8**, 293-305.
Flinn, A. M., and Pate, J. S. (1968). *Ann. Bot.* **32**, 479-495.
Flinn, A. M., and Pate, J. S. (1970). *J. Exp. Bot.* **21**, 71-82.
Flinn, A. M., Atkins, C. A., and Pate, J. S. (1977). *Plant Physiol.* **60**, 412-418.
Gunning, B. E. S., and Pate, J. S. (1974). *In* 'Dynamic Aspects of Plant Ultrastructure' (A. W. Robards, ed.), pp. 441-480. McGraw Hill, London.
Gunning, B. E. S., Pate, J. S., and Green, L. W. (1970). *Protoplasma* **71**, 147-171.
Herridge, D. F., and Pate, J. S. (1977). *Plant Physiol.* **60**, 759-764.
Herridge, D. F., Atkins, C. A., Pate, J. S., and Rainbird, R. M. (1978). *Plant Physiol.* **62**, 495-498.
Hocking, P. J., Pate, J. S., Atkins, C. A., and Sharkey, P. J. (1978). *Ann. Bot.* **42**, 1277-1290.
Kuo, J., Pate, J. S., Rainbird, R. M., and Atkins, C. A. (1980). *Protoplasma* **104**, 181-185.
Layzell, D. B., and LaRue, T. A. G. (1982). *Plant Physiol.* **70**, 1290-1298.
Layzell, D. B., Pate, J. S., Atkins, C. A., and Canvin, D. T. (1981). *Plant Physiol.* **67**, 30-36.
Lewis, O. A. M., and Pate, J. S. (1973). *J. Exp. Bot.* **24**, 596-606.
McNeil, D. L., Atkins, C. A., and Pate, J. S. (1979). *Plant Physiol.* **63**, 1076-1081.
O'Neill, T. B. (1961). *Bot. Gaz.* **123**, 1-9.
Pate, J. S. (1976). *In* 'Transport and Transfer Processes in Plants' (I. F. Wardlaw and J. B. Passioura, eds.), pp. 253-281. Academic Press, New York.
Pate, J. S. (1979). *J. Royal Soc. West. Aust.* **62**, 83-94.
Pate, J. S. (1980). *Annu. Rev. Plant Physiol.* **31**, 313-340.

Pate, J. S., and Atkins, C. A. (1983). *Plant Physiol.* **71**, 835-840.
Pate, J. S., and Farrington, P. (1981). *Aust. J. Plant Physiol.* **8**, 307-318.
Pate, J. S., and Flinn, A. M. (1973). *J. Exp. Bot.* **24**, 1090-1099.
Pate, J. S. and Herridge, D. F. (1978). *J. Exp. Bot.* **29**, 401-412.
Pate, J. S., and Hocking, P. J. (1978). *Ann. Bot.* **42**, 911-921.
Pate, J. S., and Kuo, J. (1981). *In* 'Advances in Legume Systematics' (R. M. Polhill and P. H. Raven, eds.), Part 2, pp. 903-912. Royal Botanic Gardens, Kew.
Pate, J. S., and Layzell, D. B. (1981). *In* 'Nitrogen and Carbon Metabolism' (J. D. Bewley, ed.), pp. 94-134. Junk, The Hague.
Pate, J. S., and Minchin, F. R. (1980). *In* 'Advances in Legume Science' (R. J. Summerfield and A. H. Bunting, eds.), pp. 105-114. Royal Botanic Gardens, Kew.
Pate, J. S., and O'Brien, T. P. (1968). *Planta* **78**, 60-71.
Pate, J. S., and Wallace, W. (1964). *Ann. Bot.* **28**, 80-99.
Pate, J. S., Walker, J., and Wallace, W. (1965). *Ann. Bot.* **29**, 475-493.
Pate, J. S., Gunning, B. E. S., and Milliken, F. F. (1970). *Protoplasma* **71**, 313-334.
Pate, J. S., Sharkey, P. J., and Lewis, O. A. M. (1974a, b). *Planta* **120**, 229-243.
Pate, J. S., Sharkey, P. J., and Lewis, O. A. M. (1975). *Planta* **122**, 11-26.
Pate, J. S., Sharkey, P. J., and Atkins, C. A. (1977). *Plant Physiol.* **59**, 506-510.
Pate, J. S., Kuo, J., and Hocking, P. J. (1978). *Aust. J. Plant Physiol.* **5**, 321-326.
Pate, J. S., Layzell, D. B., and McNeil, D. M. (1979a). *Plant Physiol.* **63**, 730-738.
Pate, J. S., Atkins, C. A., Hamel, K., McNeil, D. L., and Layzell, D. B. (1979b). *Plant Physiol.* **63**, 1083-1088.
Pate, J. S., Layzell, D. B., and Atkins, C. A. (1979c). *Plant Physiol.* **64**, 1083-1088.
Pate, J. S., Atkins, C. A., and Perry, M. W. (1980a). *Aust. J. Plant Physiol.* **7**, 283-297.
Pate, J. S., Atkins, C. A., White, S. T., Rainbird, R. M., and Woo, K. C. (1980b). *Plant Physiol.* **65**, 961-965.
Pate, J. S., Layzell, D. B., and Atkins, C. A. (1980c). *Ber. Deutsch. Bot. Ges.* **93**, 243-255.
Pate, J. S., Atkins, C. A., Herridge, D. F., and Layzell, D. B. (1981). *Plant Physiol.* **67**, 37-42.
Pate, J. S., Peoples, M. B., and Atkins, C. A. (1983). *J. Exp. Bot.* **34**, 544-562.
Pate, J. S., Atkins, C. A., and Peoples, M. B. (1984a). *Plant Physiol.* **74**, 499-505.
Pate, J. S., Williams, W., and Farrington, P. (1984b). *In* 'Grain Legume Crops' (R. J. Summerfield and E. H. Roberts, eds.) (in press) Granada, England.
Peoples, M. B., Pate, J. S., and Atkins, C. A. (1983). *J. Exp. Bot.* **34**, 563-578.
Sharkey, P. J., and Pate, J. S. (1975). *Planta* **127**, 251-262.
Sharkey, P. J., and Pate, J. S. (1976). *Planta* **128**, 63-72.
Wallace, W., and Pate, J. S. (1965). *Ann. Bot.* **29**, 654-671.

CHAPTER 3

Accumulation of Seed Reserves of Nitrogen

DAVID R. MURRAY

I.   Albumins 84
II. Lectins 86
     A. Properties and Relationships 88
     B. Intracellular Locations 95
     C. Functions 95
III. Seed Proteinase Inhibitors 98
     A. Properties and Relationships 98
     B. Activities and Intracellular Locations 101
     C. Functions 102
IV. Timing of Reserve Protein Synthesis 104
V.   Protein Accumulation in Protein Bodies 107
     A. The Endoplasmic Reticulum 108
     B. The Vacuole 109
     C. The Dictyosome (Golgi Apparatus) 114
VI. Processing of Seed Reserve Proteins 117
     A. Removal of the Signal Sequence 118
     B. Proteolytic Activity of Protein Bodies 118
     C. Glycosylation 125
VII. Conclusions 126
     References 127

The main nitrogenous reserves of seed are proteins, and in certain species, various non-protein amino acids. Non-protein amino acids are toxic and are considered in this context (Volume 1, Chapter 7). It has been argued that storage proteins of seeds are only those major proteins localized in protein bodies (Millerd, 1975). This view would exclude most albumins (Section I), some lectins (Section II) and proteinase inhibitors (Section III).

I have taken a functional approach. Reserve proteins in seeds are those the storage tissue is genetically capable of manufacturing in amounts sufficient to constitute a useful reserve of nitrogen and sulphur. Any protein may function as a reserve if it is degraded and its component amino acids reutilized, regardless of its initial location within the cell.

I. ALBUMINS

For almost 30 years the predominant view of albumins has been that they are enzymes or 'metabolic proteins'. This view became established following the work of Danielsson (1956) on pea (*Pisum sativum*). It has gradually become evident that enzymes in seed tissues are not confined to the albumin fraction. Furthermore, it has been shown that the albumins of pea cotyledons are used as a reserve during and after germination, whether the albumin fraction is regarded as the proteins soluble at 70% of saturation with ammonium sulphate (Basha and Beevers, 1975; Collier and Murray, 1977) or the proteins soluble in water (Murray, 1979a). Studies on the protein constituents of oil seeds (Youle and Huang, 1978a, b; 1979; 1980) also lead to the conclusion that albumins constitute a widespread and neglected category of reserve protein.

The albumins of castor bean (*Ricinus communis*) endosperm are identical with the allergens from this species, are localized in the matrix of the protein bodies, and are utilized after germination (Youle and Huang, 1978a, b). The albumin fraction of pea seeds includes allergens (Malley *et al.*, 1975) and proteins that are distinct antigens (Simola, 1969; Guldager, 1978), but the albumins from legumes are not localized inside protein bodies (Varner and Schidovsky, 1963; Millerd *et al.*, 1978). Albumins stored outside protein bodies may be among the first proteins degraded (Volume 2, Chapter 7).

Albumins frequently account for 30% to 40% of extractable seed protein, but up to 50% has been reported for pea (Murray and Vairinhos, 1982a) and 62% for sunflower (*Helianthus annuus*) (Youle and Huang, 1981). A number of wild species yield the major proportion of seed protein as albumins, for example, *Lupinus cosentinii* (Varasundharosoth, 1982).

The potential of the albumin fraction for improving the quality of seed protein is related to its higher content of the essential sulphur-containing amino acid methionine and its substitute cysteine, compared to major globulins from the same species. Bajaj *et al.* (1971) found that the protein efficiency ratio for pea cultivars ranged from 0.46 to 2.20 gram weight gain per gram of protein eaten by young rats. This ratio was strongly correlated with seed albumin content, and linked to its sulphur amino acid content

3. *Nitrogenous Reserves* 85

rather than to lysine. In total albumins, higher contents of either methionine or cysteine or both sulphur amino acids compared to globulins have been reported often for pea (Goa and Strid, 1959; Hurich *et al.*, 1977; Murray, 1979a), and also for cowpea (*Vigna unguiculata*) (Carasco *et al.*, 1978), white lupin (*Lupinus albus*) (Duranti and Cerletti, 1979), chickpea (*Cicer arietinum*; Table VIII of Chapter 1) and a range of oil seeds (Youle and Huang, 1978a, 1981).

The heterogeneity of the albumin fractions is well illustrated in Table I, which shows the amino acid compositions of seven of the sixteen different albumins purified from seeds of *Lupinus angustifolius* by Varasundharosoth (1982). Nine of these albumins had zero or trace methionine content, with other values ranging from 0.16 to 2.90 (albumin 4). Three had no cysteine, with other values ranging from 0.17 to 6.09 (e.g. albumin 1). Of the three albumins with no cysteine, two also had no methionine (e.g., albumin 5). Albumin 1, with the highest cysteine content, is devoid of methionine and deficient in most of the essential amino acids, resembling the double-headed

Table I. Amino acid composition of seven albumins isolated from seeds of *Lupinus angustifolius* cv. Uniwhite. (mole %).

Reference no. MW × 10^{-3} dal. Mobility	1 24.8 high	2 n.d.	3 42.1 medium	4 40.5	5 142.7	6 53.8 low	7 394
Amino acid							
Aspartic	8.26	6.69	10.67	9.93	13.28	11.56	8.40
Glutamic	34.20	11.60	11.36	10.30	19.75	20.25	10.42
Threonine[a]	0.43	5.20	6.70	6.58	3.55	3.82	6.21
Serine	7.04	3.51	3.66	4.83	6.92	5.29	5.18
Glycerine	10.25	11.30	15.59	11.86	7.71	12.73	11.45
Alanine	1.96	8.05	7.84	6.21	5.01	5.40	9.09
Valine[a]	1.38	8.58	7.01	9.15	4.50	4.23	7.04
Cysteine	6.09	0.40	0.35	0.48	0.00	1.28	2.06
Methionine[a]	0.00	1.95	0.00	2.90	0.00	0.16	2.42
Isoleucine[a]	3.06	3.31	4.35	3.72	5.63	3.67	3.32
Leucine[a]	8.08	6.39	8.14	7.03	7.88	7.14	7.04
Tyrosine[a]	0.95	tr	tr	3.42	4.39	2.02	4.05
Phenylalanine[a]	1.56	4.14	5.31	3.61	4.67	3.89	4.35
Tryptophane[a]	0.08	n.d.	n.d.	n.d.	0.00	0.74	0.72
Histidine	1.99	5.63	0.87	3.20	1.24	1.54	2.09
Arginine	8.08	2.72	4.75	3.90	6.92	7.54	4.18
Lysine[a]	2.57	14.68	7.05	7.14	4.11	3.23	6.07
Proline	4.13	5.80	6.36	5.73	4.45	6.25	6.64

[a] Considered essential for adults.
Data of Varasundharosoth (1982).

type of proteinase inhibitor. The albumins offering the best internal balance of essential amino acids are 7 and possibly 4 (Table I).

Jakubek and Przybylska (1979) separated four main fractions of albumins from five distinct lines of pea. Their S3 fraction (components of MW 17 000 daltons) was the richest in cysteine (3.38% to 4.75%). Their suggestion that this fraction may contain proteinase inhibitors is consistent with the known properties of double-headed trypsin inhibitors from pea (Weder and Hory, 1972a, b). Dimers of this type of inhibitor are very stable, even in sodium dodecyl sulphate (SDS) (e.g., Gennis and Cantor, 1976a) and the MW 8000-17 000 dalton series observed in pea albumins (Table II) correspond to the sizes expected for monomer and dimer respectively. The 23 000-25 000 dalton polypeptides would be homologous with the Kunitz inhibitor (Section III).

A consequence of sulphur deficiency in peas is the diminished synthesis of albumins and legumin compared to vicilin (Randall *et al.*, 1979). Legumin is substantially richer in sulphur (S) amino acid content than vicilin (Jackson *et al*, 1969). The slower synthesis of legumin in S-deficient pea seeds is clearly linked to a much reduced synthesis of the corresponding mRNA species (Chandler *et al.*, 1983). However, even under ideal conditions, there appear to be genetic restrictions on the incorporation of sulphur amino acids into reserve proteins in pea. Schroeder (1982) found that the seed albumin content was inversely related to the content of legumin (Fig. 1). Davies (1980) has related low legumin content to the $r_a r_a$ genotype for wrinkled seed (distinguished in Fig. 1).

Plant breeders may ultimately have to accept that for any species there is an optimal mixture of albumins that will coexist with a set of globulins or other proteins but it would be premature to dismiss the possibility of selecting for one or a few nutritionally well-balanced polypeptides. It may be possible to alter genotypes in favour of increasing the duration of periods of maximum net synthesis displayed by desirable albumins during seed development (Section IV).

II. LECTINS

The two groups of proteins stored in seeds that have excited most interest and received the most detailed study are the lectins and the proteinase inhibitors. Representatives of these two groups were the first reserve proteins from seeds to be fully sequenced by direct techniques (Tan and Stevens, 1971a, b; Edelman *et al.*, 1972), before the faster techniques of DNA and mRNA base sequencing were established. Much more is known

Table II. Albumins corresponding in estimated molecular weight to proteinase inhibitors.

Species	MW estimate ($\times 10^{-3}$ daltons)		References
Pisum sativum	23	19, 18	Murray (1979a, c)
	23	18	Murray and Vairinhos (1982a)
	25	15.5	Grant *et al.* (1976)
	23	17	Jakubek and Przybylska (1979)
	22	8	Schroeder (1982)
Cicer arietinum	24	21, 18	Vairinhos and Murray (1983)
Vigna unguiculata	22.5		Khan *et al.* (1980)
	25	15	Murray *et al.* (1983)
Edwardsia microphylla	24	20,19,17	Murray (1979b)

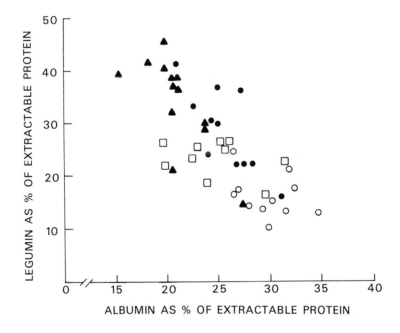

Fig. 1. The relationship between seed content of albumin and legumin in 45 lines of pea: ▲, wild forms; □, field peas; ●, round-seeded garden peas; ○, wrinkled-seeded garden peas. Correlation coefficient $r = -0.757$ (from Schroeder, 1982).

about their properties than about any other categories of seed protein. Accordingly, they are a useful source of information about the evolutionary origins of reserve proteins.

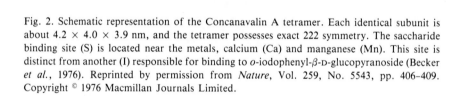

Fig. 2. Schematic representation of the Concanavalin A tetramer. Each identical subunit is about 4.2 × 4.0 × 3.9 nm, and the tetramer possesses exact 222 symmetry. The saccharide binding site (S) is located near the metals, calcium (Ca) and manganese (Mn). This site is distinct from another (I) responsible for binding to o-iodophenyl-β-D-glucopyranoside (Becker et al., 1976). Reprinted by permission from *Nature*, Vol. 259, No. 5543, pp. 406–409. Copyright © 1976 Macmillan Journals Limited.

A. Properties and Relationships

The term 'lectin' was devised by Boyd and Shapleigh (1954) to describe haemagglutinins, but current usage is broader and does not imply that the compound referred to is a haemagglutinin. Lectins are proteins or glycoproteins characterized by their ability to bind particular sugar residues that belong to polysaccharide moieties of glycoproteins, glycolipids, polysaccharides, or simple glycosides. Their most spectacular effects, the agglutination of erythrocytes, or the stimulation of mitosis in resting lymphocytes are attributed to their selective binding properties.

Agglutination of red blood cells is achieved by dimeric or tetrameric association of polypeptide subunits, generally with one binding site per subunit, as shown for concanavalin A (Con A) from jack bean (*Canavalia*

3. *Nitrogenous Reserves* 89

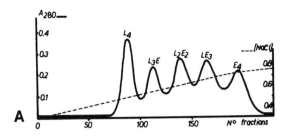

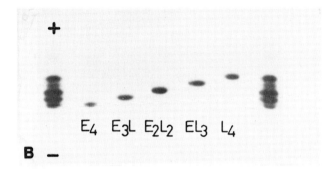

Fig. 3. Five isolectins from seeds of bean (*Phaseolus vulgaris* cv. Contender) separated by ion exchange chromatography on SP-Sephadex using 0.05 M acetate buffer, pH 3.6, with increasing content of NaCl as shown (A). Each fraction yields a single band of distinct mobility on electrophoresis at pH 4.5 (B). The two subunit types are designated E for erythrocyte agglutination and L for lymphocyte stimulation (from Manen and Pusztai, 1982).

ensiformis) (Fig. 2). Tetramers of non-identical subunits are also known. This subunit heterogeneity was first established for a set of isolectins from bean (*Phaseolus vulgaris*) (Miller et al., 1973, 1975). Mitogenic activity resides in a distinct type of subunit that lacks agglutinating activity. Conversely, an agglutinating subunit is not mitogenic. Thus five kinds of tetramer are formed, each with different proportions of the two subunit types (Fig. 3). A similar set of isolectins has been found in *Bandeiraea simplicifolia* (type I, Hayes and Goldstein, 1974; Murphy and Goldstein, 1977).

The progress of the characterization of lectins and their application in typing of blood cells and affinity chromatography may be followed in reviews by Boyd (1963), Liener (1963; 1976), Lis and Sharon (1973; 1982), Toms and Western (1971) and Toms (1981).

```
 1                            10                                    20
Ala-Asp-Thr-Ile-Val-Ala-Val-Glu-Leu-Asp-Thr-Tyr-Pro-Asn-Thr-Asp-Ile-Gly-Asp-Pro-

21                            30                                    40
Ser-Tyr-Pro-His-Ile-Gly-Ile-Asp-Ile-Lys-Ser-Val-Arg-Ser-Lys-Lys-Thr-Ala-Lys-Trp-

41                            50                                    60
Asn-Met-Gln-Asp-Gly-Lys-Val-Gly-Thr-Ala-His-Ile-Ile-Tyr-Asn-Ser-Val-Asp-Lys-Arg-

61                            70                                    80
Leu-Ser-Ala-Val-Val-Ser-Tyr-Pro-Asn-Ala-Asp-Ala-Thr-Ser-Val-Ser-Tyr-Asp-Val-Asp-

81                            90                                   100
Leu-Asn-Asp-Val-Leu-Pro-Glu-Trp-Val-Arg-Val-Gly-Leu-Ser-Ala-Ser-Thr-Gly-Leu-Tyr-

101                           110                                  120
Lys-Glu-Thr-Asn-Thr-Ile-Leu-Ser-Trp-Ser-Phe-Thr-Ser-Lys-Leu-Lys-Ser-Asn-Ser-Thr-

121                           130                                  140
His-Gln-Thr-Asp-Ala-Leu-His-Phe-Met-Phe-Asn-Gln-Phe-Ser-Lys-Asp-Gln-Lys-Asp-Leu-

141                           150                                  160
Ile-Leu-Gln-Gly-Asp-Ala-Thr-Thr-Gly-Thr-Asp-Gly-Asn-Leu-Glu-Leu-Thr-Arg-Val-Ser-

161                           170                                  180
Ser-Asn-Gly-Ser-Pro-Glu-Gly-Ser-Ser-Val-Gly-Arg-Ala-Leu-Phe-Tyr-Ala-Pro-Val-His-

181                           190                                  200
Ile-Trp-Glu-Ser-Ser-Ala-Thr-Val-Ser-Ala-Phe-Glu-Ala-Thr-Phe-Ala-Phe-Leu-Ile-Lys-

201                           210                                  220
Ser-Pro-Asp-Ser-His-Pro-Ala-Asp-Gly-Ile-Ala-Phe-Phe-Ile-Ser-Asn-Ile-Asp-Ser-Ser-

221                           230                          237
Ile-Pro-Ser-Gly-Ser-Thr-Gly-Arg-Leu-Leu-Gly-Leu-Phe-Pro-Asp-Ala-Asn
```

Fig. 4. The amino acid sequence of Concanavalin A (from Cunningham *et al.*, 1975).

1. Concanavalin A

This lectin was purified as one of a number of globulins from jack bean seed (Sumner, 1919) and only later identified as a haemagglutinin (Sumner and Howell, 1936). Its physical and chemical properties have now been studied in such detail that Con A provides a reference for all comparative studies. The amino acid sequence has been determined in full (Edelman *et al.*, 1972; Wang *et al.*, 1975; Cunningham *et al.*, 1975). The polypeptide comprises 237 amino acid residues (Fig. 4). Complementary X-ray crystallographic studies (Becker *et al.*, 1975, 1976; Reeke *et al.*, 1975) have revealed the shape of the molecule. A feature of the polypeptide chain is a pair of antiparallel β-pleated sheets. Amino acids important for the non-covalent association of two or four identical subunits (Fig. 2) have been identified in the regions 117 to 139 and 175 to 178 (dimer) and at positions 114, 116

3. Nitrogenous Reserves 91

and 192 (tetramer). Con A contains no cysteine, in agreement with earlier observations (Olsen and Liener, 1967). The number of methionines also varies with species: Con A from *Canavalia maritima* lacks the methionine at position 129, retaining that at position 42 (Hague, 1975).

The metal ions that are essential for specific saccharide binding are held by acidic residues close to the N-terminus (positions 7-25, Becker *et al.*, 1976; Reeke *et al.*, 1978). This specific site (S in Fig. 2) is about 1.3 nm from the manganese ion (Becker *et al.*, 1976) and is distinct from a second binding site for *o*-iodo-phenyl-β-D-glucopyranoside in hydrophobic pockets (I in Fig. 2). The specificity of binding at the metal-dependent site has been reviewed by Goldstein *et al.* (1972). Terminal non-reducing α-D-mannopyranosyl and α-D-glucopyranosyl residues and internal 2-o linked mannosyl residues are preferred.

2. Have Lectins Evolved from Glycosidases ?

Despite the variety of carbohydrate binding specificities displayed by seed lectins (Table III), there are several types of evidence that indicate they are

Table III. Major saccharide binding specificity of seed lectins.

Sugar residue	Lectin source and identity	Reference
Mannose or glucose	*Canavalia ensiformis* (Con A)	Goldstein *et al.* (1973)
	Lens culinaris	Howard and Sage (1969)
	Pisum sativum	Bures *et al.* (1973)
	Vicia faba (favin)	Wang *et al.* (1974)
	Lathyrus odoratus	Sletten *et al.* (1983)
	Lathyrus sativus	Kolberg and Sletten (1982)
N-acetyl-glucosamine	*Bandeiraea simplicifolia* (II)	Shankar Iyer *et al.* (1976)
	Datura stramonium	Kilpatrick and Yeoman (1978)
	Triticum aestivum	Nagata and Burger (1974)
	Ulex europaeus (II)	Pereira *et al.* (1979)
Galactose	*Abrus precatorius*	Olsnes *et al.* (1974b)
	Arachis hypogaea	Baues and Gray (1977)
	Bandeiraea simplicifolia (I)	Hayes and Goldstein (1974)
	Phaseolus vulgaris (E$_4$)	Hankins *et al.* (1979)
	Ricinus communis	Nicholson *et al.* (1974)
	Styphnolobium japonica	Poretz *et al.* (1974)
N-acetyl-galactosamine	*Dolichos biflorus*	Etzler and Kabat (1970)
	Glycine max (soyin)	Lis *et al.* (1970)
	Phaseolus lunatus	Galbraith and Goldstein (1972)
	Wistaria floribunda	Kurokawa *et al.* (1976)
L-fucose	*Cytisus sessilifolius*	Matsumoto and Osawa (1974)
	Lotus tetragonolobus	Blumberg *et al.* (1972)
	Ulex europaeus (I)	Pereira *et al.* (1979)
	Vicia cracca	Sundberg *et al.* (1970)

Table IV. Seed lectins displaying glycosidase activity.

Species and activity	MW^a ($\times$ 10^{-3})	Reference
α-Mannosidase:		
Phaseolus vulgaris	220	Paus and Steen (1978)
α-Galactosidase:		
Vigna radiata	160	Hankins and Shannon (1978)
Phaseolus lunatus	180	Hankins et al. (1980)
Lupinus arboreus	190	Hankins et al. (1980)
Thermopsis caroliniana	180	Hankins et al. (1980)
Pueraria thunbergiana	150	Hankins et al. (1980)
Glycine max	160	del Campillo and Shannon (1982)
Vicia faba	160	Dey et al. (1982)

^{a} Values are estimates from gel filtration for tetrameric forms except for α-mannosidase: types I and II are both dimers.

a family of related globular proteins. Immunochemical evidence for close structural relationships among the seed lectins of Solanaceae has been obtained by using purified antibodies specific for the lectin from common thornapple (*Datura stramonium*) (Kilpatrick et al., 1980). Immunochemical studies involving most of the other lectins listed in Table III have been undertaken by Howard et al. (1979) and Hankins et al. (1979). Purified antibodies specific for eight different lectins were used to map the cross-reactivity of either purified lectins or crude extracts from seeds of 16 species. The levels of partial or complete identity obtained indicate close relationships between the lectins that bind galactose, N-acetyl galactosamine or L-fucose, even across plant families, whereas lectins from the mannose/glucose group did not cross-react with antibodies specific for lectins in the galactose grouping. This suggests a major structural difference in the polypeptide types comprising these two major classes of lectins.

A lectin from mung bean (*Vigna radiata*) with α-galactosidase activity (Hankins and Shannon, 1978; del Campillo et al., 1981) is structurally related to classic lectins that lack glycosidase activity (Hankins et al., 1979). Seed lectins shown to possess glycosidase activity are listed in Table IV. Those with α-galactosidase activity show considerable diversity in their properties, and the α-galactosidase lectin from soybean (*Glycine max*) (del Campillo and Shannon, 1982) is immunochemically distinct from soyin (Liener, 1953), the non-enzymic agglutinin from soybean (Fig. 5).

3. Sequence Homology with Concanavalin A
The lectins from several members of Vicieae all possess similar saccharide specificity to Con A (Table III), but consist of small α chains (MW 5700

Fig. 5. Double diffusion tests for the α-galactosidase agglutinin from soybean (wells 3 and 4) against antisera raised against the α-galactosidase agglutinin from mung bean (well 1) and soyin, the classic soybean agglutinin (well 2). Homologous reactions were shown by placing the antigens in well 5 (mung bean α-galactosidase agglutinin) and 6 (soyin) (from del Campillo and Shannon, 1982).

to 6000 daltons) and β chains (MW 20 000 daltons or less) associating non-covalently as $(\alpha\beta)_2$. Partial N-terminal amino acid sequences of lectin polypeptides from several different species, including some with α and β chains, first revealed the homology of the β chains with part of the Con A sequence, beginning unexpectedly at residue 123 (Foriers et al., 1977a, b; 1978). The α chain from lentil (*Lens culinaris*), the first to be fully sequenced, was found to comprise 52 amino acids and to correspond to a section of the Con A sequence fitting in before the N-terminus of the β subunit (Foriers et al., 1978). The α chain from favin, the lectin from broad bean (*Vicia faba*), comprises 51 amino acids (Hemperly et al., 1979). Compared to Con A, these α subunits are homologous with positions 70 to 120 (favin) and 70 to 121 (lentil) (Cunningham et al., 1979; Foriers et al., 1981). The β subunit of favin corresponds not only to the carboxyterminal portion of the Con A chain, but extends to match residues 1 to 69 of the N-terminal region as well (Cunningham et al., 1979; Fig. 6). The β chain of favin contains 182 amino acid residues and is glycosylated at the asparagine in position β 168 (Hopp et al., 1982). The β subunit of the lentil lectin comprises only 159 residues and terminates with the residue corresponding to position 45 of Con A (Foriers et al., 1981).

During the evolution of these lectins from a common ancestral form, there has obviously been rearrangement of sequences within the gene. It is by no means certain that Con A is closer to the ancestral type.

The lectins with the subunit structure $(\alpha\beta)_2$ do not form associations equivalent to the Con A tetramer. This is because the lysines in positions that correspond to 114 and 116 in Con A have been replaced by glutamine and glycine in the lentil α chain, by glutamate and threonine in the favin

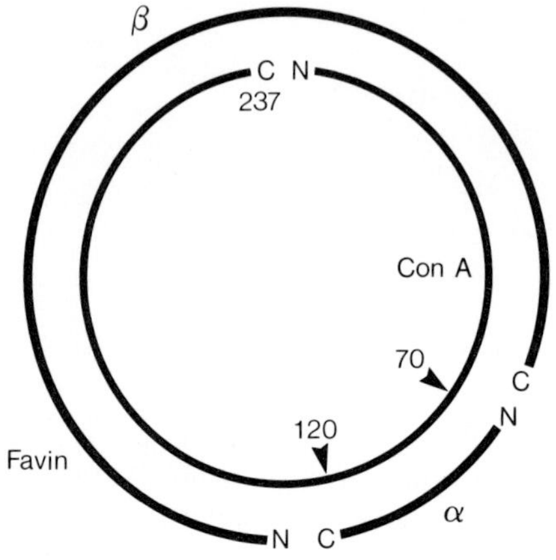

Fig. 6. Alignment of the α and β subunits of favin against Concanavalin A (after Hopp et al., 1982).

Table V. Estimates of the proportion of seed protein accounted for by lectins.

Species and source	Percentage	Reference
(i) from endosperm		
Mucuna flagellipes	14	Mbadiwe and Agogbua (1978)
Ricinus communis	5	Tully and Beevers (1976)
(ii) embryo (cotyledons)		
Canavalia ensiformis	23	Hague (1975)
C. gladiata	28	Hague (1975)
C. maritima	32	Hague (1975)
Abrus precatorius	30	Murray and Vairinhos (1982c)
Phaseolus vulgaris	15	Bollini and Chrispeels (1978)
P. vulgaris	10	Pusztai and Watt (1974)
Styphnolobium japonica	8.5	Poretz et al. (1974)
Psophocarpus tetragonolobus	6	Kortt (1984)
Vicia faba	1	Weber et al. (1978)

α chain, and by glutamate and serine in the pea lectin α chain. The glutamate at position 192 of Con A that would normally interact with these lysine residues on other chains (above pH 5.6) has also been replaced by threonine at position 65 of the favin β chain (Hopp et al., 1982) and by valine in the β chain of both the pea and lentil lectins.

The lectins that have been fully sequenced, as well as the isolectins of *Phaseolus vulgaris* do not contain cysteine, whereas many others do. Wheat germ agglutinin has a high content of cysteine, involved in intramolecular disulphide bridges (Nagata and Burger, 1974). The main lectins of *Abrus precatorius* and castor bean consist of disulphide-linked pairs of similar subunits (Olsnes et al., 1974b) containing up to 10% cysteine (Nicolson et al., 1974). A free thiol group is essential for the activity of enzymic lectins (Hankins and Shannon, 1978). This diversity of subunit types often revealed within a single species is evidence for gene duplication.

B. Intracellular Locations

The proportion of seed protein accounted for by lectins varies from very minor to at least 30% (Table V). There are many reports of lectins in protein body preparations, but whether their associations with protein bodies are intrinsic or adventitious has been unclear. Mialonier et al. (1973) used a histochemical method that indicated a cytoplasmic location for lectins in bean cotyledon cells. Clarke et al. (1975), using immunofluorescent techniques, found that the main lectins from many legumes were cytoplasmic and prominent in bounding membranes of protein bodies and starch grains. Similar procedures indicate an association of lectin with plasmalemma and cytoplasm in cotyledon cells of common thornapple (Kilpatrick et al., 1979). If ethylene diamine tetra acetate (EDTA) is used during extraction, the proportion of lectin associated with pelletable material is much reduced, consistent with a main location in membranes of cytoplasmic organelles or cytoplasm (Kilpatrick et al., 1979).

By comparison, the lectins of castor bean have been identified among the major components of the matrix portion of protein bodies (Tully and Beevers, 1976). Ultrastructural studies of the localization of soyin show that this lectin is present mainly inside protein bodies (Horisberger and Vonlanthen, 1980). Similarly, Manen and Pusztai (1982) have shown that isolectins of bean (Fig. 3) are mainly localized inside protein bodies of cotyledon storage parenchyma (Fig. 7), but in some vascular cells and in parenchyma from the embryonic axis, these lectins are cytoplasmic.

C. Functions

What support is there for the claim that lectins are reserve proteins ? Firstly, they may occupy a considerable proportion of total seed protein (Table V). Secondly, they often occur inside protein bodies. Finally and conclusively, their main fate following germination is to be degraded and their constituent amino acids reutilized, as seen in lentil (Rougé, 1974) and bean (Bollini and Chrispeels, 1978; Murray, 1982).

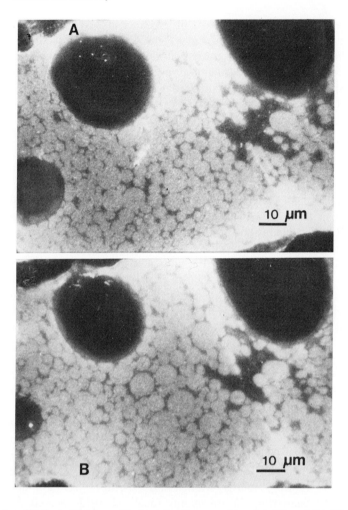

Fig. 7. Localization of lectins with the subunit composition L4 (A) and E4 (B) in bean (*Phaseolus vulgaris*) cotyledons cells by indirect immunofluorescence. Sections were first incubated with the respective IgG antibodies, followed by fluorescent anti-IgG (tetramethyl rhodamine isothiocyanate-labelled goat anti-rabbit IgG) (from Manen and Pusztai, 1982).

Are there no more subtle functions for a class of proteins with such striking properties? They could well participate in the incorporation of glycoproteins into protein bodies with their capacity for cross-linking. Their location inside protein bodies and their demonstrated interactions with seed glycoproteins from the same species (Basha and Roberts, 1981) would seem to enhance this possibility.

However, it has been established for bean (Brücher *et al.*, 1969; Klozová and Turková, 1978), *Pisum elatius* W1293 (Guldager, 1978) and soybean that individual plants lacking seed lectins can survive and reproduce successfully. Inheritance of soyin, the main seed lectin in soybean, is Mendelian: the double recessive condition at a single locus (*le le*) prevents synthesis of the lectin polypeptide (Orf *et al.*, 1978). Protein bodies form readily in the cotyledons of cowpea (Harris and Boulter, 1976), a species with no detectable lectin (Janzen *et al.*, 1976), indicating clearly that reserve glycoproteins are capable of independent assembly.

In some instances, lectins may be responsible for the interaction of seedling root cells with *Rhizobium* (Clarke and Knox, 1978; Toms, 1981). This cannot be generally true, as soybeans lacking soyin are still capable of nodulation with *R. japonicum* (Pull *et al.*, 1978). Cowpea also nodulates effectively. The wide (but not universal) distribution of seed lectins suggests a role such as defence, appropriate against general or specific insect predators (Janzen *et al.*, 1976), or pathogenic fungi (Mirelman *et al.*, 1975). A secondary storage function following germination would be simply economic and entirely compatible with a protective role.

The occurrence of phytotoxins that share a subunit with the major agglutinins from the same sources lends weight to the idea that a major role of lectins is defence. The most potent phytotoxins known are abrin from *Abrus precatorius* (LD_{50} for mice 40 ng) and ricin from castor bean (LD_{50} for mice 100 ng, Olsnes *et al.*, 1974a). Each toxin consists of a toxic A chain linked by covalent (disulphide) bonds to a galactose-binding B chain (Olsnes and Phil, 1973; Olsnes *et al.*, 1974a). The B chain is the common subunit with the main lectins from these species (Olsnes *et al.*, 1974b). The A chain differs from the B chain in many respects (Table VI) and lacks the capacity

Table VI. Properties of abrin, ricin, and their constituent polypeptides.

Toxin	Estimated MW (daltons)	pI	Carbohydrate (% w/w)
Abrin	65 000	6.1	
Abrin A chain	30 000	4.6	0
Abrin B chain	35 000	7.2	7.4
Ricin	65 000	7.1	
Ricin A chain	32 000	7.5	2.4
Ricin B chain	34 000	4.8	6.5

Data from Olsnes *et al.* (1974a). Reprinted by permission from *Nature*, Vol. 249, No. 5458, pp.627–631. Copyright © 1974 Macmillan Journals Limited.

to bind sugars. It is such a strong inhibitor of protein synthesis *in vitro*, however, that it is believed to act catalytically against ribosomes.

III. SEED PROTEINASE INHIBITORS

The evolution of the Angiosperms has been accompanied by a remarkable proliferation of seed proteins capable of inhibiting proteolytic enzymes from diverse alien sources. Weder (1981) lists 41 enzyme activities inhibited by legume inhibitors alone. The enzymes inhibited are mostly endopeptidases that depend upon a serine residue in the active site and include not only trypsin and chymotrypsin from mammals, but the trypsin-like enzymes of insect larvae and proteinases from fungi (e.g., *Aspergillus oryzae* in Kirsi and Mikola, 1971; *Colletotrichum lindemuthianum* in Mosolov *et al.*, 1979).

A. Properties and Relationships

Studies that led to the isolation of two distinct kinds of trypsin inhibitor from soybean seeds are described by Kunitz (1945; 1946; 1947) and Bowman (1944; 1946). Many subsequent studies are reviewed by Birk (1976a).

The Kunitz inhibitor of MW 25 000 daltons has a single active site that involves an arginine residue (Bidlingmeyer *et al.*, 1972) located at position 63. There are four cysteine residues forming two disulphide bridges within the whole chain (positions 39–86 and 136–145), which comprises a total of 181 amino acids (Koide *et al.*, 1972). It is possible that many of the inhibitors from cereals (Birk, 1976c) are related to the Kunitz inhibitor, but for most species crucial information is lacking. Homologous inhibitors are known from silk tree (*Albizzia julibrissin*) (Odani *et al.*, 1979, 1980), winged bean (*Psophocarpus tetragonolobus*) (Kortt, 1979, 1980, 1981) and cedar wattle (*Acacia elata*) (Kortt and Jermyn, 1981) and appear to be widely distributed (Weder, 1978; Weder and Murray, 1981). The chymotrypsin inhibitor of winged bean is specific for chymotrypsin at two sites, not one (Kortt, 1980, 1981).

Another inhibitor from lentil is similar to the Kunitz inhibitor in size, but has two binding sites, one specific for trypsin, the other for chymotrypsin (Weder *et al.*, 1983). The cysteine content of this inhibitor is exceptionally high, 30 residues per molecule.

The inhibitors from Mimosoideae can be separated into two polypeptide chains on reduction of disulphide linkages, suggesting the evolution of a peptide site susceptible to cleavage during processing (Section VI.B).

The inhibitor described by Bowman (1946) and Birk (1961) is

commonly referred to as the Bowman–Birk inhibitor. It has a smaller subunit size, about MW 8000 daltons. The amino acid sequence obtained by Odani and Ikenaka (1972) is included in Figure 8. Tryptophane is absent, and the proportions of aspartate, serine and cysteine are high. The cysteine residues are all combined in disulphide linkages within the molecule. There are two distinct inhibitory sites per molecule (Birk *et al.*, 1967; Seidl and Liener, 1971) on small loops formed by two of the disulphide bridges (Fig. 8). The first site is specific for trypsin and depends on a lysine residue, the second is specific for chymotrypsin and depends on a leucine residue. These sites are identical in the related Lima bean (*Phaseolus lunatus*) inhibitors I and IV (Fig. 8), but inhibitor IV' has phenylalanine substituting in the second site, like inhibitor 3 (PVI 3) from bean (Belitz and Fuchs, 1973).

The independence of these inhibitory sites was fully established in studies with active fragments from various *Phaseolus* inhibitors (Krahn and Stevens, 1970; Weder, 1973) and from the Bowman–Birk inhibitor (Odani and Ikenaka, 1973; 1978b). However, the correlation of site 1 for trypsin, with site 2 for chymotrypsin was not upheld in the type II inhibitors from *P. vulgaris* (GB II and GB II') of Wilson and Laskowski Sr. (1973). These inhibitors did not inhibit chymotrypsin, but inhibited elastase as well as trypsin (Wilson and Laskowski Sr., 1975). The trypsin inhibitory site proved to be not the first, but the second, which contained arginine, whereas in the first (elastase specific) site, alanine replaced lysine. Arginine also appeared to be involved in the active site or sites of the inhibitors from peanut (*Arachis hypogaea*) (Hochstrasser *et al.*, 1970; Birk, 1976b).

Gerstenberg *et al.* (1980) have reviewed the properties of the isoinhibitors of bean. Besides these differences in active site specificity, differences are mainly evident in the N-terminal and C-terminal chain lengths. The number of amino acid residues per chain ranges from 74 to 86, with MW in the range 8100 to 9400 daltons. There is always no tryptophane and sometimes no valine, methionine or phenylalanine. There are always 14 cysteine residues, a number commonly found in the Papilionaceae (Fig. 8; Gennis and Cantor, 1976b).

The discovery of trypsin inhibitors from soybean with arginine in both sites (Odani and Ikenaka, 1976, 1978a; Hwang *et al.*, 1977) has special significance. It provides tangible evidence for the long-held suspicion that duplication of an ancestral gene gave rise to the first double-headed inhibitor, which diversified with subsequent mutations. To reconstruct what might have happened in terms of single base changes, CGG may be nominated as the likely codon for arginine (Table VII). Since the first two bases for the four possible alanine codons are GC, a minimum of two base changes must be invoked for the transition from arginine (or lysine) to alanine in site 1. Despite this difficulty, the gene duplication hypothesis is attractive, offering a distant link with the Kunitz inhibitor and its relatives (Odani *et al.*, 1980).

Since most species studied possess sets of isoinhibitors, it seems very likely that gene reduplication has also occurred, thereby providing a mechanism for perpetuating the now evident diversity in substrate specificities.

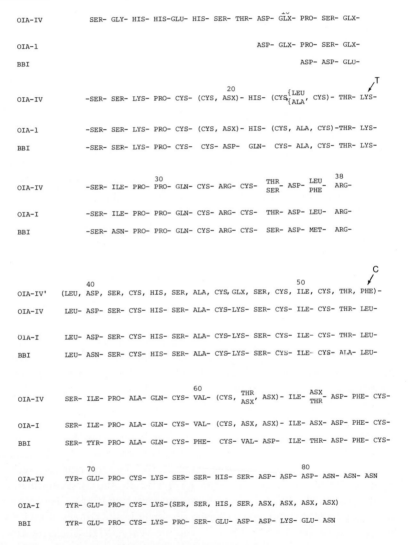

Fig. 8. Amino acid sequences of double-headed proteinase inhibitors from Lima bean (OIA-I, OIA-IV, OIA-IV′) elucidated by Tan and Stevens (1971a, b) and Stevens *et al.* (1974) compared with the Bowman–Birk inhibitor (BBI) from soybean, determined by Odani and Ikenaka (1972). T, site of trypsin binding; C, site of chymotrypsin binding.

Table VII. Single base mutations in mRNA codons accounting for most of the observed alternatives to arginine in the active sites of double-headed proteinase inhibitors.

	Site 1	Site 2	
arginine	CGG	CGG	arginine
arginine	AGG	CUG	leucine
lysine	AAG	UUG or CUC	leucine
		UUC	phenylalanine
		UAC	tyrosine

B. Activities and Intracellular Locations

Proteinase inhibitors are found in the cotyledons and axis of embryos adapted for storage, and in both the embryo and endosperm of cereal grains (Mikola and Enari, 1970; Kirsi and Mikola, 1971). Estimates of the proportion of total seed protein accounted for by the inhibitors are rarely given, but do not exceed 10% (barley (*Hordeum vulgare*), Mikola and Kirsi, 1972). For common bean, an estimate of 2.2 mg/g seed weight has been made (Pusztai, 1972).

Some soybean lines possess a double recessive condition (*ti ti*) resulting in the absence of the Kunitz inhibitor (Orf and Hymowitz, 1979). In such instances, 50% to 70% of the normal trypsin inhibitory activity remains, which can be attributed to all the double-headed inhibitors. Double-headed inhibitors total 6.3% of protein extractable from oil-free meal (Hwang *et al.*, 1978), with the Bowman-Birk inhibitor accounting for more than half this value. In contrast to soybean, the winged bean is a species with most of its proteinase inhibitors of the Kunitz type, specific either for trypsin (Kortt, 1979) or chymotrypsin (Kortt, 1980, 1981). Together these inhibitors comprise 4.5% of seed protein.

In surveys of all three sub-families of legumes, Weder (1978, 1981) has measured inhibitory activities against trypsin and chymotrypsin. The highest activities against trypsin (as mg inhibited per g of seed meal) are those of *Machaerium* sp. (31.9), *Robinia pseudoacacia* (27.3), *Elephantorrhiza burkei* (25.2) and *Lathyrus sativus* (21.5). Similarly, the highest recorded activities against chymotrypsin are those of *Robinia pseudoacacia* (15.2), winged bean (13.4) and *Tamarindus indica* (13.2).

As judged by activity against trypsin, proteinase inhibitors from legumes may be distributed mainly in the globulin fraction (e.g., *Cordeauxia edulis*) or mainly in the albumin fraction (e.g., hyacinth bean (*Lablab purpureus*), Miège and Miège, 1978). They are evidently not components of protein bodies, at least in those species for which this has been studied. Hobday *et al.* (1973) concluded that the association of trypsin inhibitors with protein bodies from pea cotyledons was largely adventitious, because the inhibitory activity was selectively removed during washing of the isolated protein bodies, and could be re-attached. Using rhodamine-labelled antibodies, Chrispeels and Baumgartner (1978) showed that the main location of a low molecular weight trypsin inhibitor in mung bean cotyledons is the cytoplasm. For this species, less than 5% of the total trypsin inhibitory activity was ever associated with the protein body-mitochondrial band in sucrose gradients. The characteristics of trypsin inhibitors of cultivated legumes generally indicate that they are cytoplasmic (Miège et al., 1976; Pusztai et al., 1977), although the proteolytic cleavage of inhibitors from *Albizzia* and *Acacia* that are otherwise homologous with the Kunitz inhibitor (Section III,A) suggests that they will be found in vacuolar protein bodies (Section VI). A major proteinase inhibitor from leaf cells accumulates in vacuolar protein bodies (Shumway et al., 1976).

C. Functions

Seed proteinase inhibitors have been considered to fulfil three possible functions: as storage proteins; as regulators of endogenous proteolytic enzymes; or as deterrents towards insect larvae or pathogenic microorganisms (Ryan, 1973). There is abundant evidence that they are mobilized after germination. They disappear from the endosperm of barley (Kirsi and Mikola, 1971; Kirsi, 1974) and maize (*Zea mays*) (Melville and Scandalios, 1972), many of them within the first few days. Pusztai (1972) reported some complex changes for individual inhibitors in the cotyledons of common bean, with some increasing in concentration before declining. The appearance of new forms during early growth of mung bean seedlings results from limited proteolysis of peptide bonds adjacent to aspartyl residues (Wilson and Chen, 1983). Compared to values obtained for fully imbibed cotyledons, the total content of proteinase inhibitors declines as proteins are mobilized from the cotyledons of pea (Hobday et al., 1973), serido bean (*Vigna unguiculata*) (Xavier-Filho, 1973), broad bean (Bhatty, 1977), soybean (Hwang et al., 1978) and mung bean (Chrispeels and Baumgartner, 1978). Slower rates of decline occur in the cotyledons of pumpkin (*Cucurbita pepo*) (Hara and Matsubara, 1980) and jojoba bean (*Simmondsia chinensis*) (Samac and Storey, 1981).

Quantitatively, there is little support for the view that proteinase inhibitors might regulate the activities of the endopeptidases that are responsible for reserve protein processing or breakdown in Angiosperm seeds. A number of possibilities warrant further investigation; for example, the relationship of a trypsin-like enzyme activity and endogenous inhibitors in lettuce (*Lactuca sativa*) seed (Shain and Mayer, 1965, 1968). Removal of trypsin inhibitors enhances the caseolytic activity of the albumin fraction from cowpea cotyledons (Royer, 1975). Gennis and Cantor (1976a) have described the release of an alkaline proteinase from an inhibitor complex from the same species. However, the major endopeptidase involved in reserve protein breakdown is a thiol-dependent enzyme with acidic pH optimum (Harris *et al.*, 1975), as in mung bean (Chrispeels and Boulter, 1975). The proteolytic activity of mung bean cotyledon extracts is not enhanced by the complete removal of trypsin inhibitors (Chrispeels and Baumgartner, 1978).

In Gymnosperms, an endogenous regulatory function may still be discerned. Proteinase inhibitors from the storage tissue (megagametophyte) of seeds of Scots pine (*Pinus sylvestris*) have been separated into fractions of MW 24 000, 14 600, 14 000 and 9000 daltons (Salmia, 1980). They do not inhibit trypsin or chymotrypsin, but are evidently specific for pine seed proteinases I and II, which are thiol-dependent enzymes formed *de novo* in the reserve tissue during germination and early seedling growth (Fig. 3 of Chapter 5, Volume 2). The major proportion of proteinase activity is initially the pepstatin-sensitive type that is unaffected by the endogenous inhibitors. With time, this proteinase activity occupies a decreasing proportion of total activity compared to proteinases I and II, which presumably become increasingly effective as the inhibitory activity declines.

Seed proteinase inhibitors of Angiosperms may have evolved originally from endogenous inhibitors with regulatory functions that also proved to have adaptive value as deterrents. There is considerable support for the idea that the Angiosperm inhibitors proliferated in response to insect predation and co-evolution (Applebaum, 1964; Ryan, 1973). Specificity for larval proteinases is evident; for example, the trypsin-like enzyme of *Tenebrio molitor* larvae is unaffected by a soybean inhibitor active against *Tribolium* proteinase (Birk *et al.*, 1963), but is inhibited by both the Kunitz and Bowman–Birk inhibitors (Applebaum *et al.*, 1964). Bruchid beetles such as *Callosobruchus maculatus*, the subjects of experiments by Janzen *et al.* (1976, 1977), have adapted to feeding on seeds rich in proteinase inhibitors (Applebaum, 1964).

It may well be that the effects of seed proteinase inhibitors on mammalian and avian proteinases are largely coincidental. When ingested by mammals, their anti-nutritional effects are often overshadowed by other factors (Liener and Kakade, 1980; Liener, 1980). Indeed, some soybean inhibitors have little or no effect on human trypsin and chymotrypsin (Coan and Travis, 1971).

As with lectins, it may be concluded that the proteinase inhibitors of seed storage tissues of most species fulfil a dual role, primarily in defence against insect predators and fungi, and secondly, as a reserve of both nitrogen and sulphur.

IV. TIMING OF RESERVE PROTEIN SYNTHESIS

In embryonic seed tissues destined for conversion to storage parenchyma, mitotic cell division ceases relatively early and gives way to endoreduplication of the nuclear DNA. Levels of DNA commonly found in cotyledons are 8C (peanut, Dhillon and Miksche, 1982), 16C (broad bean, Millerd and Whitfeld, 1973) and 64C (pea, Smith, 1973; Scharpé and Van Parijs, 1973). Endoreduplication is clearly distinguished from selective amplification as seen in suspensor cells (Nagl, 1974) and is not universal in differentiating plant tissues (Evans and Van't Hof, 1975; Nagl, 1979). The level of endoreduplication is correlated with cell size, but not protein content (Smith, 1973). Madison et al. (1976) compared cotyledons from pea, peanut, soybean and bean and showed that the final protein content is unrelated either to DNA or RNA content. Endosperm cells may also undergo endoreduplication (Section 1.I,C).

Once endoreduplication has begun, an early peak in nucleolar volume follows (Scharpé and Van Parijs, 1973). This is correlated with the capacity of isolated nuclei to direct the synthesis of RNA, much of it ribosomal (Millerd and Spencer, 1974). The onset of rapid reserve protein synthesis is preceded by a marked increase in the number of ribosomes (Bain and Mercer, 1966) and by an accumulation of free amino acids and amides in fruit wall, seed coats, embryo sac liquid, and the embryo itself (Murray, 1979d; 1983a).

It is rare to find that rapid synthesis of reserve proteins is confined to a distinct phase of development sandwiched neatly between cell enlargement and desiccation ('Phase 3' of Bain and Mercer, 1966). Reserve protein accumulation generally accompanies cell expansion and frequently continues throughout the final period of desiccation and consequent cell shrinkage, despite declining respiration rates (Loewenberg, 1955; Kollöffel, 1970).

The times of first appearance of individual reserve proteins may differ considerably, as may the periods of most rapid synthesis of their component polypeptides. Many of these differences, examples of which follow, are likely to depend upon genetic regulation of mRNA availability, whereas others can be related to site of accumulation and processing of precursor polypeptides (Section VI).

In developing soybean embryos, the β-conglycinins show a changing spectrum of three constituent polypeptides with increasing age (Gayler and

Sykes, 1981; Spielmann *et al.*, 1982). In castor bean endosperm, the crystalloid interior of the protein bodies is composed largely of a characteristic globulin (Youle and Huang, 1976; Tully and Beevers, 1976). During endosperm development, each protein body acquires the crystalloid proteins first (Fig. 9), then the matrix proteins (primarily albumins and lectins) are added with no subsequent increase in the crystalloid material (Gifford *et al.*, 1982). A similar distribution of protein types occurs in protein bodies of pumpkin cotyledons (Hara-Nishimura *et al.*, 1982).

Such a clear-cut disjunction in the programme for reserve protein synthesis is remarkable. Although it is often assumed or stated that albumin synthesis occurs in a distinct period before globulin synthesis in developing embryos of legumes, generally the situation is complex, with the times of first appearance of some major albumins and globulins indistinguishable, as in chick pea (Vairinhos and Murray, 1982a) and cowpea (Murray *et al.*, 1983). In developing cowpea cotyledons, major albumin polypeptides show different lag times before their periods of most rapid synthesis; for example, that of MW 25 000 daltons became prominent after seven days, that of 35 000 daltons after 12 days, and that of 43 000 daltons after 16 days (Fig. 10). Of three major globulin polypeptides abundant after nine days

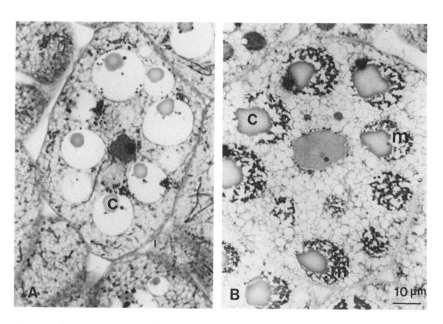

Fig. 9. Protein bodies of developing castor bean endosperm cells 25 days (A) and 30 days (B) after fertilization. Note the central nucleus in each cell, and the vacuoles that first acquire crystalloid protein (c) and globoid inclusions followed by matrix protein (m) (micrographs courtesy of J. S. Greenwood).

(MW 56 000, 52 500 and 50 000 daltons), only two are prominent at maturity; the synthesis of the smallest member of this trio is selectively curtailed (Fig. 10).

In developing embryos of broad bean (Wright and Boulter, 1972) and pea (Millerd and Spencer, 1974), vicilin has always been detected before legumin, with several of the major reserve albumins of pea appearing and accumulating later still (Guldager, 1978). It is interesting that the accumulation of legumin is not confined to an early and discrete phase of protein deposition as occurs with the comparable globulins that form the crystalloid inclusions of protein bodies from castor bean and pumpkin.

During seed development the accumulation of protein may cease relatively early, as in some cultivars of soybean (Hill and Breidenbach, 1974) and castor bean (Gifford et al., 1982). Usually, a considerable proportion of the total reserve protein accumulates while the concentration of water in the storage cells is decreasing, and while there is an absolute loss of water from the seed. Examples include peanut (Aldana et al., 1972), field pea (*Pisum sativum*) (Smith, 1973), garden pea (Murray, 1979d), chick pea (Vairinhos and Murray, 1982a), soybean (Gayler and Sykes, 1981), *Mucuna utilis* (Janardhanan, 1982), almond (Hawker and Buttrose, 1980) and wheat (Fig. 11).

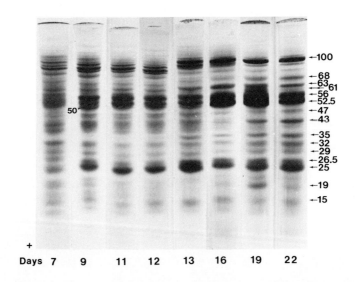

Fig. 10. Polypeptides of developing embryos or cotyledons of cowpea (*Vigna unguiculata* subsp. *unguiculata* cv. Vita 3) dissociated with sodium dodecyl sulphate (SDS) plus 2-mercaptoethanol and separated by electrophoresis on 10% polyacrylamide disc gels. MW × 10^{-3} daltons (from Murray et al., 1983).

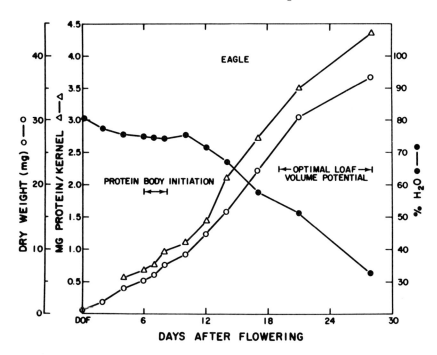

Fig. 11. Changes in dry matter, water content and protein content estimated from total N values during caryopsis development of the hard red winter wheat (*Triticum aestivum* cv. Eagle) (data of Bechtel *et al.*, 1982a).

Ribosomes clearly operate throughout desiccation, and remain functional even in species of cultivars where the accumulation of protein ceases well before maturity (Bewley and Larsen, 1979; Bewley, 1981). Since messenger RNA is subject to a higher rate of turnover than ribosomal RNA (Püchel *et al.*, 1979), it is likely that in all species the rate of reserve protein synthesis is finally restricted by declining rates of synthesis of the corresponding mRNA molecules, rather than by critical water content. This has been demonstrated for developing cotton (*Gossypium hirsutum*) (Dure and Galau, 1981) and is also clearly true of pea cotyledons (Chandler *et al.*, 1983; Higgins *et al.*, 1983b).

V. PROTEIN ACCUMULATION IN PROTEIN BODIES

The major proportion of the accumulated reserves of protein in most species is found in membrane-bounded protein bodies or aggregates derived from them. Among species, the wide variation in size, shape, and the nature of their inclusions has been reviewed by Lott (1980; see also Chapter 4). How they form has been the subject of debate since the time of Sachs (1875), who attributed their formation to dehydration.

A. The Endoplasmic Reticulum

The main site of synthesis of reserve protein in seed tissues is the rough endoplasmic reticulum (RER), the visible association of functional polyribosomes with endoplasmic reticulum (ER) (e.g., Briarty et al., 1969; Bailey et al., 1970). This is hardly surprising. A feature of all growing cells is RER (Shore and Tata, 1977). With improved fixation and use of high voltage electron microscopes, ultrastructural features of the ER have been clarified (Harris, 1979; Briarty, 1980). The RER is classified as cisternal, distinct from interconnecting smooth tubular ER. Both forms become prominent during protein deposition, with cisternal ER declining in favour of reticular ER as the rate of reserve protein synthesis declines (Harris, 1979).

Many studies *in vitro* have shown that the products of translation on free ribosomes can be distinguished by size or immunochemical behaviour from the polypeptides whose synthesis is directed by the mRNA species bound within the polysomes of RER. Examples representing both endosperm and embryo include maize (Larkins et al., 1976a), wheat (Greene, 1981; Donovan et al., 1982) common bean (Bollini and Chrispeels, 1979) and pea (Hurkman and Beevers, 1982). Cisternal RER is able to inject newly synthesized polypeptides into the lumen, thus channelling them either to protein bodies formed *in situ*, or to the dictyosome (Section V.C), The RER also participates in the initial glycosylation of some reserve polypeptides (Section VI.C).

The type of protein body that forms as a deposit in the lumen of RER was recognized in maize endosperm by Khoo and Wolf (1970). There is now abundant evidence that the alcohol-soluble reserve proteins of maize, the zeins, are localized in protein bodies formed by accumulation of polypeptides discharged throughout the RER, and not solely by those adjacent to the developing protein body (e.g., Larkins et al., 1976a, b; 1979; Burr et al., 1978). According to Miflin and Shewry (1979, p.262) and Miflin and Burgess (1982), this is the only way protein bodies can form in cereal endosperm.

The RER deposition specifically involves prolamins and is confined to prolamin-producing endosperm cells. Prolamins generally account for no more than 50%-60% of total protein in cereal grains, and in certain instances much less: 7%-10% in oats (*Avena sativa*) (Peterson, 1976) and

5% in rice (*Oryza sativa*) (Juliano, 1972). The protein bodies of aleurone layers of the endosperm in cereal grains form differently, indeed it has long been known that they have a vacuolar origin. This is true of wheat (Buttrose, 1963; Morrison *et al.*, 1975), barley (Buttrose, 1971), rice (Bechtel and Pomeranz, 1977), maize (Kyle and Styles, 1977), oats, and other species (Fulcher *et al.*, 1972, 1981). A developing protein body from the aleurone layer of wheat is shown in Figure 3 of Chapter 4. There is clearly ample scope for the operation of more than one mechanism of protein body formation in endosperm, even in species where prolamins are prominent in the total spectrum of reserve proteins.

The heterogeneity of protein bodies within the starchy endosperm of rice grains provides further evidence that more than one mechanism can operate in the one tissue. Harris and Juliano (1977) distinguished spherical bodies from irregular crystalline bodies in developing endosperm. Three types have been recognized in the mature grain: large spherical (Fig. 12A), small spherical (Fig. 12B) and crystalline (Fig. 12C).

All types are found in the sub-aleurone region, but only large spherical protein bodies occur in the protein-deficient central endosperm (Bechtel and Pomeranz, 1978). Differences in time of first appearance (Fig. 13) and in constituent polypeptides have been determined. As in maize, the spherical bodies arise as deposits in the lumen of RER (Fig. 12B), and are the sole repository of the prolamins, of estimated MW 13 000 daltons (Tanaka *et al.*, 1980; Yamagata *et al.*, 1982a, b). The crystalline protein aggregates include the major glutelin polypeptides (Section VI.B) and a globulin of 26 000 daltons. The formation of the crystalline protein bodies occurs by filling of vacuoles (Bechtel and Juliano, 1980).

B. The Vacuole

The functions of vacuoles in plant cells have been discussed by Matile (1976, 1978), Buvat and Robert (1979) and Marty *et al.* (1980). A reversible relationship between protein bodies and vacuoles in seed storage tissues has been demonstrated conclusively since the proposals of Matile (1968) and Yatsu and Jacks (1968), based on their content of acid hydrolases (Table VIII).

Many studies have related the onset of rapid reserve protein synthesis in legumes and oilseeds to the appearance in the central vacuole of material presumed to be protein (e.g. Briarty, 1978; Lott, 1980). The appearance of these peripheral deposits inside vacuoles is not an artifact of preparation; that is, it is not induced by the fixative employed (Goodchild and Craig, 1982). In developing pea cotyledons, studies with labelled antibodies have

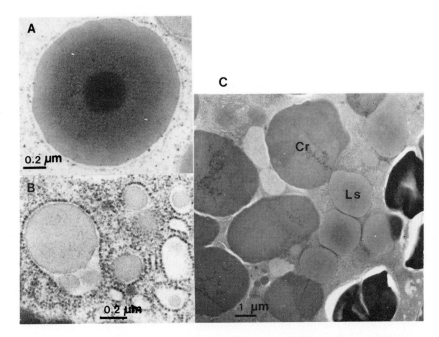

Fig. 12. Protein bodies from the starchy endosperm of rice grains (*Oryza sativa* L.). (A) Large spherical protein body with dense centre and boundary from an endosperm cell six days after flowering. (B) Small spherical protein bodies enclosed singly or in groups within RER vesicles of an endosperm cell. (C) Crystalline (Cr) and large spherical (Ls) protein bodies in a rice endosperm cell 14 days after flowering. (From Bechtel and Juliano, 1980).

established that the reserve globulins legumin and vicilin are deposited in central vacuoles (Craig et al., 1979b, 1980b; Craig and Millerd, 1981). Subdivision of vacuoles to form numerous small protein bodies has been indicated in pea (Craig et al., 1979a, 1980a) and in soybean (Yoo and Chrispeels, 1980). In castor bean endosperm, the number of vacuoles formed initially is greater, with apparently no subdivision during or after the acquisition of protein. The crystalloid proteins form an obvious deposit in these vacuoles, surrounded later by the matrix proteins (Fig. 9).

Both the storage and lytic functions of vacuolar protein bodies are demonstrated following germination, when protein bodies are seen to degrade their contents while maintaining membrane integrity and gradually fuse to reform one or a few central vacuoles (Horner and Arnott, 1965; Öpik, 1966; Table VIII, references). The few studies on cereals indicate that it is usual to find a similar range of hydrolytic enzymes associated with protein bodies from all parts of the grain (Adams and Novellie, 1975a, b; Adams et al., 1975).

3. *Nitrogenous Reserves* 111

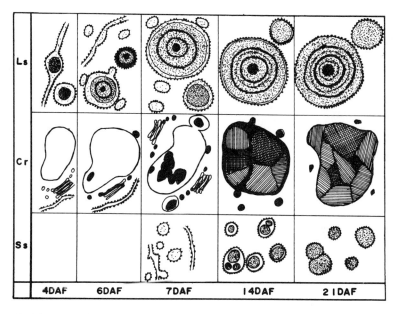

Fig. 13. Diagrammatic scheme of protein body formation in rice subaleurone starchy endosperm cells. Ls, large spherical protein body; Cr, crystalline protein body; Ss, small spherical protein body; DAF, days after flowering (from Bechtel and Juliano, 1980).

Enzyme activities that would serve as markers for the tonoplast or protein body membrane have been sought, but so far with little success. In yeast cells α-mannosidase is localized in the tonoplast (Wiemken, 1975), but this enzyme activity is contained entirely within the vacuole in several types of higher plant cell (Boller and Kende, 1979), and in protein bodies from mung bean cotyledons (Table VIII). In pea cotyledons, α-mannosidase activity increases during development and is attributed mainly to a single isoenzyme (Murray, 1983b). In developing bean cotyledons, two isoenzymes are associated with the protein bodies: isoenzyme I is selectively removed from protein bodies with detergent but is not associated with the protein body membranes (Van der Wilden and Chrispeels, 1983). It is possible that these two isoenzymes correspond to the enzymic lectins of Paus and Steen (1978). Two isoenzymes of α-mannosidase have also been isolated from protein bodies of lupin (Plant and Moore, 1982). Membranes from protein bodies from castor bean endosperm display several enzyme activities, but none that is not also associated with both the matrix and crystalloid protein compartments (Mettler and Beevers, 1979).

Whether protein bodies form in the vacuoles of starchy endosperm cells

Table VIII. Enzyme activities of protein bodies isolated from cotyledons or endosperm* of dicotyledons.

Enzyme activity	Pisum sativum	Vicia faba	Gossypium hirsutum	Vigna radiata	Ricinus communis*
Acid phosphatase	A	B	C	F	H, I
Phytase		B^a			H^a
Acid proteinase	A	B	C	E^a, G^a	H^a, J^a
Carboxypeptidase				D, F	H^a
Aminopeptidase					I
Ribonuclease	A			F, G	H
Phosphodiesterase	K			F	I, H
Carboxyesterase	A				
β-Amylase	A				
α-Glucosidase	A				
α-Mannosidase				F	
N-acetyl-β-glucosaminidase	K			D, F	
β-Glucosidase					H
Acid lipase					I
Phospholipase D				F	
Malate dehydrogenase					I

^{a} acquired following imbibition.
A, Matile (1968); B, Morris et al. (1970); C, Yatsu and Jacks (1968); D, Harris and Chrispeels (1975); E, Baumgartner et al. (1978); F, Van der Wilden et al. (1980b); G, Van der Wilden et al. (1980a); H, Nishimura and Beevers (1978); I, Mettler and Beevers (1979); J, Alpi and Beevers (1981); K, Miflin et al. (1981).

of cereals, particularly wheat and barley, has been questioned (Miflin and Shewry, 1979; Miflin et al., 1981). Miflin and Burgess (1982) have claimed that the eventual lack of a membrane around protein aggregates in endosperm cells of barley and wheat precludes there being any contribution from vacuolar protein deposits. These authors have discounted evidence to the contrary for both barley (Ingversen, 1975; Bechtel and Pomeranz, 1979) and wheat.

Campbell et al. (1981) considered that the bounding membrane of protein deposits in wheat endosperm might arise in three ways: from ER, from vacuole or from dictyosome. Other studies of bread wheat (Parker, 1980; Bechtel et al., 1982a, b) and the wild tetraploid wheat (*Triticum dicoccoides*) (Parker, 1982) have provided ample evidence that the vacuole receives and concentrates protein-containing material from a variety of sources (Figs 14, 15). Protein granules in the vacuoles enlarge and fuse with other newly deposited protein granules (Fig. 15), and with one another (Fig. 16), a process that accelerates about halfway through development (Fig. 11). The irregularly shaped protein masses are eventually released

3. *Nitrogenous Reserves* 113

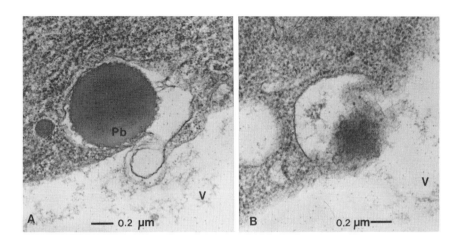

Fig. 14. Two stages in the fusion of protein body (Pb) and vacuole (V) in starchy wheat endosperm cells 10 DAF (refer Fig. 11) (from Bechtel *et al.*, 1982a).

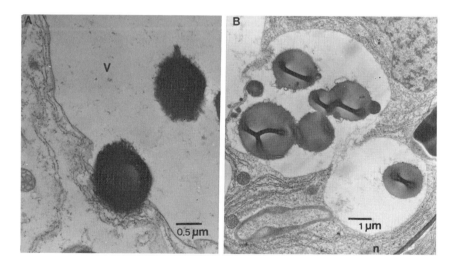

Fig. 15. Acquisition of protein granules by the vacuole (V) is accompanied by fusion of the vacuoles in starchy wheat endosperm cells. A, 7 DAF; B, 10 DAF (from Bechtel *et al.*, 1982a).

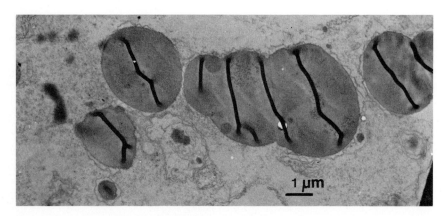

Fig. 16. Further amalgamation of vacuolar protein deposits in starchy wheat endosperm cells 17 DAF (from Bechtel et al., 1982a).

from their bounding membranes and condensed with residual cytoplasmic material to form the proteinaceous matrix of the mature endosperm. This matrix contains a combination of gliadins (wheat prolamins) and glutenins (wheat glutelins), the proportions of which determine the ability of the resulting flour to form doughs suitable for bread making (Wall, 1979; Wrigley et al., 1980, 1981, 1982).

C. The Dictyosome (Golgi Apparatus)

Many reviews emphasize the importance of this organelle in the sorting process necessary to direct proteins and other macromolecules formed in the ER to their various destinations both inside and outside the cell (Morré and Mollenhauer, 1974; Mollenhauer and Morré, 1980; Juniper et al., 1982). Rothman (1981) attributes distinct functions to the cis and trans faces. Thus proteins belonging to the ER membranes are returned to the ER after circulating in the cis stacks, whereas proteins for export proceed through the trans stacks, concentrating into different types of vesicles that depart from the outermost face. The dictyosome has the ability to retain its own structural components, including a number of glycosyl transferases (Chrispeels, 1976, 1980; Mollenhauer and Morré, 1980).

In developing storage tissues of seeds, at first only the electron-lucent dictyosome vesicles are observed. The vacuole itself is probably derived from such vesicles (Marty et al., 1980). Then, vesicles develop that have contents indistinguishable from protein in their intensity. A precursor

3. Nitrogenous Reserves 115

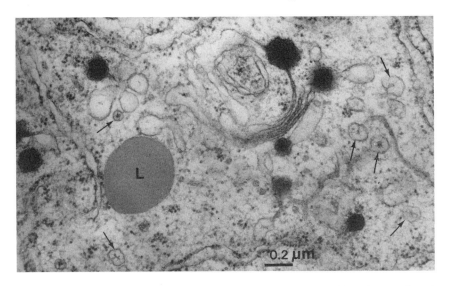

Fig. 17. Dictyosomes in wheat endosperm cell (12 DAF) producing vesicles with star-shaped inclusions (arrows) whose destination is the cell wall, and vesicles with dense inclusions, believed to be the protein bodies acquired by the vacuoles. L, lipid (from Bechtel et al., 1982b).

relationship between dictyosome vesicles and developing protein bodies has often been suggested, for cereals (Buttrose, 1963) and dicotyledons alike (Bain and Mercer, 1966; Engleman, 1966).

The precise role of the dictyosome in protein synthesis has been difficult to establish, despite good circumstantial evidence that it might transport proteins from RER to vacuolar protein bodies in legumes and oilseeds (Dieckert and Dieckert, 1976; Briarty, 1978, 1980). Even this correlation has been lacking in most studies of cereal endosperm. Briarty et al. (1979) concluded that the dictyosome was not involved in the pathway taken by proteins to protein bodies in wheat endosperm because no dictyosomes could be detected beyond 12 days after flowering.

In more recent studies of wheat, Bechtel et al. (1982a, b) observed the first formed protein granules close to dictyosomes 6-7 days after flowering. The dictyosomes were prominent in both early and mid-term stages of protein accumulation (Fig. 17) and persisted throughout most of the final period to maturity (Fig. 11). Persistence of dictyosomes in wheat endosperm has also been observed by Campbell et al. (1981), Parker (1980, 1982) and Parker and Hawes (1982). A number of vesicle types may be distinguished, including one with star-shaped contents (Fig. 17) which are identifiable in the cell-wall material deposited outside the cells (Bechtel et al., 1982b).

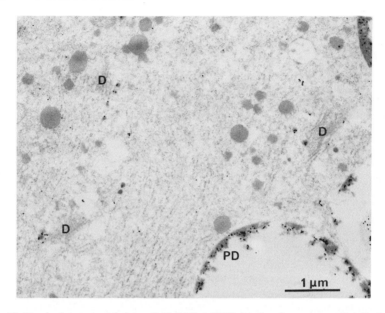

Fig. 18. Developing pea cotyledon cell 15 DAF at 25°C showing the presence of vicilin in the vacuolar protein deposits (PD) and in the electron dense vesicles associated with the dictyosomes (D). The tissue was fixed with glutaraldehyde and osmium tetroxide, embedded in epoxy resin, then immunolabelled by an indirect method employing colloidal gold as the marker (courtesy of S. Craig and D. J. Goodchild).

Certain types of electron-dense dictyosome vesicle have been shown to consist largely of protein, since their locations are vacant following treatment of unfixed thin sections with proteolytic enzymes (Dieckert and Dieckert, 1976; Bechtel and Gaines, 1982; Parker, 1982). Furthermore, in developing pea cotyledons, dictyosome vesicles have been shown to contain proteins identifiable as the reserve proteins that are ultimately localized in vacuolar protein bodies (Fig. 18). It has also proved possible to redirect protein-containing dictyosome vesicles by treatment with ionophores specific for potassium ions (nigericin), sodium ions (monensin), or simply EDTA (e.g., Fig. 19). Deposits of reserve protein then appeared between the plasmalemma and cell wall. This response was stopped by cycloheximide, indicating that only newly synthesized reserve polypeptides were involved in the vesicle-mediated transport to the plasmalemma. This is the first convincing demonstration for any seed storage tissue that the dictyosome functions in internal protein secretion.

The situation in wheat endosperm is more complex than in the pea cotyledon. The protein accumulating in vacuoles appears to come partly

3. Nitrogenous Reserves 117

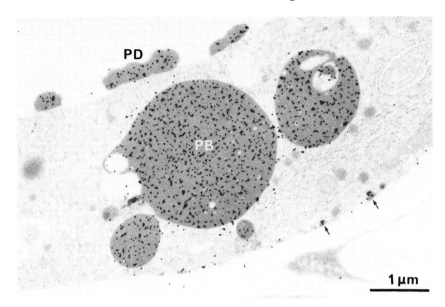

Fig. 19. Developing pea cotyledon cell treated with nigericin before immunochemical localization of vicilin as in Fig. 18. Note exocytosis of vicilin (arrows) which normally accumulates in protein bodies (PB) derived from vacuolar protein deposits (PD) (courtesy of S. Craig and D. J. Goodchild).

from incorporation of dictyosome vesicles, from material added to dictyosome vesicles while in transit, from engulfment or phagocytosis of ER deposits perhaps corresponding to gliadins, and from engulfment of other cytoplasmic (membranous) material.

VI. PROCESSING OF SEED RESERVE PROTEINS

Processing is an economic term that includes the events of 'post-translational modification'. The number of such modifications is potentially enormous. Uy and Wold (1977) have pointed out that some 140 derivatives of the 20 common polypeptide-forming amino acids can be isolated from proteins. Major events in the processing of seed reserve proteins include glycosylation, the formation of disulphide bonds, and the limited cleavage of peptide bonds. The allocation of these events to different cellular compartments is an area of rapid and continuing progress.

A. Removal of the Signal Sequence

With improved precautions against ribonuclease activity and the widespread adoption of a wheat-germ translation system (Marcu and Dudock, 1974), it has proved possible to translate mRNA species derived from RER in the absence of their endogenous ER membranes, and among the products identify precursor polypeptides that include an N-terminal leader or signal sequence (Table IX). By analogy with the synthesis of secreted proteins in many other organisms (Blobel et al., 1979), it is probable that the relatively hydrophobic signal sequence assists the penetration of the lumen of the RER by the newly synthesized polypeptide *in vivo*. The peptide bond between this leader sequence and the balance of the polypeptide chain is often hydrolysed during passage through the ER membrane and before the chain is complete; that is, its removal may be co-translational.

The lectin leader sequences contain a high proportion of hydrophobic amino acids, mainly leucine, isoleucine and phenylalanine, as seen in many secretory proteins (Table X).

B. Proteolytic Activity of Protein Bodies

Once the signal sequence of a reserve protein precursor is removed on passage through RER membrane, the residual precursor is stable: experimentally, protection from subsequent cleavage by added endopeptidases can be demonstrated as long as the ER vesicles are intact. The addition of detergents abolishes this protection and allows further proteolysis to occur (e.g., Cameron-Mills et al., 1978; Higgins and Spencer, 1981). *In vivo*, pulse-chase labelling experiments have established the precursor–product

Table IX. Apparent molecular weight differences indicating an N-terminal signal sequence for precursor polypeptides of reserve proteins.

Species	Proteins	MW difference	References
Zea mays	zeins	ca. 2000	Larkins and Hurkman (1978)
		1100 to 2000	Burr et al. (1978)
	A zeins	900	Melcher (1979)
	B zeins	1300	Melcher (1979)
Triticum	gliadins	ca. 1000	Greene (1981)
aestivum		5000 to 1000	Donovan et al. (1982)
Avena sativa	globulin	2000	Brinegar and Peterson (1982b)
Glycine max	glycinin	1000 to 2000	Tumer et al. (1982)
Gossypium hirsutum		2000	Dure and Galau (1981)
Vicia faba	favin	2700	Hemperly et al. (1982)
Ricinus communis	lectins	1500 to 2000	Roberts and Lord (1981)

Table X. The signal sequences removed cotranslationally from the pea lectin (I) and favin (II)

	MET	ALA	SER	LEU	GLU	THR	GLU	MET	ILE	SER	PHE	TYR	ALA	ILE	PHE
I.	MET	ALA	SER	LEU	GLU	THR	GLU	MET	ILE	SER	PHE	TYR	ALA	ILE	PHE
II.	—	—	—	ILE	—	—	—	MET	LEU	—	PHE	—	—	ILE	PHE
I.	LEU	SER	ILE	LEU	LEU	THR	THR	ILE	LEU	PHE	PHE	LYS	VAL	ASN	SER
II.	LEU	—	ILE	LEU	LEU	—	—	ILE	LEU	PHE	PHE	—	—	—	—

The sequences begin from the N-terminus. I is derived indirectly by base sequence analysis (T.J.V. Higgins et al., 1983a). II is derived from direct analysis of favin synthesized in vitro (J. J. Hemperly et al., 1982). A dash signifies that the residue was not identified.

relationship for polypeptides from several species. Subsequent proteolytic changes occur relatively slowly, only after precursor polypeptides have arrived inside developing vacuolar protein bodies.

1. Lectins

Cleavage of peptide bonds is necessary for the conversion of the 'prolectins' of MW 23 000 daltons to the β and α subunits of the mature lectin. This has been demonstrated for favin (Hemperly *et al.*, 1982) and for the pea lectin (Fig. 20). By obtaining the mRNA base sequence, Higgins *et al.* (1983a) have established that there is no stop codon between the regions of RNA coding for each subunit; rather there is a link peptide, which has the composition SER LEU GLU GLU GLU ASN in the pea prolectin. Thus there are two proteolytic cleavage sites between the β and α subunits, and another site which results in the loss of the peptide ALA ALA ASP ALA from the C-terminus of the α subunit (Higgins *et al.*, 1983a). All of these events occur after arrival of the prolectin inside the protein bodies (Higgins *et al.*, 1983b) and compared to the processing of the major globulins, they occur relatively slowly (Fig. 20).

2. Legumin

Three precursor forms of legumin with estimated MW 60 000 to 65 000 daltons appeared to have a lifetime of 1 to 2 hours before complete division into disulphide-bridged polypeptides of MW 40 000 and 19 000 daltons in developing pea cotyledons (Spencer and Higgins, 1980). Croy *et al.* (1980) reported the presence of only a single legumin precursor of MW 60 000 daltons in another cultivar of pea (cv. Feltham First). These authors proposed a single peptide cleavage site between the two disulphide-linked polypeptides of mature legumin. Vairinhos and Murray (1982b) concluded that disulphide bonds must form within the polypeptide chain of legumin precursors before cleavage of connecting peptide bonds, because in several species the associations between large and small polypeptides of differing apparent molecular weight are not random.

The association of legumin subunits to form a hexamer (MW 350 000 ± 10 000 daltons, Blagrove *et al.*, 1980) probably occurs before or during final proteolytic processing, although the precursor polypeptides appear to migrate from the ER as tetramers (Chrispeels *et al.*, 1982).

3. Glycinin

Glycinin of soybean is homologous with legumin (Gilroy *et al.*, 1979) and consists of disulphide-linked pairs of acidic (large) and basic (small) polypeptides that differ remarkably in their content of methionine (Table XI).

3. *Nitrogenous Reserves* 121

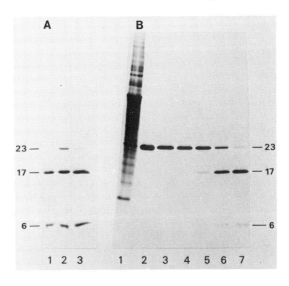

Fig. 20. (A) SDS-polyacrylamide gel electrophoresis of pea lectin isolated by saccharide affinity chromatography from 25-day immature seeds (lane 2) and mature seeds (lanes 1 and 3). Estimated MW × 10^{-3}. (B) Pulse-chase labelling of pea lectin polypeptides: cotyledons from 15-day seeds were labelled with ^{14}C-amino acids for 1.5 h then incubated normally for up to 22 h longer. Gel electrophoresis and fluorography were used to detect the conversion of the pro-lectin (MW 23 000) to mature subunits of MW 17 000 and 6 000 daltons. Lanes 1, 2 and 3 show respectively total cotyledon extract, lectin purified as in (A), or by immunoaffinity, all after 1.5 h labelling. Lanes 4, 5, 6 and 7 show lectin purified after chase intervals of 2, 4, 11 and 22 hours respectively. (From Higgins *et al.*, 1983a).

Table XI. The methionine content of glycinin subunits.

Identification	Group	Estimated MW[a] (daltons)	Number of methionines[b]
$A_{1a}B_2$	I	57 000	5 or 6
$A_{1b}B_{1b}$	I	57 000	5 or 6
A_2B_{1a}	I	57 000	7 or 8
A_3B_4	II	62 000	3
$A_5A_4B_3$	II	69 000	3

[a] Sum of MW determinations for component polypeptides following electrophoresis in polyacrylamide gels with SDS and 2-mercaptoethanol present. These are likely to be overestimates, since amino acid sequencing indicated A_2 = 31 600 and B_{1a} = 19 900 daltons respectively.
[b] Variation in this number reflects heterogeneity in amino acid sequences.
Data of P. E. Staswick *et al.* (1981; 1983).

The methionine positions differ widely within each type of polypeptide (Moreira et al., 1981a, b), supporting the conclusion that the mature subunits arise as products of distinct genes. A variety of techniques including cloned DNA sequencing (Moreira et al., 1981a; Tumer et al., 1982) have been employed to determine the positions of the large (acidic) and small (basic) polypeptides within most of these precursors as follows:

(N) signal—large—link polypeptide—small (C)

This arrangement is the same as that proposed for four legumin precursors in chick pea (Fig. 10 of Chapter 1). There is one disulphide bridge per polypeptide pair in glycinin (N. C. Nielsen, personal communication, 1983). These bonds must form before excision of the link polypeptides, which in glycinin vary between MW 1000 and 4000 daltons (Tumer et al., 1982). A small peptide is cleaved from the C-terminus of the basic polypeptide in at least one instance, A_2B_{1a}, as occurs with the α subunit of the pea lectin.

4. Castor Bean Crystalloid Protein

The crystalloid protein from protein bodies of castor bean endosperm (Fig. 9) is a hexamer of subunits that display considerable charge heterogeneity. Eight main forms have been distinguished, each comprising a disulphide-linked pair of polypeptides of approximate molecular weight (MW) 30 000 and 20 000 daltons (Gifford and Bewley, 1983). These are as varied in their methionine content as the glycinin polypeptides.

5. Cotton Seed Proteins

Two principal sets of polypeptides of estimated MW 52 000 and 48 000 daltons are formed in developing cotton embryos (Dure and Chlan, 1981). Each group of size isomers is comprised of polypeptides that vary in net charge and hence isoelectric point. The set of MW 52 000 polypeptides is glycosylated (Section VI.C). Its precursors of initial MW 60 000 daltons are modified co-translationally to MW 70 000 daltons, the net result of glycosylation and removal of signal sequence (Dure and Galau, 1981). The other major precursors are not glycosylated, and these lose a signal sequence of about MW 2000 daltons (Table IX), reducing them from MW 69 000 to 67 000 daltons. These polypeptides are subsequently cleaved to those of MW 48 000 and 19 000 daltons (Dure and Galau, 1981).

The final cleavage of both sets of precursors is relatively slow, but the glycosylated set is processed more slowly. Immuno-precipitation of the glycosylated polypeptides is more difficult than for the non-glycosylated polypeptides, but antigenic determinants are shared, which indicates a close relationship between the two groups of polypeptides and therefore gene duplication (Dure and Chlan, 1981).

6. Helianthin

Helianthin, the predominant globulin of sunflower cotyledons, comprises disulphide-linked polypeptide pairs of two main types, with estimated MW 37 000 + 22 000 = 59 000 daltons, and 29 000 + 22 000 = 51 000 daltons. At least one type of subunit is glycosylated, which allows interaction with the jack bean lectin Con A (Murray and Vairinhos, 1982b).

7. Oat Globulin

Oat is unusual among the well-known cereals in possessing a globulin as its major reserve protein (e.g., Peterson, 1978). Small polypeptides of MW 22 000–24 000 daltons and large of MW 32 500–37 500 daltons are combined in disulphide-linked pairs of MW 53 000–58 000 daltons (Brinegar and Peterson, 1982a). *In vitro* studies have distinguished two stages of processing: the co-translational removal of signal sequences (Table IX), then later conversion of precursors of MW 58 000–62 000 daltons to the mature disulphide-linked pairs of MW 53 000–58 000 daltons (Luthe and Peterson, 1977; Brinegar and Peterson, 1982b). The second processing stage is correlated with the accumulation of this globulin in vacuolar protein bodies (Bechtel and Pomeranz, 1981).

8. Rice Glutelin

There is a striking resemblance between the major globulin of oat and the major glutelin of rice (Tanaka *et al.*, 1980; Yamagata *et al.*, 1982a, b). Cleavage of precursor polypeptides of MW 57 000 into subunits of 22 000–23 000 and 37 000–39 000 daltons occurs slowly (Fig. 21) and corresponds with the deposition of the precursor polypeptides in vacuolar protein bodies (Fig. 12C). The precursors of the rice glutelin subunits are salt-soluble, that is, globulins (Yamagata *et al.*, 1982b).

Since the oat globulin strongly resembles both legumin and glycinin (Brinegar and Peterson, 1982a, b), it appears both it and the rice glutelin are more of the 'homologous counterparts' of legumin sought by Derbyshire *et al.* (1976). As noted with lectins (Section II.A), content of sulphur amino acids, the presence or absence of disulphide-bridges, and whether glycosylation occurs or not, are all highly variable characteristics. Legumin counterparts may also show wide variation in the same properties and be

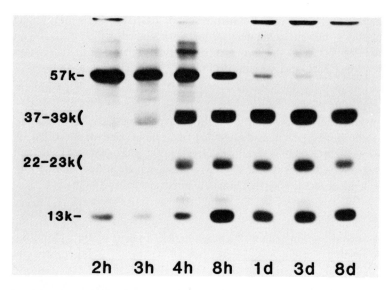

Fig. 21. Pulse-chase labelling of polypeptides in the starchy endosperm of developing rice grains. [^{14}C]leucine was supplied to cut fruiting stalks ten days after flowering for a period no longer than 30 minutes. A chase of non-radioactive leucine was provided for the period indicated beneath each gel track. Polypeptides were separated under reducing conditions by electrophoresis in 14% polyacrylamide gels and detected by fluorography (from Yamagata et al., 1982b).

widespread among both dicotyledons and monocotyledons. It may now be predicted that glutelins of the type found in rice will be found to share the amino acid sequence of homologous globulins that ensure their arrival and deposition in vacuolar protein bodies.

9. Vicilin

The diversity of subunits in vicilin with proportions determined by genotype (Thomson and Schroeder, 1978; Thomson et al., 1978) is now explicable in terms of partial processing of the group of precursors of about MW 50 000 daltons after their arrival inside vacuolar protein bodies. The products of proteolytic processing remain associated with uncleaved precursors by non-covalent forces, giving rise to the final spectrum of subunits found on isolation and analysis of the holoprotein (Higgins and Spencer, 1977; 1981; Gatehouse et al., 1981). It is possible that oligomers form inside ER cisternae while the proteins are en route to the dictyosome (Chrispeels et al., 1982). The vacuolar processing of vicilin takes from 6 to 20 hours, in contrast to the 2 hours usually sufficient for the complete proteolytic processing of legumin (Chrispeels et al., 1982).

C. Glycosylation

Glycoproteins consist of polypeptide chains modified by the covalent attachments of oligosaccharides to asparagine, serine and threonine residues (Elbein, 1979; Gibson *et al.*, 1980). The subunits of many reserve globulins of seeds have proved to be glycoproteins with one or more asparagine-linked oligosaccharides that consist mainly of mannose and glucosamine. The oliogosaccharide of soybean lectin (soyin) consists of eight variously branched α-linked mannose residues linked to a core of β-(1-4)-linked mannose, N-acetyl glucosamine, and N-acetyl glucosamine (Lis and Sharon, 1978). This type of linked oligosaccharide is of universal occurrence, as is the metabolic pathway leading to the formation of the glycosyl donor, glycosyl-phosphoryl-polyprenol, where the polyprenol is of the dolichol category. Analytical evidence for the operation of this pathway in developing bean cotyledons has been presented by Ericson and Delmer (1977) and Delmer *et al.* (1978).

The subcellular location of the glycosyl transferases that are necessary to build and transfer the oligosaccharide is the RER. This has been determined for pea cotyledons (Beevers and Mense, 1977; Nagahashi and Beevers, 1978; Nagahashi *et al.*, 1980) and for castor bean endosperm (Marriott and Tanner, 1979). *In vitro* studies confirm that glycosylation is co-translational; for example, favin (Hemperly *et al.*, 1982).

Tunicamycin has been used specifically to block the formation of the dolichol phosphate glycosyl donor in many organisms including plants (Ericson *et al.*, 1977). In mammals, the conformation of the resulting 'naked' protein is often not that attained by the glycosylated protein; it may not be fully functional, and may not arrive at its correct destination (Gibson *et al.*, 1980).

These functional considerations are certainly valid in specific instances, but when applied to the reserve proteins of seeds they become irrelevant. The glycosylation of lectins, for example, is determined purely by local amino acid sequence. For an asparagine residue to be eligible for glycosylation, it must fall in the sequence —asparagine—X—threonine (or serine). Such a sequence occurs in positions 12 to 14 of both types of polypeptide subunit comprising the isolectins of bean (—asparagine—glutamate—threonine) with the result that both are glycosylated at position 12 (Miller *et al.*, 1975). The same positions in homologous lectin chains are not highly conserved (Foriers *et al.*, 1977a, b). The favin β chain (Section II.A) is glycosylated at a completely different site, residue 168. Many other lectins are not glycosylated; for example, Con A and the lectins from pea and lentil, yet all of these lectins attain shapes appropriate to their detectable binding functions and intracellular locations (Section II).

The capacity for self-association shown characteristically by reserve globulin polypeptides does not depend upon glycosylation. The accumulation of major globulins in vacuolar protein bodies occurs independently of glycosylation. Legumin from pea cotyledons is not glycosylated (Casey, 1979; Hurkman and Beevers, 1980). Of the polypeptides comprising vicilin, only a minor one of MW 14 000 daltons is glycosylated, and even in this instance, an alternative non-glycosylated form coexists (Davey et al., 1981). The major crystalloid globulin of protein bodies of castor bean endosperm is not glycosylated, and neither are the albumins from the matrix of the same protein bodies (Gifford et al., 1982).

Thus, in many instances the pathway from RER lumen to dictyosome vesicle to vacuole can be traversed without the aid of any glycosyl attachments. Both the transport and the assembly of reserve globulins can be governed by amino acid sequence alone.

VII. CONCLUSIONS

The acquisition of reserve proteins by vacuoles is ubiquitous as a mechanism for the formation of protein bodies in the storage tissues of seeds. Dictyosome-mediated transport of reserve proteins to vacuoles, shown conclusively for one species (Figs 18, 19), may prove to be an integral component of this storage mechanism. Proteolytic processing of reserve protein precursors occurs relatively slowly in vacuolar protein bodies under the control of enzymes that remain to be characterized. Reserve proteins share proteolysis with vacuolar enzymes that require cleavage for activation (Kauss et al., 1978, 1979).

The reserve proteins that fill vacuoles share this destination with many enzymes (Table VIII). Their translocation from ER lumen to vacuole depends upon amino acid sequence in some part or parts of the polypeptide chain, and it is therefore reasonable to predict that the reserve proteins in protein bodies, when fully sequenced, will prove to be homologous with vacuolar hydrolases. This is already evident in the structural relationships of canavalin (Smith et al., 1982) and certain lectins to α-mannosidase and α-galactosidase (Section II.A,2).

A second mechanism of protein body formation occurs in the endosperm of cereals and involves the deposition of prolamins in the lumen of RER. Prolamins secreted into the lumen of RER have removable signal sequences, and cannot be distinguished in this respect from polypeptides destined for inclusion in vacuolar protein bodies.

Proteins that accumulate in the cytoplasm outside protein bodies (Section I) may or may not be translated on free ribosomes. Because of the higher contents of sulphur amino acids in many albumins and proteinase inhibitors, genotypes should be sought that will maximize their representation in the seed of species intended for human consumption.

The size heterogeneity of polypeptide subunits of proteinase inhibitors, lectins, legumin and many other reserve proteins indicates clearly that gene duplication has occurred often. Beyond this, there are many instances of complex heterogeneity that have not been satisfactorily explained; for example, for the subunits of glycinin (Table XI) and the crystalloid protein from castor bean (Gifford and Bewley, 1983). The extensive heterogeneity of the prolamins has long been recognized (Wall, 1979). In the case of zein, at least 20 forms occur that result neither from glycosylation, from deamidation (Rhigetti *et al.*, 1977), nor from a correspondingly large number of totally distinct mRNA types (Park *et al.*, 1980; Burr and Burr, 1982).

Does each form of subunit arise from a distinct gene copy, or is the number of complete gene copies relatively small? It seems possible that some species could have developed a coding system similar to the one governing the hypervariable region of immunoglobulin light chains (Leder *et al.*, 1980). In this way many species of mRNA with very slight differences could be generated from a few gene copies for a particular type of polypeptide. Techniques now available for direct genotype analysis should soon resolve these questions concerning gene organization and expression as they relate to protein synthesis.

REFERENCES

Adams, C. A., and Novellie, L. (1975a). *Plant Physiol.* **55**, 1-6.
Adams, C. A., and Novellie, L. (1975b). *Plant Physiol.* **55**, 7-11.
Adams, C. A., Watson, T. G., and Novellie, L. (1975). *Phytochemistry* **14**, 953-956.
Aldana, A. B., Fites, R. C., and Pattee, H. E. (1972). *Plant Cell Physiol.* **13**, 515-520.
Alpi, A., and Beevers, H. (1981). *Plant Physiol.* **68**, 851-853.
Applebaum, S. W. (1964). *J. Insect Physiol.* **10**, 783-788.
Applebaum, S. W., Birk, Y., Harpaz, I., and Bondi, A. (1964). *Comp. Biochem. Physiol.* **11**, 85-103.
Bailey, C. J., Cobb, A., and Boulter, D. (1970). *Planta* **95**, 103-108.
Bain, J. M., and Mercer, F. V. (1966). *Aust. J. Biol. Sci.* **19**, 49-67.
Bajaj, S., Mickelsen, O., Lillevik, H. A., Baker, L. R., Bergen, W. G., and Gill, J. L. (1971). *Crop Sci.* **11**, 813-815.
Basha, S. M. M., and Beevers, L. (1975). *Planta* **124**, 77-87.
Basha, S. M. M., and Roberts, R. M. (1981). *Plant Physiol.* **67**, 936-939.
Baues, R. J., and Gray, G. R. (1977). *J. Biol. Chem.* **252**, 57-60.
Baumgartner, B., Tokuyasu, K. T., and Chrispeels, M. J. (1978). *J. Cell Biol.* **79**, 10-19.
Bechtel, D. B., and Gaines, R. L. (1982). *Am. J. Bot.* **69**, 880-884.
Bechtel, D. B., and Juliano, B. O. (1980). *Ann. Bot.* **45**, 503-509.
Bechtel, D. B., and Pomeranz, Y. (1977). *Am. J. Bot.* **64**, 966-973.

Bechtel, D. B., and Pomeranz, Y. (1978). *Am. J. Bot.* **65**, 684–691.
Bechtel, D. B., and Pomeranz, Y. (1979). *Cereal Chem.* **56**, 446–452.
Bechtel, D. B., and Pomeranz, Y. (1981). *Cereal Chem.* **58**, 61–69.
Bechtel, D. B., Gaines, R. L., and Pomeranz, Y. (1982a). *Cereal Chem.* **59**, 336–343.
Bechtel, D. B., Gaines, R. L., and Pomeranz, Y. (1982b). *Ann. Bot.* **50**, 507–518.
Becker, J. W., Reeke, G. N. Jr., Wang, J. L., Cunningham, B. A., and Edelman, G. M. (1975). *J. Biol. Chem.* **250**, 1513–1524.
Becker, J. W., Reeke, G. N. Jr., Cunningham, B. A., and Edelman, G. M. (1976). *Nature* **259**, 406–409.
Beevers, L., and Mense, R. M. (1977). *Plant Physiol.* **60**, 703–708.
Belitz, H.-D., and Fuchs, A. (1973). *Z. Lebensm. Unters.-Forsch.* **152**, 129–133.
Bewley, J. D. (1981). In 'The Physiology and Biochemistry of Drought Resistance in Plants' (L. G. Paleg and D. Aspinall, eds.), pp. 261–282. Academic Press, Sydney.
Bewley, J. D., and Larsen, K. (1979). *Phytochemistry* **18**, 1617–1619.
Bhatty, R. S. (1977). *Can. J. Plant Sci.* **57**, 979–982.
Bidlingmeyer, U. de V., Leary, T. R., and Laskowski, M. Jr. (1972). *Biochm.* **11**, 3303–3310.
Birk, Y. (1961). *Biochim. Biophys. Acta* **54**, 378–381.
Birk, Y. (1976a). In 'Methods in Enzymology XLV, Proteolytic Enzymes Part B' (L. Lorand, ed.), pp. 700–707. Academic Press, New York.
Birk, Y. (1976b). In 'Methods in Enzymology XLV, Proteolytic Enzymes Part B' (L. Lorand, ed.), pp. 716–722. Academic Press, New York.
Birk, Y. (1976c). In 'Methods of Enzymology XLV, Proteolytic Enzymes Part B' (L. Lorand, ed.), pp. 723–728. Academic Press, New York.
Birk, Y., Gertler, A., and Khalef, S. (1963). *Biochim. Biophys. Acta* **67**, 326–328.
Birk, Y., Gertler, A., and Khalef, S. (1967). *Biochim. Biophys. Acta* **147**, 402–404.
Blagrove, R. J., Lilley, G. G., and Davey, R. (1980). *Aust. J. Plant Physiol.* **7**, 221–225.
Blobel, G., Walter, P., Chang, C. N., Goldman, B. M., Erickson, A. H., and Lingappa, V. R. (1979). *Symp. Soc. Exp. Biol.* **33**, 9–36.
Blumberg, S., Hildesheim, J., Yariv, J., and Wilson, K. J. (1972). *Biochim. Biophys. Acta* **264**, 171–176.
Boller, Th., and Kende, H. (1979). *Plant Physiol.* **63**, 1123–1132.
Bollini, R., and Chrispeels, M. J. (1978). *Planta* **142**, 291–298.
Bollini, R., and Chrispeels, M. J. (1979). *Planta* **146**, 487–501.
Bowman, D. E. (1944). *Proc. Soc. Exp. Biol. Med.* **57**, 139–140.
Bowman, D. E. (1946). *Proc. Soc. Exp. Biol. Med.* **63**, 547–550.
Boyd, W. C. (1963). *Vox Sang.* **8**, 1–32.
Boyd, W. C., and Shapleigh, E. (1954). *Science* **119**, 419.
Briarty, L. G. (1978). In 'Plant Proteins' (G. Norton, ed.), pp. 81–106. Butterworths, London.
Briarty, L. G. (1980). *J. Exp. Bot.* **31**, 1387–1398.
Briarty, L. G., Coult, D. A., and Boulter, D. (1969). *J. Exp. Bot.* **20**, 358–372.
Briarty, L. G., Hughes, C. E., and Evers, A. D. (1979). *Ann. Bot.* **44**, 641–658.
Brinegar, A. C., and Peterson, D. M. (1982a). *Arch. Biochem. Biophys.* **219**, 71–79.
Brinegar, A. C., and Peterson, D. M. (1982b). *Plant Physiol.* **70**, 1767–1769.
Brücher, O., Wecksler, M., Levy, A., Palozzo, A., and Jaffé, W. G. (1969). *Phytochemistry* **8**, 1739–1743.
Bures, L., Entlicher, G., Tichá, M., and Kocourek, J. (1973). *Experientia* **29**, 1546–1547.
Burr, B., Burr, F. A., Rubenstein, I., and Simon, M. N. (1978). *Proc. Nat. Acad. Sci. U.S.A.* **75**, 696–700.
Burr, F. A., and Burr, B. (1982). *J. Cell Biol.* **94**, 201–206.

Buttrose, M. S. (1963). *Aust. J. Biol. Sci.* **16**, 305-317.
Buttrose, M. S. (1971). *Planta* **96**, 13-26.
Buvat, R., and Robert, G. (1979). *Am. J. Bot.* **66**, 1219-1237.
Cameron-Mills, V., Ingversen, J., and Brandt, A. (1978). *Carlsberg Res. Commun.* **43**, 91-102.
Campbell, W. P., Lee, J. W., O'Brien, T. P., and Smart, M. G. (1981). *Aust. J. Plant Physiol.* **8**, 5-19.
Carasco, J. F., Croy, R., Derbyshire, E., and Boulter, D. (1978). *J. Exp. Bot.* **29**, 309-323.
Casey, R. (1979). *Biochem. J.* **177**, 509-520.
Chandler, P. M., Higgins, T. J. V., Randall, P. J., and Spencer, D. (1983). *Plant Physiol.* **71**, 47-54.
Chrispeels, M. J. (1976). *Annu. Rev. Plant Physiol.* **27**, 19-38.
Chrispeels, M. J. (1980). *In* 'The Biochemistry of Plants Volume 1. The Plant Cell' (N. E. Tolbert, ed.), pp. 389-412. Academic Press, New York.
Chrispeels, M. J., and Baumgartner, B. (1978). *Plant Physiol.* **61**, 617-623.
Chrispeels, M. J., and Boulter, D. (1975). *Plant Physiol.* **55**, 1031-1037.
Chrispeels, M. J., Higgins, T. J. V., and Spencer, D. (1982). *J. Cell Biol.* **93**, 306-313.
Clarke, A. E., and Knox, R. B. (1978). *Quart. Rev. Biol.* **53**, 3-28.
Clarke, A. E., Knox, R. B., and Jermyn, M. A. (1975). *J. Cell Sci.* **19**, 157-167.
Coan, M. H., and Travis, J. (1971). *In* 'Proceedings of the 1st International Conference on Proteinase Inhibitors' (H. Fritz and H. Tschesche, eds.), pp. 294-298. W. de Gruyter, Berlin.
Collier, M. D., and Murray, D. R. (1977). *Aust. J. Plant Physiol.* **4**, 571-582.
Craig, S., and Millerd, A. (1981). *Protoplasma* **105**, 333-339.
Craig, S., Goodchild, D. J., and Hardham, A. (1979a). *Aust. J. Plant Physiol.* **6**, 81-98.
Craig, S., Goodchild, D. J., and Millerd, A. (1979b). *J. Histochem. Cytochem.* **27**, 1312-1316.
Craig, S., Goodchild, D. J., and Miller, C. (1980a). *Aust. J. Plant Physiol.* **7**, 329-337.
Craig, S., Millerd, A., and Goodchild, D. J. (1980b). *Aust. J. Plant Physiol.* **7**, 339-351.
Croy, R. R. D., Gatehouse, J. A., Evans, I. M., and Boulter, D. (1980). *Planta* **148**, 49-56.
Cunningham, B. A., Wang, J. L., Waxdal, M. J., and Edelman, G. M. (1975). *J. Biol. Chem.* **250**, 1503-1512.
Cunningham, B. A., Hemperly, J. J., Hopp, T. P., and Edelman, G. M. (1979). *Proc. Nat. Acad. Sci. U.S.A.* **76**, 3651-3655.
Danielsson, C.-E. (1956). *Annu. Rev. Plant Physiol.* **7**, 215-236.
Davey, R. A., Higgins, T. J. V., and Spencer, D. (1981). *Biochem. Int.* **3**, 595-602.
Davies, D. R. (1980). *Biochem. Genet.* **18**, 1207-1219.
del Campillo, E., and Shannon, L. M. (1982). *Plant Physiol.* **69**, 628-631.
del Campillo, E., Shannon, L. M., and Hankins, C. N. (1981). *J. Biol. Chem.* **256**, 7177-7180.
Delmer, D. P., Kulow, C., and Ericson, M. C. (1978). *Plant Physiol.* **61**, 25-29.
Derbyshire, E., Wright, D. J., and Boulter, D. (1976). *Phytochemistry* **15**, 3-24.
Dey, P. M., Pridham, J. B., and Sumar, N. (1982). *Phytochemistry* **21**, 2195-2199.
Dhillon, S. S., and Miksche, J. P. (1982). *Am. J. Bot.* **69**, 219-226.
Dieckert, J. W., and Dieckert, M. C. (1976). *In* 'Genetic Improvements of Seed Proteins' pp. 18-56. Nat. Acad. Sci. Washington D.C., U.S.A.
Donovan, G. R., Lee, J. W., and Longhurst, T. J. (1982). *Aust. J. Plant Physiol.* **9**, 59-68.
Duranti, M., and Cerletti, P. (1979). *J. Agric. Food Chem.* **27**, 977-978.

Dure, L. III, and Chlan, C. (1981). *Plant Physiol.* **68**, 180-186.
Dure, L. III, and Galau, G. A. (1981). *Plant Physiol.* **68**, 187-194.
Edelman, G. M., Cunningham, B. A., Reeke, G. N. Jr., Becker, J. W., Waxdal, M. J., and Wang, J. L. (1972). *Proc. Nat. Acad. Sci. U.S.A.* **69**, 2580-2584.
Elbein, A. D. (1979). *Annu. Rev. Plant Physiol.* **30**, 239-272.
Engleman, E. M. (1966). *Am. J. Bot.* **53**, 231-237.
Ericson, M. C., and Delmer, D. P. (1977). *Plant Physiol.* **59**, 341-347.
Ericson, M. C., Gafford, J. T., and Elbein, A. D. (1977). *J. Biol. Chem.* **252**, 7431-7433.
Etzler, M. E., and Kabat, E. A. (1970). *Biochemistry* **9**, 869-877.
Evans, L. S., and Van't Hof, J. (1975). *Am. J. Bot.* **62**, 1060-1064.
Foriers, A., Van Driessche, E., de Neve, R., Kanarek, L., Strosberg, A. D., and Wuilmart, C. (1977a). *FEBS Lett.* **75**, 237-240.
Foriers, A., Wuilmart, C., Sharon, N., and Strosberg, A. D. (1977b). *Biochem. Biophys. Res. Commun.* **75**, 980-986.
Foriers, A., de Neve, R., Kanarek, L., and Strosberg, A. D. (1978). *Proc. Nat. Acad. Sci. U.S.A.* **75**, 1136-1139.
Foriers, A., Lebrun, E., Van Rapenbusch, R., de Neve, R., and Strosberg, A. D. (1981). *J. Biol. Chem.* **256**, 5550-5560.
Fulcher, R. G., O'Brien, T. P., and Simmonds, D. H. (1972). *Aust. J. Biol. Sci.* **25**, 487-497.
Fulcher, R. G., O'Brien, T. P., and Wong, S. I. (1981). *Cereal Chem.* **58**, 130-135.
Galbraith, W., and Goldstein, I. J. (1972). *Biochem.* **11**, 3976-3984.
Gatehouse, J. A., Croy, R. R. D., Morton, H., Tyler, M., and Boulter, D. (1981). *Eur. J. Biochem.* **118**, 627-633.
Gayler, K. R., and Sykes, G. E. (1981). *Plant Physiol.* **67**, 958-961.
Gennis, L. S., and Cantor, C. R. (1976a). *J. Biol. Chem.* **251**, 734-740.
Gennis, L. S., and Canton, C. R. (1976b). *J. Biol. Chem.* **251**, 741-746.
Gerstenberg, H., Belitz, H.-D., and Weder, J. K. P. (1980). *Lebensm. Unters.-Forsch.* **171**, 28-34.
Gibson, R., Kornfeld, S., and Schlesinger, S. (1980). *TIBS* **5**, 290-293.
Gifford, D. J., and Bewley, J. D. (1983). *Plant Physiol.* **72**, 376-381.
Gifford, D. J., Greenwood, J. S., and Bewley, J. D. (1982). *Plant Physiol.* **69**, 1471-1478.
Gilroy, J., Wright, D. J., and Boulter, D. (1979). *Phytochemistry* **18**, 315-316.
Goa, J., and Strid, L. (1959). *Arch. Mikrobiol.* **33**, 253-259.
Goldstein, I. J., Reichert, C. M., Misaki, A., and Gorin, P. A. J. (1973). *Biochim. Biophys. Acta* **317**, 500-504.
Goodchild, D. J., and Craig, S. (1982). *Aust. J. Plant Physiol.* **9**, 689-704.
Grant, D. R., Sumner, A. K., and Johnson, J. (1976). *Can. Inst. Food Sci. Technol.* **9**, 84-91.
Greene, F. C. (1981). *Plant Physiol.* **68**, 778-783.
Guldager, P. (1978). *Theor. Appl. Genet.* **53**, 241-250.
Hague, D. R. (1975). *Plant Physiol.* **55**, 636-642.
Hankins, C. N., and Shannon, L. M. (1978). *J. Biol. Chem.* **253**, 7791-7797.
Hankins, C. N., Kindinger, J. I., and Shannon, L. M. (1979). *Plant Physiol.* **64**, 104-107.
Hankins, C. N., Kindinger, J. I., and Shannon, L. M. (1980). *Plant Physiol.* **65**, 618-622.
Hara, I., and Matsubara, H. (1980). *Plant Cell. Physiol.* **21**, 233-245.
Hara-Nishimura, I., Nishimura, M., Matsubara, H., and Akazawa, T. (1982). *Plant Physiol.* **70**, 699-703.
Harris, N. (1979). *Planta* **146**, 63-69.

Harris, N., and Boulter, D. (1976). *Ann. Bot.* **40**, 739-744.
Harris, N., and Chrispeels, M. J. (1975). *Plant Physiol.* **56**, 292-299.
Harris, N., and Juliano, B. (1977). *Ann. Bot.* **41**, 1-5.
Harris, N., Chrispeels, M. J., and Boulter, D. (1975). *J. Exp. Bot.* **26**, 544-554.
Hawker, J. S., and Buttrose, M. S. (1980). *Ann. Bot.* **46**, 313-321.
Hayes, C. E., and Goldstein, I. J. (1974). *J. Biol. Chem.* **249**, 1904-1914.
Hemperly, J. J., Hopp, T. P., Becker, J. W., and Cunningham, B. A. (1979). *J. Biol. Chem.* **254**, 6803-6810.
Hemperly, J. J., Mostov, K. E., and Cunningham, B. A. (1982). *J. Biol. Chem.* **257**, 7903-7909.
Higgins, T. J. V., and Spencer, D. (1977). *Plant Physiol.* **60**, 655-661.
Higgins, T. J. V., and Spencer, D. (1981). *Plant Physiol.* **67**, 205-211.
Higgins, T. J. V., Chandler, P. M., Zurawski, G., Button, S. C., and Spencer, D. (1983a). *J. Biol. Chem.* **258**, 9544-9549.
Higgins, T. J. V., Chrispeels, M. J., Chandler, P. M., and Spencer, D. (1983b). *J. Biol. Chem.* **258**, 9550-9552.
Hill, J. E., and Breidenbach, R. W. (1974). *Plant Physiol.* **53**, 747-751.
Hobday, S. M., Thurman, D. A., and Barber, D. J. (1973). *Phytochemistry* **12**, 1041-1046.
Hochstrasser, K., Illchman, K., and Werle, E. (1970). *Hoppe-Seyler's Z. Physiol. Chem.* **351**, 1503-1512.
Hopp, T. P., Hemperly, J. J., and Cunningham, B. A. (1982). *J. Biol. Chem.* **257**, 4473-4483.
Horisberger, M., and Vonlanthen, M. (1980). *Histochemie* **65**, 181-186.
Horner, H. T. Jr., and Arnott, H. J. (1965). *Am. J. Bot.* **52**, 1027-1038.
Howard, I. K., and Sage, H. J. (1969). *Biochem.* **8**, 2436-2441.
Howard, J., Kindinger, J., and Shannon, L. M. (1979). *Arch. Biochem. Biophys.* **192**, 457-465.
Hurich, J., Parzysz, H., and Przybylska, J. (1977). *Genetica Polonica* **18**, 241-252.
Hurkman, W. J., and Beevers, L. (1980). *Planta* **150**, 82-88.
Hurkman, W. J., and Beevers, L. (1982). *Plant Physiol.* **69**, 1414-1417.
Hwang, D. L.-R., Lin, K.-T. D., Yang, W.-K., and Foard, D. E. (1977). *Biochim. Biophys. Acta* **495**, 369-382.
Hwang, D. L., Yang, W.-K., Foard, D. E., and Lin, K.-T. D. (1978). *Plant Physiol.* **61**, 30-34.
Ingversen, J. (1975). *Hereditas* **81**, 69-76.
Irimura, T., and Osawa, T. (1972). *Arch. Biochem. Biophys.* **151**, 475-482.
Jackson, P., Boulter, D., and Thurman, D. A. (1969). *New Phytol.* **68**, 25-33.
Jakubek, M., and Przybylska, J. (1979). *Genetica Polonica* **20**, 369-380.
Janardhanan, K. (1982). 'Studies on Seed Development and Germination in *Mucuna utilis* Wall. ex Wt. (Papilionaceae);. Ph.D. Thesis, University of Madras, India.
Janzen, D. H., Juster, H. B., and Liener, I. E. (1976). *Science* **192**, 795-796.
Janzen, D. H., Juster, H. B., and Bell, E. A. (1977). *Phytochemistry* **16**, 223-227.
Juliano, B. O. (1972). *In* 'Rice Chemistry and Technology' (D. F. Houston, ed.), pp. 16-74. American Association of Cereal Chemists Inc., St. Paul, Minnesota.
Juniper, B. E., Hawes, C. R., and Horne, J. C. (1982). *Bot. Gaz.* **143**, 135-145.
Kauss, H., Thomson, K. S., Tetour, M., and Jeblick, W. (1978). *Plant Physiol.* **61**, 35-37.
Kauss, H., Thomson, K. S., Thomson, M., and Jeblick, W. (1979). *Plant Physiol.* **63**, 455-459.

Khan, M. R. I., Gatehouse, J. A., and Boulter, D. (1980). *J. Exp. Bot.* **31**, 1599-1611.
Khoo, U., and Wolf, M. J. (1970). *Am. J. Bot.* **57**, 1042-1050.
Kilpatrick, D. C., and Yeoman, M. M. (1978). *Biochem. J.* **175**, 1151-1153.
Kilpatrick, D. C., Yeoman, M. M., and Gould, A. R. (1979). *Biochem. J.* **184**, 215-219.
Kilpatrick, D. C., Jeffree, C. E., Lockhart, C. M., and Yeoman, M. M. (1980). *FEBS Lett.* **113**, 129-133.
Kirsi, M. (1974). *Physiol. Plant* **32**, 89-93.
Kirsi, M., and Mikola, J. (1971). *Planta* **96**, 281-291.
Klozova, E., and Turková, V. (1978). *Biol. Plant.* **20**, 129-134.
Koide, T., Tsunasawa, S., and Ikenaka, T. (1972). *J. Biochem.* **71**, 165-167.
Kolberg, J., and Sletten, K. (1982). *Biochim. Biophys. Acta* **704**, 26-30.
Kollöffel, C. (1970). *Planta* **91**, 321-328.
Kortt, A. A. (1979). *Biochim. Biophys. Acta* **577**, 371-382.
Kortt, A. A. (1980). *Biochim. Biophys. Acta* **624**, 237-248.
Kortt, A. A. (1981). *Biochim. Biophys. Acta* **657**, 212-221.
Kortt, A. A. (1984). *Eur. J. Biochem.* **138**, 519-525.
Kortt, A. A., and Jermyn, M. A. (1981). *Eur. J. Biochem.* **115**, 551-557.
Krahn, J., and Stevens, F. C. (1970). *Biochem.* **9**, 2646-2652.
Kunitz, M. (1945). *Science* **101**, 668-669.
Kunitz, M. (1946). *J. Gen. Physiol.* **29**, 149-154.
Kunitz, M. (1947). *J. Gen. Physiol.* **30**, 291-310.
Kurokawa, T., Tsuda, M., and Sugino, Y. (1976). *J. Biol. Chem.* **251**, 5686-5693.
Kyle, D. J., and Styles, E. D. (1977). *Planta* **137**, 185-193.
Larkins, B. A., and Hurkman, W. J. (1978). *Plant Physiol.* **62**, 256-263.
Larkins, B. A., Bracker, C. E., and Tsai, C. Y. (1976a). *Plant Physiol.* **57**, 740-745.
Larkins, B. A., Jones, R. A., and Tsai, C. Y. (1976b). *Biochem.* **15**, 5506-5511.
Larkins, B. A., Pearlmutter, N. L., and Hurkman, W. J. (1979). In 'The Plant Seed. Development, Preservation, Germination' (I. Rubenstein, R. L. Phillips, C. E. Green, and B. C. Gengenbach, eds.), pp. 49-65. Academic Press, New York.
Leder, P., Max, E. E., and Seidman, J. G. (1980). In 'Immunology 80' (M. Fougereau and J. Dausset, eds.), Vol. 1, pp. 34-50. Academic Press, London.
Liener, I. E. (1953). *J. Nutr.* **49**, 527-539.
Liener, I. E. (1963). *Econ. Bot.* **18**, 27-33.
Liener, I. E. (1976). *Annu. Rev. Plant Physiol.* **27**, 291-319.
Liener, I. E. (1980). In 'Advances in Legume Science' (R. J. Summerfield and A. H. Bunting, eds.), pp. 157-170. Royal Botanic Gardens, Kew.
Liener, I. E., and Kakade, M. L. (1980). In 'Toxic Constituents of Plant Foodstuffs' (I. E. Liener, ed.), pp. 7-71. Academic Press, New York.
Lis, H., and Sharon, N. (1973). *Annu. Rev. Biochem.* **42**, 541-574.
Lis, H., and Sharon, N. (1978). *J. Biol. Chem.* **253**, 3468-3476.
Lis, H., and Sharon, N. (1981). In 'The Biochemistry of Plants. Vol. 6. Proteins and Nucleic Acids' (A. Marcus, ed.), pp. 371-447. Academic Press, New York.
Lis, H., Sela, B., Sachs, L., and Sharon, N. (1970). *Biochim. Biophys. Acta* **211**, 582-585.
Loewenberg, J. R. (1955). *Plant Physiol.* **30**, 244-250.
Lott, J. N. A. (1980). In 'The Biochemistry of Plants. Vol. 1. The Plant Cell' (N. E. Tolbert, ed.), pp. 589-623. Academic Press, New York.
Luthe, D. S., and Peterson, D. M. (1977). *Plant Physiol.* **59**, 836-841.
Madison, J. T., Thompson, J. F., and Muenster, A.-M. E. (1976). *Ann. Bot.* **40**, 745-756.
Malley, A., Baecher, L., Mackler, B., and Perlman, F. (1975). *J. Allergy Clin. Immunol.* **56**, 282-290.

Manen, J. F., and Pusztai, A. (1982). *Planta* **155**, 328-334.
Marcu, K., and Dudock, B. (1974). *Nucleic Acids Res.* **1**, 1385-1397.
Marriott, K. M., and Tanner, W. (1979). *Plant Physiol.* **64**, 445-449.
Marty, F., Branton, D., and Leigh, R. A. (1980). In 'The Biochemistry of Plants. Vol. 1. The Plant Cell' (N. E. Tolbert, ed.), pp. 625-658. Academic Press, New York.
Matile, Ph. (1968). *Z. Pflanzephysiol.* **58**, 365-368.
Matile, Ph. (1976). In 'Plant Biochemistry' (J. Bonner and J. E. Varner, eds.), pp. 189-224. Academic Press, New York.
Matile, Ph. (1978). *Annu. Rev. Plant Physiol.* **29**, 193-213.
Matsumoto, I., and Osawa, T. (1974). *Biochem.* **13**, 582-588.
Mbadiwe, E. I., and Agogbua, S. I. O. (1978). *Phytochemistry* **17**, 1057-1058.
Melcher, U. (1979). *Plant Physiol.* **63**, 354-358.
Melville, J. C., and Scandalios, J. G. (1972). *Biochem. Genet.* **7**, 15-31.
Mettler, I. J., and Beevers, H. (1979). *Plant Physiol.* **64**, 506-511.
Mialonier, G., Privat, J.-P., Monsigny, M., Kahlem, G., and Durand, R. (1973). *Physiol. Vég.* **11**, 519-537.
Miège, J., and Miège, M.-N. (1978). *Econ. Bot.* **32**, 336-345.
Miège, M.-N., Mascherpa, J.-M. Royer-Spierer, A., Grange, A., and Miège, J. (1976). *Planta* **131**, 81-86.
Miflin, B. J., and Burgess, S. R. (1982). *J. Exp. Bot.* **33**, 251-260.
Miflin, B. J., and Shewry, P. R. (1979). In 'Recent Advances in the Biochemistry of Cereals' (D. L. Laidman and R. G. Wyn Jones, eds.), pp. 239-273. Academic Press, London.
Miflin, B. J., Burgess, S. R., and Shewry, P. R. (1981). *J. Exp. Bot.* **32**, 199-219.
Mikola, J., and Enari, T.-M. (1970). *J. Inst. Brew.* **76**, 182-188.
Mikola, J., and Kirsi, M. (1972). *Acta. Chem. Scand.* **26**, 787-795.
Miller, J. B., Noyes, C., Heinrikson, R., Kingdon, H. S., and Yachnin, S. (1973). *J. Exp. Med.* **138**, 939-951.
Miller, J. B., Hsu, R., Heinrikson, R., and Yachnin, S. (1975). *Proc. Nat. Acad. Sci. U.S.A.* **72**, 1388-1391.
Millerd, A. (1975). *Annu. Rev. Plant Physiol.* **26**, 53-72.
Millerd, A., and Spencer, D. (1974). *Aust. J. Plant Physiol.* **1**, 331-341.
Millerd, A., and Whitfeld, P. R. (1973). *Plant Physiol.* **51**, 1005-1010.
Millerd, A., Thomson, J. A., and Schroeder, H. E. (1978). *Aust. J. Plant Physiol.* **5**, 519-534.
Mirelman, D., Galun, E., Sharon, N., and Lotan, R. (1975). *Nature* **256**, 414-416.
Mollenhauer, H. H., and Morré, D. J. (1976). *Cytobiology* **13**, 297-306.
Mollenhauer, H. H., and Morré, D. J. (1980). In 'The Biochemistry of Plant Cells. Vol. 1. The Plant Cell' (N. E. Tolbert, ed.), pp. 437-488. Academic Press, New York.
Moreira, M. A., Hermodson, M. A., Larkins, B. A., and Nielsen, N. C. (1981a). *Arch. Biochem. Biophys.* **210**, 633-642.
Moreira, M. A., Mahoney, W. C., Larkins, B. A., and Nielsen, N. C. (1981b). *Arch. Biochem. Biophys.* **210**, 643-646.
Morré, D. J., and Mollenhauer, H. H. (1974). In 'Dynamic Aspects of Plant Ultrastructure' (A. W. Robards, ed.), pp. 84-137. McGraw-Hill, New York.
Morris, G. F. I.,Thurman, D. A., and Boulter, D. (1970). *Phytochemistry* **9**, 1707-1714.
Morrison, I. N., Kuo, J., and O'Brien, T. P. (1975). *Planta* **123**, 105-116.
Mosolov, V. V., Loginova, M. D., Malova, E. L., and Benken, I. I. (1979). *Planta* **144**, 265-269.
Murphy, L. A., and Goldstein I. J. (1977). *J. Biol. Chem.* **252**, 4739-4742.
Murray, D. R. (1979a). *Plant,Cell Environ.* **2**, 223-226.
Murray, D. R. (1979b). *Z. Pflanzenphysiol.* **93**, 423-428.

Murray, D. R. (1979c). *Planta* **147**, 117-121.
Murray, D. R. (1979d). *Plant Physiol.* **64**, 763-769.
Murray, D. R. (1982). *Z. Pflanzenphysiol.* **108**, 17-25.
Murray, D. R. (1983a). *New Phytol.* **93**, 33-41.
Murray, D. R. (1983b). *Z. Pflanzenphysiol.* **110**, 7-15.
Murray, D. R., and Vairinhos, F. (1982a). *Z. Pflanzenphysiol.* **106**, 465-468.
Murray, D. R., and Vairinhos, F. (1982b). *Z. Pflanzenphysiol.* **108**, 181-185.
Murray, D. R., and Vairinhos, F. (1982c). *Z. Pflanzenphysiol.* **108**, 471-476.
Murray, D. R., Mackenzie, K. F., Vairinhos, F., Peoples, M. B., Atkins, C. A., and Pate, J. S. (1983). *Z. Pflanzenphysiol.* **109**, 363-370.
Nagahashi, J., and Beevers, L. (1978). *Plant Physiol.* **61**, 451-459.
Nagahashi, J., Browder, S. K., and Beevers, L. (1980). *Plant Physiol.* **65**, 648-657.
Nagata, Y., and Burger, M. M. (1974). *J. Biol. Chem.* **249**, 3116-3122.
Nagl, W. (1974). *Z. Pflanzenphysiol.* **73**, 1-44.
Nagl, W. (1979). *Z. Pflanzenphysiol.* **95**, 283-314.
Nicolson, G. L., Blaudstein, J., and Etzler, M. E. (1974). *Biochem.* **13**, 196-204.
Nishimura, M., and Beevers, H. (1978). *Plant Physiol.* **62**, 44-48.
Odani, S., and Ikenaka, T. (1972). *J. Biochem.* **71**, 839-848.
Odani, S., and Ikenaka, T. (1973). *J. Biochem.* **74**, 857-860.
Odani, S., and Ikenaka, T. (1976). *J. Biochem.* **80**, 641-643.
Odani, S., and Ikenaka, T. (1978a). *J. Biochem.* **83**, 737-745.
Odani, S., and Ikenaka, T. (1978b). *J. Biochem.* **83**, 747-753.
Odani, S., Odani, S., Ono, T., and Ikenaka, T. (1979). *J. Biochem.* **86**, 1795-1805.
Odani, S., Ono, T., and Ikenka, T. (1980). *J. Biochem.* **88**, 297-301.
Olsnes, S., and Pihl, A. (1973). *Eur. J. Biochem.* **35**, 179-185.
Olsnes, S., Refsnes, K., and Pihl, A. (1974a). *Nature* **249**, 627-631.
Olsnes, S., Saltvedt, E., and Pihl, A. (1974b). *J. Biol. Chem.* **249**, 803-810.
Olson, M. O. J., and Liener, I. E. (1967). *Biochem.* **6**, 105-111.
Öpik, H. (1966). *J. Exp. Bot.* **17**, 427-439.
Orf, J. H., and Hymowitz, T. (1979). *Crop Sci.* **19**, 107-109.
Orf, J. H., Hymowitz, T., Pull, S. P., and Pueppke, S. G. (1978). *Crop Sci.* **18**, 899-900.
Park, W. D., Lewis, E. D., and Rubinstein, I. (1980). *Plant Physiol.* **65**, 98-106.
Parker, M. L. (1980). *Ann. Bot.* **46**, 29-36.
Parker, M. L. (1982). *Plant, Cell Environ.* **5**, 37-43.
Parker, M. L., and Hawes, C. R. (1982). *Planta* **154**, 277-283.
Paus, E., and Steen, H. B. (1978). *Nature* **272**, 452-454.
Pereira, M. E. A., Gruezo, F., and Kabat, E. A. (1979). *Arch. Biochem. Biophys.* **194**, 511-525.
Peterson, D. M. (1976). *Crop Sci.* **16**, 663-666.
Peterson, D. M. (1978). *Plant Physiol.* **62**, 506-509.
Plant, A. R., and Moore, K. G. (1982). *Phytochemistry* **21**, 985-989.
Poretz, R. D., Riss, H., Timberlake, J. W., and Chien, S. (1974). *Biochem.* **13**, 250-256.
Püchel, M., Müntz, K., Parthier, B., Aurich, O., Bassüner, R., Manteuffel, R., and Schmidt, P. (1979). *Eur. J. Biochem.* **96**, 321-329.
Pull, S. P., Pueppke, S. G., Hymowitz, T., and Orf, J. G. (1978). *Science* **200**, 1277-1279.
Pusztai, A. (1972). *Planta* **107**, 121-129.
Pusztai, A., and Watt, W. B. (1974). *Biochim. Biophys. Acta* **365**, 57-71.
Pusztai, A., Croy, R. R. D., Grant, G., and Watt, W. B. (1977). *New Phytol.* **79**, 61-71.
Randall, P. J., Thomson, J. A., and Schroeder, H. E. (1979). *Aust. J. Plant Physiol.* **6**, 11-24.
Reeke, G. N. Jr., Becker, J. W., and Edelman, G. M. (1975). *J. Biol. Chem.* **250**, 1525-1547.

Reeke, G. N. Jr., Becker, J. W., and Edelman, G. M. (1978). *Proc. Nat. Acad. Sci. U.S.A.* **75**, 2286-2290.
Righetti, P. G., Gianazza, E., Viotti, A., and Soave, C. (1977). *Planta* **136**, 115-123.
Roberts, L. M., and Lord, J. M. (1981). *Eur. J. Biochem.* **119**, 31-41.
Rothman, J. E. (1981). *Science* **213**, 1212-1219.
Rougé, P. (1974). *C. R. Acad. Sci. Paris* **278D**, 449-452.
Royer, A. (1975). *Phytochemistry* **14**, 915-919.
Ryan, C. A. (1973). *Annu. Rev. Plant Physiol.* **24**, 173-196.
Sachs, J. (1875). 'Textbook of Botany — Morphological and Physiological' (A. W. Bennett and W. T. T. Dyer, translators). The University of Oxford (Clarendon Press), MacMillan and Co., London.
Salmia, M. A. (1980). *Physiol. Plant.* **48**, 266-270.
Samac, D., and Storey, R. (1981). *Plant Physiol.* **68**, 1339-1344.
Scharpé, A., and Van Parijs, R. (1973). *J. Exp. Bot.* **24**, 216-222.
Schroeder, H. E. (1982). *J. Sci. Food Agric.* **33**, 623-633.
Seidl, D. S., and Liener, I. E. (1971). *Biochim. Biophys. Acta* **251**, 83-93.
Shain, Y., and Mayer, A. M. (1965). *Physiol. Plant.* **18**, 853-859.
Shain, Y., and Mayer, A. M. (1968). *Phytochemistry* **7**, 1491-1498.
Shankar Iyer, P. N., Wilkinson, K. D., and Goldstein, I. J. (1976). *Arch. Biochem. Biophys.* **177**, 330-333.
Shore, G. C., and Tata, J. R. (1977). *Biochim. Biophys. Acta* **472**, 197-236.
Shumway, L. K., Yang, V. V., and Ryan, C. A. (1976). *Planta* **129**, 161-165.
Simola, L. K. (1969). *Flora* B **158**, 645-658.
Sletten, K., Kolberg, J., and Michaelsen, T. E. (1983). *FEBS Lett.* **156**, 253-256.
Smith, D. L. (1973). *Ann. Bot.* **37**, 795-804.
Smith, S. C., Johnson, S., Andrews, J., and McPherson, A. (1982). *Plant Physiol.* **70**, 1199-1209.
Spencer, D., and Higgins, T. J. V. (1980). *Biochem. Int.* **1**, 502-509.
Spencer, D., Higgins, T. J. V., Button, S. C., and Davey, R. A. (1980). *Plant Physiol.* **66**, 510-515.
Spielmann, A., Schürmann, P., and Stutz, E. (1982). *Plant Sci. Lett.* **24**, 137-145.
Staswick, P. E., Hermodson, M. A., and Nielsen, N. C. (1981). *J. Biol. Chem.* **256**, 8752-8755.
Staswick, P. E., Broué, P. and Nielsen, N. C. (1983). *Plant Physiol.* **72**, 1114-1118.
Stevens, F. C., Wuerz, S., and Krahn, J. (1974). *In* 'Proteinase Inhibitors' (H. Fritz, H. Tschesche, L. J. Greene and E. Truscheit, eds.), pp. 344-354. Springer-Verlag, Berlin.
Sumner, J. B. (1919). *J. Biol. Chem.* **37**, 137-141.
Sumner, J. B., and Howell, S. F. (1936). *J. Biol.Chem.* **115**, 583-588.
Sundberg, L., Porath, J., and Aspberg, K. (1970). *Biochim. Biophys. Acta* **221**, 394-395.
Tan, C. G. L., and Stevens, F. C. (1971a). *Eur. J. Biochem.* **18**, 503-514.
Tan, C. G. L., and Stevens, F. C. (1971b). *Eur. J. Biochem.* **18**, 515-523.
Tanaka, K., Sugimoto, T., Ogawa, M., and Kasai, Z. (1980). *Agric. Biol. Chem.* **44**, 1633-1639.
Thomson, J. A., and Schroeder, H. E. (1978). *Aust. J. Plant Physiol.* **5**, 281-294.
Thomson, J. A., Schroeder, H. E., and Dudman, W. F. (1978). *Aust. J. Plant Physiol.* **5**, 263-279.
Toms, G. C. (1981). *In* 'Advances in Legume Systematics' (R. M. Polhill and P. H. Raven, eds.), Part 2, pp. 561-577. Royal Botanic Gardens, Kew, U.K.
Toms, G. C., and Western, A. (1971). *In* 'Chemotaxonomy of the Leguminosae' (J. B. Harborne, D. Boulter and B. L. Turner, eds.), pp. 367-462. Academic Press, London.
Tully, R. E., and Beevers, H. (1976). *Plant Physiol.* **58**, 710-716.

Turner, N. E., Richter, J. D., and Nielsen, N. C. (1982). *J. Biol. Chem.* **257**, 4016-4018.
Uy, R., and Wolf, F. (1977). *Science* **196**, 890-896.
Vairinhos, F., and Murray, D. R. (1982a). *Z. Pflanzenphysiol.* **106**, 447-452.
Vairinhos, F., and Murray, D. R. (1982b). *Z. Pflanzenphysiol.* **107**, 25-32.
Vairinhos, F., and Murray, D. R. (1983). *Plant System Evol.* **142**, 11-22.
Van der Wilden, W., and Chrispeels, M. J. (1983). *Plant Physiol.* **71**, 82-87.
Van der Wilden, W., Gilkes, N. R., and Chrispeels, J. M. (1980a). *Plant Physiol.* **66**, 390-394.
Van der Wilden, W., Herman, E. M., and Chrispeels, M. J. (1980b). *Proc. Nat. Acad. Sci. U.S.A.* **77**, 428-432.
Varasundharosoth, D. (1982). 'The Fractionation and Analysis of Lupin Seed Proteins with a Detailed Investigation of the Albumins'. Ph.D. Thesis, The University of Canterbury, N.Z.
Varner, J. E., and Schidlovsky, G. (1963). *Plant Physiol.* **38**, 139-144.
Wall, J. S. (1979). In 'Recent Advances in the Biochemistry of Cereals' (D. L. Laidman and R. G. Wyn Jones, eds.), pp. 275-311. Academic Press, London.
Wang, J. L., Becker, J. W., Reeke, G. N. Jr., and Edelman, G. M. (1974). *J. Mol. Biol.* **88**, 259-262.
Wang, J. L., Cunningham, B. A., Waxdal, M. J., and Edelman, G. M. (1975). *J. Biol. Chem.* **250**, 1490-1502.
Weber, E., Manteuffel, R., and Neumann, D. (1978). *Biochem. Physiol. Pflanzen* **172**, 597-614.
Weder, J. K. P. (1973). *Z. Lebensm. Unters.-Forsch.* **153**, 83-86.
Weder, J. K. P. (1978). *Z. Pflanzenphysiol.* **90**, 285-291.
Weder, J. K. P. (1981). In 'Advances in Legume Systematics' (R. M. Polhill and P. H. Raven, eds.), Part 2, pp. 533-560. Royal Botanic Gardens, Kew.
Weder, J. K. P., and Hory, H.-D. (1972a). *Lebensm. Wiss. u. Technol.* **5**, 54-63.
Weder, J. K. P., and Hory, H.-D. (1972b). *Lebensm. Wiss. u. Technol.* **5**, 86-90.
Weder, J. K. P., and Murray, D. R. (1981). *Z. Pflanzenphysiol.* **103**, 317-322.
Weder, J. K. P., Hegarty, M. P., Holzner, M., and Kern-Dirndorfer, M.-L. (1983). *Z. Lebensm. Unters. Forsch.* **177**, 109-113.
Wiemken, A. (1975). In 'Methods in Cell Biology' (D. M. Prescott, ed.), XII, pp. 99-109. Academic Press, New York.
Wilson, K. A., and Chen, J. C. (1983). *Plant Physiol.* **71**, 341-349.
Wilson, K. A., and Laskowski, M. Sr. (1973). *J. Biol. Chem.* **248**, 756-762.
Wilson, K. A., and Laskowski, M. Sr. (1975). *J. Biol. Chem.* **250**, 4261-4267.
Wrigley, C. W., du Cros, D. L., Archer, M. J., Downie, P. G., and Roxburgh, C. M. (1980). *Aust. J. Plant Physiol.* **7**, 755-766.
Wrigley, C. W., Robinson, P. J., and Williams, W. T. (1981). *J. Sci. Food Agric.* **32**, 433-442.
Wrigley, C. W., Lawrence, G. J., and Shepherd, K. W. (1982). *Aust. J. Plant Physiol.* **9**, 15-30.
Xavier-Filho, J. (1973). *Physiol. Plant* **28**, 149-154.
Yamagata, H., Tanaka, K., and Kasai, Z. (1982a). *Agric. Biol. Chem.* **46**, 321-322.
Yamagata, H., Sugimoto, T., Tanaka, K., and Kasai, Z. (1982b). *Plant Physiol.* **70**, 1094-1100.
Yatsu, L. Y., and Jacks, T. J. (1968). *Arch. Biochem. Biophys.* **124**, 466-471.
Yoo, B. Y., and Chrispeels, M. J. (1980). *Protoplasma* **103**, 201-204.
Youle, R. J., and Huang, A. H. C. (1976). *Plant Physiol.* **58**, 703-709.
Youle, R. J., and Huang, A. H. C. (1978a). *Plant Physiol.* **61**, 13-16.

Youle, R. J., and Huang, A. H. C. (1978b). *Plant Physiol.* **61**, 1040-1042.
Youle, R. J., and Huang, A. H. C. (1979). *J. Agric. Food Chem.* **27**, 500-503.
Youle, R. J., and Huang, A. H. C. (1981). *Am. J. Bot.* **68**, 44-48.

CHAPTER 4

Accumulation of Seed Reserves of Phosphorus and Other Minerals

J. N. A. LOTT

I.	Introduction	139
II.	Phosphorus Reserves in Seeds	140
	A. Phytin	140
	B. Phytic Acid Synthesis	141
	C. Protein Bodies and the Formation of Globoid Crystals	142
	D. Nutritional Importance of Phytin to Man	147
III.	Distribution of Mineral Reserves within Seeds	147
	A. Concentration of Minerals in Certain Tissues	147
	B. Cellular and Intracellular Localization of Specific Elements	149
	C. Differences at Species, Cell and Protein Body Levels	156
IV.	Influence of External Soil Conditions on Mineral Levels in Seeds	157
V.	Calcium-rich Crystals	159
VI.	Procedures for Studying Minerals in Seeds	159
	A. Chemical Analysis	159
	B. Light Microscopy	160
	C. Energy Dispersive X-Ray Analysis	160
	D. Phytic Acid Determinations	161
VII.	Future Research	162
	References	163

I. INTRODUCTION

Seeds are an essential part of the life cycle of most higher plants. The developing seed accumulates reserves of nitrogen (Chapter 3), carbohydrates (Chapter 5), lipids (Chapter 6) and minerals. This chapter is concerned with the storage of minerals in seed tissues. A variety of minerals may be present in the cell walls and organelles, but the major proportions of phosphorus and several mineral cations are found together as the compound phytin. Phytin is the main constituent of the globoid inclusions present inside protein bodies. Protein bodies are, therefore, not simply sites for the storage of protein. Following imbibition and germination, most protein bodies revert to vacuoles, as their contents of protein and phytin alike are degraded (Poux, 1963; Matile, 1975; Van der Wilden *et al.*, 1980; Herman *et al.*, 1981).

II. PHOSPHORUS RESERVES IN SEEDS

A. Phytin

Most of the phosphorus present in seeds is in the form of phytin (Table I). Phytin is a salt of *myo*-inositol hexaphosphoric acid (phytic acid). Phytin is thus a store for phosphorus, carbohydrate and a variety of cations. The molecular structure of phytate was determined by Johnson and Tate (1969). Phytin frequently constitutes from one to several per cent of the seed's dry weight (Makower, 1969; Toma *et al.*, 1979). In rare instances, phytin content may reach eight per cent by weight (Mukhamedova and Akramov, 1977).

As already indicated, phytin reserves are mainly located inside protein bodies, and are especially concentrated in the globoid crystals. Phytin is the largest single component of globoid crystals. For example, Lui and Altschul (1967) found that isolated cotton seed (*Gossypium hirsutum*) globoids were low in protein, fat and carbohydrates but contained 14.2% phosphorus (P), 60% phytic acid and about 10% other minerals including potassium (K), magnesium (Mg) and calcium (Ca). They found that in cotton globoids about 97.5% of the total P was organic P, with phytic acid being the only organic P compound. Isolated castor bean (*Ricinus communis*) globoids were reported to be 77.5% phytin (Sobolev, 1966). Globoids isolated from rice (*Oryza sativa*) grains contained 66.68% phytic acid, 13.4% K, 11.9% Mg and 0.8% Ca (Ogawa *et al.*, 1975).

Although minerals are mainly concentrated in the globoid crystals, some mineral storage may take place in the proteins of protein bodies. Mineral cations may be bound directly to reserve proteins; for example, a manganese-containing protein has been isolated from protein bodies of the peanut (*Arachis hypogaea*) (Rozacky, 1968). Lectins are also metalloproteins, and may contain manganese (Mn) and Ca (Section 3.II,A). Minerals in the proteinaceous region of protein bodies may also form part

4. Mineral Reserves

Table I. The contribution of phytate to total seed phosphorus.

Species	Phytic acid P as % of total P	References
Monocotyledons (grains)		
Avena sativa, oats	48.7–70.9	Ashton and Williams (1958)
	53	Hall and Hodges (1966)
	56.7–65.4	Lolas *et al.* (1976)
Hordeum vulgare, barley	66.1–69.6	Lolas *et al.* (1976)
Oryza sativa, rice	80.6[c]	O'Dell *et al.* (1972)
Triticum aestivum, wheat	76.2[c]	O'Dell *et al.* (1972)
	61.7–79.9	Lolas *et al.* (1976)
	55–83	Nahapetian and Bassiri (1976)
	38–84	Bassiri and Nahapetian (1977)
Zea mays, maize	83.3[c]	O'Dell *et al.* (1972)
	77	Amoa and Muller (1976)
	79.3	Deosthale (1979)
opaque (high lysine)	87.1[c]	O'Dell *et al.* (1972)
Dicotyledons (usually embryo)		
Arachis hypogaea, peanut	57	Paul and Southgate (1978)
Bertholletia excelsa	86	Paul and Southgate (1978)
Glycine max, soybean	70	Smith and Rackis (1957)
	51.4–57.1[a]	Lolas *et al.* (1976)
Gossypium hirsutum, cotton	81.5–83.4	Ergle and Guinn (1959)
Phaseolus vulgaris, bean	53.6–81.6[a]	Lolas and Markakis (1975)
Pisum sativum, pea	53[b,c]	Guardiola and Sutcliffe (1971)
Prunus dulcis, almond	82	Paul and Southgate (1978)
Vicia faba, broad bean	39.5–47.2[a]	Griffiths and Thomas (1981)
cv. Minor (field)	39.5	Griffiths and Thomas (1981)
cv. Minor (glasshouse)	57.7	Griffiths and Thomas (1981)

[a] Analyses on whole seeds.
[b] Analyses on cotyledons.
[c] Calculated from data in the reference cited.

of a protein–phytin complex (Table II). Protein–phytin complexes are believed to occur naturally, but may also be formed during extraction.

B. Phytic Acid Synthesis

Synthesis of phytic acid appears to take place in the tissue in which it will be stored. For example, in developing rice grains, Ogawa *et al.* (1979c) showed that phytic acid was synthesized in the aleurone layer and the scutellum, but not in the starchy endosperm. The synthesis of phytic acid

Table II. The occurrence of protein-phytin complexes in seeds.

Species	References
Monocotyledons	
Hordeum vulgare, barley	Tluczkiewicz and Berendt (1977)
Secale cereale, rye	Tluczkiewicz and Berendt (1977)
Triticum aestivum, wheat	Tluczkiewicz and Berendt (1977)
Dicotyledons	
Arachis hypogaea, peanut	Fontaine et al. (1946)
Brassica napus, rape	Schwenke et al. (1979)
Cucurbita digitata[a]	Bolley and McCormack (1952)
Cucurbita foetidissima[a]	Bolley and McCormack (1952)
Cucurbita palmata[a]	Bolley and McCormack (1952)
Glycine max, soybean	Smith and Rackis (1957)
	Tombs (1967)
	Prattley and Stanley (1982)
Gossypium hirsutum, cotton	Fontaine et al. (1946)
	Ergle and Guinn (1959)
Linum usitatissimum, flax[a]	Bolley and McCormack (1952)
Phaseolus vulgaris, bean	Bourdillon (1951)
	Lolas and Markakis (1975)
Pisum sativum, pea	Fontaine et al. (1946)

[a] The existence of a protein-phytin complex was not established, but phytin was shown to be present in a protein rich fraction.

proceeds by phosphorylation of *myo*-inositol (Asada and Kasai, 1962; Saio, 1964; Tanaka et al., 1976). The isotope [^{3}H]*myo*-inositol applied to developing rice grains was concentrated in the protein bodies (aleurone particles), indicating that these structures are indeed the sites of accumulation of phytic acid (Tanaka et al., 1974). For additional information on *myo*-inositol metabolism, see Section 5.II,A.4, Cosgrove (1966) and Loewus and Loewus (1983).

C. Protein Bodies and the Formation of Globoid Crystals

Protein bodies may contain only a structurally homogeneous proteinaceous matrix, but more commonly, there are inclusions of one or more types. The globoid crystals may be small and infrequent (Fig. 1), small and numerous (Fig. 2), or of various sizes in a single protein body (Figs 3, 4). Other types of inclusion that may be present in protein bodies are niacin deposits (Fig. 3) and crystals, both single prismatic types, and druses (Figs 5, 6). The globoid crystals of castor bean (Figs 7, 8) are comparatively large. In this species, and many others, the protein bodies acquire distinct inclusions of crystalline protein as well as the globoid crystals (Figs 9, 10). Further details of protein body structure, composition and structural similarities within taxonomic groups have been described by Buttrose and Lott (1978b), Pernollet (1978), Weber and Neumann (1980), Lott (1980, 1981), Fulcher et al. (1981) and Webb and Arnott (1982).

Although it has received little attention, the study of the formation of phytin reserves during seed development should be a most interesting field. A few studies indicate that some exchange of bound cations occurs during seed development. Microprobe analyses of developing rice grains showed that P and Mg began to concentrate in the aleurone layer about the 12th day after flowering, whereas the K concentration did not begin to increase until about 19 days after flowering (Ogawa et al., 1979a). More sensitive energy dispersive X-ray (EDX) analyses of globoids from developing rice aleurone protein bodies were also conducted by the same group (Ogawa et al., 1979b). At seven to ten days after flowering, the globoids were mainly P and Mg with minor amounts of Ca, zinc (Zn) and K. By 17 to 19 days after flowering, the globoids contained an increased amount of K along with P and Mg, but they had no detectable Ca or Zn. These results suggest that the elements Ca and Zn are available only during the early stages of globoid crystal development, and that they are exchanged for K during the later stages of aleurone layer development. A somewhat similar shift has been observed in the composition of globoid crystals in developing castor bean endosperm. The electron dense deposits of phytin formed early in seed development contain Mg, Ca and P, but little or no K (Greenwood, 1983), whereas the globoids of the mature seed contain predominantly P, Mg and K (Lott et al., 1982).

The earlist detectable structures that are likely to be phytin are located in the cytoplasm, in association with the endoplasmic reticulum (ER). A model by Greenwood (1983) illustrates the synthesis of phytin in the cisternal ER and its movement to the developing protein body (Fig. 8). Phytin particles, each surrounded by a membrane (Figs 11, 12), are thought to be released from the ER and moved to the developing protein body (Fig. 11). Once deposited inside the developing protein body, each phytin particle is no longer membrane bound. The identification of these cytoplasmically produced particles as phytin in castor bean endosperm is based upon their demonstrated solubility in dilute acetic acid (Fig. 13), their susceptibility to degradation by alkaline phosphatase (Fig. 14) and their contents of P, Mg and Ca (Greenwood, 1983).

In addition to the studies of rice and castor bean just outlined, other evidence indicates that the rates of deposition of different mineral cations

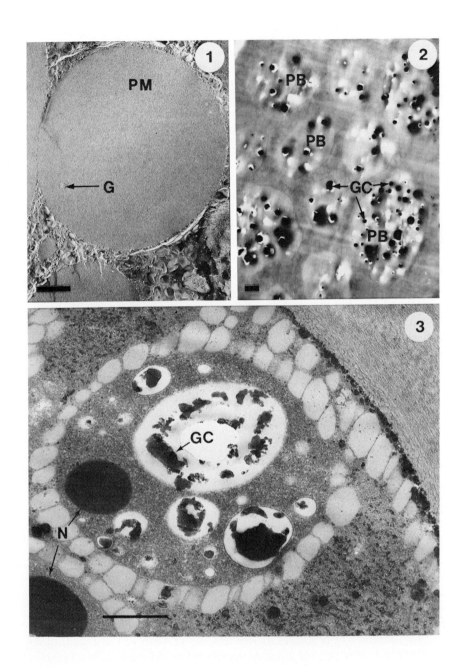

4. Mineral Reserves 145

are not the same and that further variation in composition results from organ and tissue differences within a seed. For example, in developing barley grains (*Hordeum vulgare*), the accumulation of iron (Fe) and Zn was more rapid than the accumulation of Mn (Duffus and Rosie, 1976). In the same study, it was found that Mn uptake by the developing endosperm of barley was much more gradual than Mn uptake into the embryo.

How minerals get into developing seeds is clearly a complex issue. The intakes of many minerals into fruits and seeds may be out of phase with dry matter accumulation and may occur independently (Hocking, 1982). Some of the minerals in a seed may have entered the developing seed directly from vascular connections from the stem to the ovule, whereas others may have been largely retranslocated from the fruit walls.

In the legume *Kennedia prostrata* redistribution from the pod walls provided over 50% of the P accumulated by the seeds (Hocking, 1980b) whereas in the castor bean only 13.1% of the P of the mature seed could have come from the capsule (Hocking, 1982). Studies of bean (*Phaseolus vulgaris*) have shown that most of the calcium entering the developing seed is released from the inner pod wall (endocarp) and absorbed through the testa, bypassing the vascular system (Mix and Marschner, 1976). In developing pea (*Pisum sativum*) seeds, the high acid phosphatase activities [EC 3.1.3.2.] of the seed coats seem to help regulate the supply of inorganic phosphate that is made available to the growing embryo as its sole source of P (Murray, 1980).

In pea (Ferguson and Bollard, 1976) and in other legumes (Hocking, 1980a) the seeds at each end of the pod contain less total dry matter and reduced mineral contents compared to the seeds in the middle of the fruit. Thus the entry of mineral nutrients to the developing seed may be influenced by position in the fruit. Control may be exerted at several levels,

---

Fig. 1. Freeze-fractured protein body from a soybean (*Glycine max*) cotyledon showing a small globoid (G) in a structurally homogeneous proteinaceous matrix (PM).

Fig. 2. Portion of a radish (*Raphanus sativus*) cotyledon cell showing several protein bodies (PB), each of which has numerous small globoid crystals (GC). Since this sample was fixed only with glutaraldehyde, it illustrates the natural electron density of the globoid crystals (micrograph courtesy of Douglas Holmyard).

Fig. 3. Section of glutaraldehyde–osmium tetroxide fixed wheat (*Triticum aestivum*) aleurone cell showing an aleurone grain (protein body) surrounded by lipid vesicles. The protein body contains globoid crystals (GC) and Type II inclusions (N) which are now known to be rich in niacin (from I. N. Morrison *et al.*, 1975). Bars represent 1 μm.

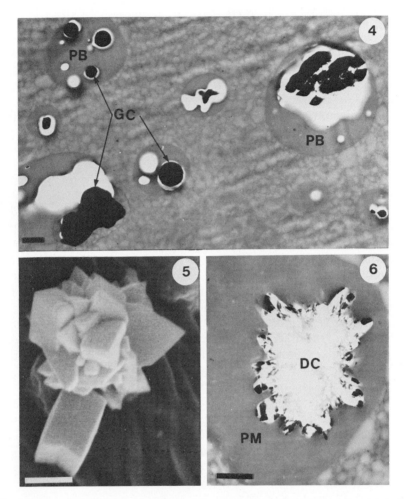

Fig. 4. Thin section of a caraway (*Carum carvi*) endosperm cell showing several protein bodies (PB). The globoid crystals (GC) in the protein bodies vary considerably in size. In umbelliferous seeds the endosperm cells either contain globoid crystals in the protein bodies as illustrated here, or they contain druse crystals of calcium oxalate (micrograph courtesy of Ernest Spitzer).

Fig. 5. Scanning electron micrograph of a druse crystal from caraway endosperm. The druse was treated with enzymes to remove materials that normally encase the druse (micrograph courtesy of Ernest Spitzer).

Fig. 6. Section of a caraway endosperm protein body that contained a druse crystal (DC). The druse, which normally is encased in proteinaceous matrix (PM), has shattered during sectioning so that only some electron dense material remains (micrograph courtesy of Ernest Spitzer). Bars represent 1 μm.

4. Mineral Reserves 147

with selective transmission of metabolites from fruit tissues to seeds (Chapter 2) and from the seed coats to the enclosed embryo (Murray, 1979, 1980).

D. Nutritional Importance of Phytin to Man

The phytin reserves are of undoubted importance to the growth of the seedling (see Chapter 5, Volume 2). Phytin is also of nutritional importance to monogastric animals such as man, swine and poultry. Because phytic acid chelates elements such as Ca, Fe, Mg, Zn and Mo, it may reduce the availability of these elements in the digestive tract.

The availability of minerals once ingested is of increasing concern as a result of the addition of seed protein to numerous processed foods. Phytate has been implicated in various health problems including anaemias, rickets, osteoporosis and bone deformities (Taylor, 1965; Zarei et al., 1972; Berlyne et al., 1973). Although some authors disagree, many researchers believe that phytic acid interferes with Ca uptake (Bruce and Callow, 1934; McCance and Widdowson, 1942). An effect of phytate on the absorption of Mg and Zn has also been found (Oberleas, 1973; Davies and Nightingale, 1975; Shah et al., 1979), but whether iron deficiencies associated with eating whole grains result from phytate is uncertain (Ranhotra et al., 1974; Welch and Van Campen, 1975; Morris and Ellis, 1976). Phytic acid has also been shown to inhibit peptic digestion of certain proteins, probably as a result of the formation of insoluble protein–phytin complexes (Barré, 1956). Phytic acid could also interfere with the proteolytic enzyme carboxypeptidase, which is a metalloenzyme.

III. DISTRIBUTION OF MINERAL RESERVES WITHIN SEEDS

A. Concentration of Minerals in Certain Tissues

Some of the earliest microscopic studies of seeds clearly revealed differences in the sequestering of reserves in various organs or tissues. For example, Pfeffer (1872) showed distinct differences in protein body inclusions such as globoid crystals or druse crystals in seeds of several species.

Cereal grains, which have received a great deal of study by microscopists, illustrate the uneven distribution of mineral reserves within the seed. In wheat (*Triticum aestivum*), globoid crystals are common in protein bodies in the aleurone layer (Fig. 3) and the embryo, yet are absent from most protein bodies in the bulk of the endosperm (Wada and Maeda, 1980).

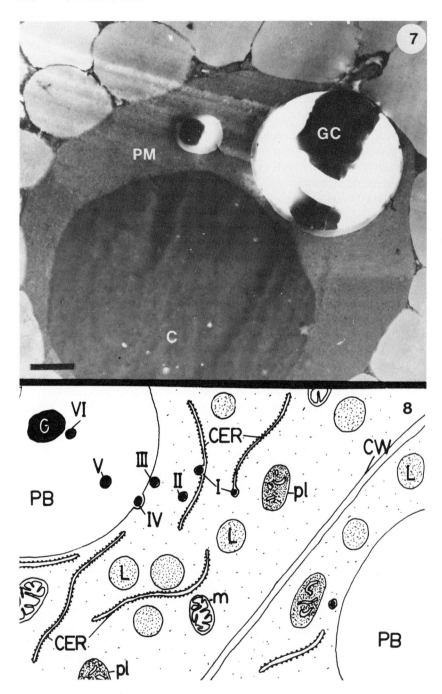

In maize (*Zea mays*) nearly 90% of the total grain phytate is in the embryo (O'Dell *et al.*, 1972). Globoid crystals are frequently smaller and less numerous in protoderm and provascular regions of embryos compared to ground meristem cells. However, even within the ground meristem globoid crystal distribution is not always uniform. For example, Kuo *et al.* (1982) found that protein bodies with globoid inclusions were confined to the central mesophyll of cotyledons of *Grevillea annulifera* and *Hakea* spp., and to the subepidermal parenchyma of cotyledons of *Xylomelum angustifolium* and *Banksia* spp.

B. Cellular and Intracellular Localization of Specific Elements

Although it has been clear for over a century that mineral reserves are not randomly distributed within seeds, we have only just begun to appreciate the degree of precision with which certain elements are sequestered. We now know that protein bodies from adjacent cells may look structurally similar and yet have distinctly different element storage. Examples are given in this section. These patterns (Figs 15-36) have been obtained using EDX analysis (Section VI,C).

1. Curcurbita
The first EDX analyses of squash (*Cucurbita maxima*) cotyledons revealed that globoid crystals usually contain P, K and Mg with occasional traces of Ca (Lott, 1975). Subsequent studies showed that Ca is more commonly found in globoid crystals from the radicle and stem than from cotyledon samples (Lott *et al.*, 1978b, 1979). In squash cotyledons, the globoid crystals containing some Ca are restricted to certain locations, such as upper and lower protoderm regions (Fig. 15) and most provascular cells (Fig. 18). In some cases the first layer of mesophyll surrounding the provascular regions also contains some Ca together with P, K and Mg (Fig. 17). Calcium is usually absent from globoid crystals of cells that will later differentiate

---

Fig. 7. Section of a castor bean (*Ricinus communis*) endosperm protein body showing the presence of a large protein crystalloid (C) and portions of fractured globoid crystals (GC) in the proteinaceous matrix (PM) (micrograph courtesy of Dr J. Greenwood). Bar represents 1 µm.

Fig. 8. Model showing the probable site of phytin synthesis in the cisternal endoplasmic reticulum (CER) and the mechanism of deposition into the developing protein bodies (PB) of castor bean endosperm. Note that the phytin particles are membrane bound (II, III) until budded into the developing protein body (IV). CW = cell wall, G = globoid, L = lipid body, m = mitochondrion, pl = plastid (courtesy of Dr J. Greenwood, 1983).

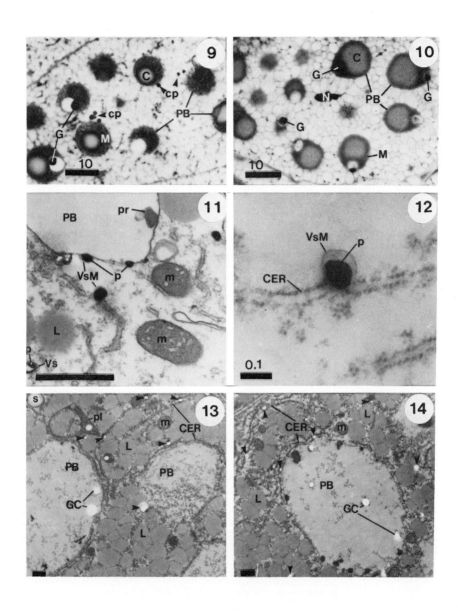

4. Mineral Reserves 151

into palisade mesophyll (Fig. 16) and spongy mesophyll (Fig. 19). A similar distribution pattern was found for winter squash, *C. mixta*, but in *C. andreana* (Fig. 20) the ground meristem cells more commonly contain Ca in their globoid crystals (Lott *et al.*, 1979). In *C. andreana* the ground meristem cells near the lower epidermis lack Ca traces in the globoid crystals. Results of studies from five different *Cucurbita* species showed a distinct correlation between seed size and the frequency with which Ca is found in globoid crystals (Lott and Vollmer, 1979b). Calcium is more frequently found in globoid crystals from the species with smaller seeds, namely *C. andreana*, *C. foetidissima* and pumpkin, *C. pepo*. In soybean (*Glycine max*) it has also been found that the Ca level increases as seed size decreases (Taira *et al.*, 1977).

2. Wheat *(*Triticum aestivum*)*
Globoid crystals containing mainly P, K and Mg are found most frequently, both in the embryo (Fig. 21) and in the aleurone, but Ca, Fe or Mn may also be present (Lott and Spitzer, 1980). Globoid crystals from the aleurone layer furthest from the embryo sometimes contain Ca (Fig. 22), whereas aleurone globoid crystals near the embryo sometimes contain Fe. The most well-defined distribution pattern occurs in the radicle. Manganese is usually found in globoid crystals from the provascular tissue in the base and mid-region of the radicle (Fig. 23). Manganese is not present in globoid crystals elsewhere in the radicle (Fig. 24) or in other parts of the grain. Throughout

---

Figs 9 and 10. Light micrographs of sections of developing castor bean (*Ricinus communis*) endosperm at 30 days after pollination and 50 days after pollination respectively. The protein bodies (PB) at both ages contain protein crystalloids (C), proteinaceous matrix (M) and some globoid crystal (G) material. Particles in the cytoplasm (cp) with the same staining characteristics as the globoid crystals are present in Fig. 9 but not in Fig. 10. Bars represent 10 μm (micrographs from J. Greenwood, 1983).

Figs 11 and 12. Electron micrographs of portions of developing castor bean endosperm cells. Phytin particles (p) surrounded by a vesicular membrane (VsM) may be associated with the cisternal endoplasmic reticulum (CER) and the membrane of the developing protein body (PB). Bar in Fig. 11 represents 1 μm; bar in Fig. 12 represents 0.1 μm (micrographs from J. Greenwood, 1983).

Fig 13 and 14. These micrographs of glycol methocrylate embedded developing castor bean endosperm show that the globoids and cytoplasmic particles are sensitive to dilute acid treatment (Fig. 13) or alkaline phosphatase treatment (Fig. 14). Since phytin is soluble in dilute acetic acid and degraded by alkaline phosphatase, these results provide clear evidence that phytin can be found in association with the cisternal endoplasmic reticulum (CER), in the cytoplasm, and in the developing protein bodies (PB). Where phytin has been removed, holes are left in the section (arrows, GC) (micrographs courtesy of Dr J. Greenwood, 1983).

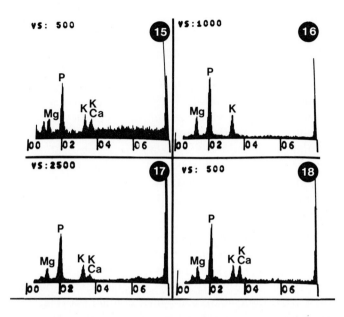

Figs 15 to 36. EDX analysis spectra of globoid crystals from protein bodies in sections of cotyledons from *Cucurbita maxima* (Figs 15 to 19) and *C. andreana* (Fig. 20). Unless otherwise stated, tissue was fixed in glutaraldehyde in distilled water, dehydrated, embedded in Spurr's low viscosity resin, and sectioned before energy dispersive X-ray (EDX) analysis. Energy levels in kiloelectronvolts (keV) are shown on the abscissa and the vertical scale (VS) is shown above each spectrum. Elements are identified on each spectrum. Elements present, energy levels in keV and principal emission lines are as follows: barium (Ba) 4.465, L_α and 4.827, $L_{\alpha 1}$ (50% of L_α peak); calcium (ca) 3.690, $K_{\alpha 1,2}$, and 4.012, K_β (10% of $K_{\alpha 1,2}$); chlorine (Cl)) 2.621, $K_{\alpha 1,2}$; chromium (Cr) 5.411, $K_{\alpha 1,2}$; copper (Cu) 0.930, L_α, and 8.040, $K_{\alpha 1,2}$; iron (Fe) 6.398, $K_{\alpha 1,2}$; magnesium (Mg) 1.253, K_α; manganese (Mn) 5.894, $K_{\alpha 1,2}$ and 6.489, $K_{\beta 1}$ (13% of $K_{\alpha 1,2}$ peak); phosphorus (P) 2.013, $K_{\alpha 1,2}$, and 2.028, $K_{\alpha 4}$ (10% of $K_{\alpha 1,2}$ peak), and 2.137, K_β (4% of $K_{\alpha 1,2}$ peak); potassium (K) 3.312, $K_{\alpha 1,2}$ and 3.589 K_β (10% of $K_{\alpha 1,2}$ peak). The Cu peak, when present, is an artifact of Cu grid usage. The Cr peak is also an artifact. Note that the major K_α peak for Ca at 3.690 keV is overlapped by the K_β peak of K at 3.589 keV. Since the potassium K_β peak is 10% of the major K_α peak for K at 3.312 keV, it can be subtracted to reveal the true Ca value when both K and Ca are present.

Fig. 15. Protoderm cell, upper surface of cotyledon. Some Ca is present in addition to P, K and Mg.

Fig. 16. Ground meristem that will become palisade mesophyll. Only P, K and Mg are present in significant amounts.

Fig. 17. Ground meristem cell immediately adjacent to a cotyledon provascular cell. Some Ca is present in addition to P, K and Mg.

Fig. 18. Provascular cell of the cotyledon. Same Ca is present in addition to P, K and Mg.

4. *Mineral Reserves* 153

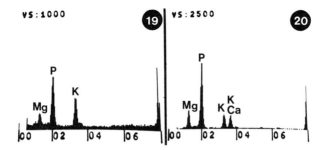

Fig. 19. Ground meristem that will become spongy mesophyll. No Ca is present.

Fig. 20. Ground meristem that will become spongy mesophyll in *C. andreana* cotyledon: note that Ca is present.

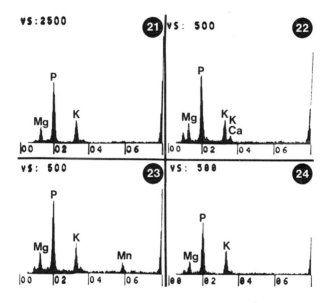

EDX analysis spectra of globoid crystals from proteins bodies in sections of wheat (*Triticum aestivum*) grain.

Fig. 21. Scutellum ground meristem cell. Only P, K and Mg are present in this globoid crystal.

Fig. 22. Aleurone layer farthest from the embryo. Some Ca is present in addition to P, K and Mg.

Fig. 23. Provascular region at the base of the radicle. Some Mn is present in globoid crystals from this region.

Fig. 24. Ground meristem of the radicle base. Only P, K and Mg are present in the globoid crystals in this region.

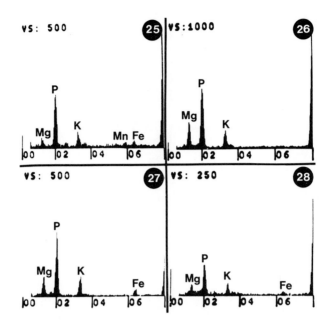

EDX analysis spectra of globoid crystals from protein bodies in section of tomato (*Lycopersicon esculentum*) embryo.

Fig. 25. Protoderm cell, radicle. In addition to P, K and Mg there are traces of Mn and Fe in this globoid crystal.

Fig. 26. Ground meristem cell away from provascular region, radicle. Only P, K and Mg are present.

Fig. 27. Ground meristem cell immediately adjacent to the provascular region, radicle. Some Fe is present in this globoid crystal in addition to P, K and Mg.

Fig. 28. Provascular cell, radicle. P, K, Mg and Fe are present in this globoid crystal.

the embryo a few globoid crystals contain Ca, but no specific patterns of Ca distribution are evident.

3. Tomato (*Lycopersicon esculentum*)

All globoid crystals contain P, K and Mg, but some may also have traces of Ca, Fe and Mn (Spitzer and Lott, 1980). Throughout the embryo and in the endosperm, the occasional cell contains Ca in its globoid crystals. The distribution of these cells appears random.

Unlike calcium-containing globoid crystals, there is a distinct distribution pattern for globoid crystals containing small amounts of Fe and Mn. Both Fe and Mn are frequently found in protodermal cell globoid crystals (Fig. 25). Iron is not present in globoid crystals from most ground

4. *Mineral Reserves* 155

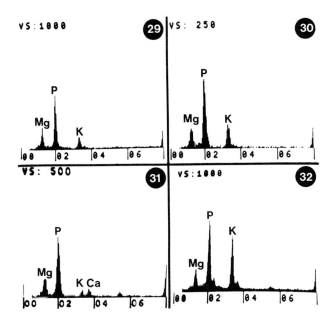

EDX analysis spectra of globoid crystals from protein bodies in castor bean (*Ricinus communis*) seeds.

Fig. 29. Ground meristem from a sector of the hypocotyl region of the embryo. Major elements present are P, K and Mg.

Fig. 30. Provascular region of the cotyledon. No calcium is present in most globoid crystals.

Fig. 31. Provascular region of the hypocotyl. Calcium is often present in globoid crystals. The Cr peak is an artifact.

Fig. 32. Globoid crystal from a freeze-dried powder of the hypocotyl region. There is considerable similarity between the elemental composition of the globoid crystals in glutaraldehyde fixed tissue and in the unfixed state. The Cr peak is an artifact.

meristem cells (Fig. 26), but globoid crystals in the first layer of large ground meristem cells surrounding the provascular region always contain some Fe (Fig. 27). Globoid crystals from the provascular cells of the radicle, hypocotyl and cotyledon often contain traces of Fe (Fig. 28).

*4. Castor Bean (*Ricinus communis*)*
Most globoid crystals of the embryo and endosperm contain only P, K and Mg (Fig. 29), but some contain Ca (Lott *et al.*, 1982). Globoid crystals in the provascular tissues of the cotyledon rarely contain Ca (Fig. 30), but globoid crystals in the provascular cells of the stem (Fig. 31) and the radicle often contain Ca.

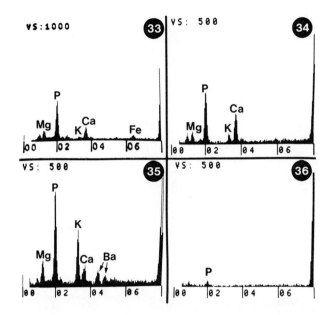

EDX spectra of globoid crystals from protein bodies of *Capsella bursapastoris* (Figs 33 and 34) and *Bertholletia excelsa* (Figs 35 and 36).

Fig. 33. Some Fe is present in globoid crystals from the cotyledon of this seed, from a plant grown at the edge of a major roadway where lead and iron levels would be elevated.

Fig. 34. Globoid crystal taken from a comparable position to that shown in the previous figure, but seed from a plant well removed from any highway. No iron traces are present.

Fig. 35. Many globoid crystals from Brazil nut embryos contain some barium.

Fig. 36. Fixation conditions can influence the preservation of mineral deposits. Use of osmium tetroxide may lead to considerable extraction of elements from globoid crystals. This Brazil nut globoid crystal was in tissue fixed with glutaraldehyde then osmium tetroxide. Compare this spectrum with the previous one.

C. Differences at Species, Cell and Protein Body Levels

As yet we cannot explain why some species store elements like Fe, Mn or Ca in globoid crystals whereas others do not. Nor can we explain why some species (e.g., squash, castor bean) show great precision in the sequestering of an element like Ca, whereas other species (e.g., tomato) have a random Ca distribution.

Although the provascular cells of the embryo differ from the ground meristem cells with regard to globoid crystal composition in many species

4. Mineral Reserves 157

that have been studied, this is not always so. For example, in the tiny embryos of umbelliferous seeds, no distinct elemental distribution patterns are found (Spitzer and Lott, 1982b).

Most EDX analytical studies have shown that all the globoid crystals within a single protein body have similar elemental composition. Not only are the globoid crystals within one protein body very similar, but so too are all the globoid crystals within one cell. This suggests that the synthesis of globoid crystal material within a cell uses biochemical pathways that are common throughout that cell. Some exceptions, however, have been found. In the coleorhiza of the wheat embryo differences in the elements present in the globoid crystals were found within one cell (Lott and Spitzer, 1980). In wheat coleorhiza the largest globoid crystals contain P, K and Mg, but lack Ca. As the globoid crystals decrease in size, the Ca content increases at the expense of K and Mg. A study of the development of globoid crystals in a tissue like this would determine whether phytin is synthesized with a different elemental content at different stages of development.

IV. INFLUENCE OF EXTERNAL SOIL CONDITIONS ON MINERAL LEVELS IN SEEDS

It is beyond the scope of this chapter to review in detail the considerable literature dealing with the influence of environmental fluctuations, various fertilizer applications or other cultural practices such as irrigation upon crop yields, seed phytin levels and seed mineral levels. However, the data of Griffiths and Thomas (1981) for broad bean (*Vicia faba*) provide a useful example. When cv. Minor was grown under glasshouse conditions compared to field conditions, the total seed P was increased by about 10%, but the proportion of seed P represented by phytate increased substantially (Table I). Even with a constant genotype, there may be variations in phytate and mineral contents of seeds from year to year (Miller *et al.*, 1980a; Nahapetian and Bassiri, 1976), with different irrigation conditions (Bassiri and Nahapetian, 1977) and with different regimes for the application of fertilizer (Srivastava *et al.*, 1955; Aulakh and Pasricha, 1978; Michael *et al.*, 1980; Miller *et al.*, 1980b; Beringer and Forster, 1981; Ahluwalia and Duffus, 1982a, b).

Although EDX analysis of globoid crystals in seeds most commonly reveals P, Mg, K and Ca, other elements that may be present include Mn (Buttrose, 1978; Fig. 23), Zn, Fe (Figs 25, 27, 28, 33), Ba (Fig. 35), Cu, Mo, Cr, Se and Na (e.g., von Hofsten, 1973; Tanaka *et al.*, 1973; Sharma and Dieckert, 1975; Singh and Reddy, 1977; Redshaw *et al.*, 1978; Deosthale, 1979). The uptake and incorporation of these other elements may well be favoured only under certain environmental conditions. For example, the Ba (Fig. 35) found in some Brazil nut (*Bertholletia excelsa*) globoid crystals (Lott and Buttrose, 1978) is most likely present because of the inability of this plant to distinguish Ba from Ca and perhaps Mg. Thus, if Ba is readily available in the soil, Brazil nut trees probably take it up with Ca and Mg, then fail to discriminate against it at any level of transport into seed tissues.

The level of Fe, which may occur naturally in some globoid crystals, can clearly be environmentally influenced. For shepherd's purse (*Capsella bursa-pastoris*) (Figs 33, 34) and tomato it has been shown that Fe will be more widespead in globoid crystals in seeds produced on plants grown on soils with high levels of Fe (Spitzer *et al.*, 1980). Similarly, some variations in Ca, Mn and Fe levels were noted in certain wheat grain tissues depending upon the availability of these elements in the soil (Spitzer *et al.*, 1981). In some preliminary experiments on the mineral uptake by seeds of plants growing in hydroponic solutions, Dr Mark Buttrose (personal communication, 1977) noted distinct variation in Zn levels in globoid crystals. Where hydroponic solutions were made from rain water stored in galvanized iron tanks, the Zn levels of globoid crystals were elevated.

Although globoid crystals may contain Ba, Fe or Zn when plants are grown under certain conditions, all indications so far suggest that other more toxic metals are excluded. Wheat grains produced on plants grown with applications of sewage sludge containing heavy metals had higher overall levels of Fe, Zn, Cu and cadmium (Cd), yet EDX analysis showed no toxic metals in the globoid crystals (Spitzer *et al.*, 1981). Studies of globoid crystals in shepherd's purse seeds taken from plants growing at various distances from a major roadway showed no tendency of the globoid crystals to concentrate toxic metals such as lead (Pb), although Fe levels (Figs 33, 34) were clearly influenced (Spitzer *et al.*, 1980). There is also no indication of selenium (Se) uptake into either globoid crystals or the proteinaceous matrix regions of protein bodies from the Se concentrating plants *Astragalus bisulcatus* or *Oxytropis lambertii* (Lott and Vollmer, 1979a).

Preliminary studies of plants growing under conditions that are severe enough to cause mineral deficiency symptoms but not severe enough to halt seed production showed that developing seeds generally obtain the minerals they require (M. Buttrose, personal communication, 1977). Thus developing seeds are effective in preventing the uptake of toxic metals such as Pb or Cd into protein bodies, yet are efficient in mobilizing the minerals normally required for globoid crystal formation.

The presence of sodium (Na) in globoid crystals of *Crambe abyssinica* was reported by von Hofsten (1973). Since Na is often volatile it may be removed from samples by vacuum treatment before EDX analysis. Detectors that are not of the window-less type will not be particularly sensitive to detection of Na. Thus it is possible that Na is more common in globoid crystals than previously determined, but our measurement and sample preparation systems do not routinely permit its detection. In some studies of seeds from halophytic plants my laboratory was unable to get any reproducible data on the presence or absence of Na in the protein bodies. The status of Na as an element commonly present in protein bodies is uncertain, although most studies have not reported its presence.

V. CALCIUM-RICH CRYSTALS

Webb and Arnott (1982) have recently reviewed what is known of the structure and composition of Ca-containing crystals in seeds and this topic will be considered here only briefly. Calcium is often stored in seeds in small amounts relative to P and K (Weber and Neumann, 1980) even though in hydroponic experiments Ca is required before several other elements. However, there are species whose seeds possess Ca rich cells or tissues, where the Ca is usually present in crystals of Ca oxalate (Figs 5, 6) or Ca carbonate. In some cases the crystals are located in protein bodies (Buttrose and Lott, 1978a; Spitzer and Lott, 1982a, c) but in other cases, such as the Rosanoffian crystals of okra (*Abelmoschus esculentus*), they are not (Webb and Arnott, 1981).

At present it is not clear whether any of the Ca in crystals is available for seedling growth; even within the same genus or family the evidence is unclear. Dwarte and Ashford (1982) recently reported that some druse crystals in celery (*Apium graveolens*) endosperm were mobilized. However, Lott *et al.* (1982) reported that in celery and a number of other umbelliferous seeds there is no evidence that Ca oxalate crystals are used extensively as a Ca source during early seedling growth. In *Eucalyptus* embryos, the Ca oxalate crystals present in protein bodies do not appear to be degraded and end up in the central vacuole of cotyledon cells in the seedling (Michael White, personal communication, 1982).

VI. PROCEDURES FOR STUDYING MINERALS IN SEEDS

This chapter would not be complete without a brief appraisal of the methods currently available for studying mineral distribution in seed tissues.

A. Chemical Analysis

The entire seed can be ashed, dissolved and assayed with various procedures that include atomic absorption spectroscopy (Hocking, 1982). This may seem an easy procedure, but a variety of complex interference effects with ash from seeds can make measurement of elements like Ca very difficult (Irene Ockenden, personal communication, 1983). It is also difficult to obtain suitable standards for many seed tissues. The National Bureau of Standards (USA) sells such standards as tomato leaves and wheat flour, but the ratios of elements found in many seeds are different from these standards. Some seeds do not ash well under procedures that will usually give an acid soluble ash with most plant materials.

Certain elements present in seeds may be measured with considerable sensitivity by neutron activation or X-ray fluorescence analysis without the need for preparative procedures such as ashing. However, obtaining adequate standards is also a major problem with these methods.

B. Light Microscopy

With bright field microscopy, Toluidine Blue O will stain globoid crystals rich in phytin a red-pink colour while the rest of the protein body is stained blue (Fulcher *et al.*, 1981). Fluorescence procedures such as the use of Acriflavine-HCl or Alizarin Red S provide an additional very useful and sensitive procedure for phytin localization (Fulcher, 1982; Yiu *et al.*, 1982), Polarizing light microscopy is useful for localizing deposits of crystalline material such as calcium oxalate, as staining procedures for calcium oxalate have not proved reliable (Spitzer and Lott, 1982c).

C. Energy Dispersive X-Ray Analysis

Localization of phytin by light microscopy, total seed assays of elemental content, and chemical measurement of phytin in bulk samples are useful, but one of the most interesting and effective procedures for studying mineral storage in seeds is energy dispersive X-ray analysis. With EDX analysis, it is possible to determine the presence of elements of atomic number 11 (Na) and greater in selected areas of a tissue. With window-less or thin-window detectors, elements with even lower atomic number can be detected. Depending upon the electron microscope used, it is possible to measure the elements present in either a large area of a seed or in a small region of a cell. Since globoid crystals are naturally electron dense, they not only make ideal subjects for EDX analysis, but are easy to locate.

EDX analysis permits simultaneous detection of all elements present in a sample. This is a distinct advantage since the elements present may not always be the ones expected. For example the discovery of Ba in globoid crystals of Brazil nuts (Lott and Buttrose, 1978) was unexpected. Another

4. Mineral Reserves   161

advantage of simultaneous measurement of all the elements present as compared to a sequential analysis which would be required for wavelength dispersive systems is that loss of mass as a result of electron beam damage is not as serious a problem.

With EDX analysis equipment mounted on a scanning electron microscope (SEM) large pieces of seed tissue can be studied. Dry seeds present minimal danger of element redistribution and thus may be studied directly. Developing seeds present more technical problems in preparation. Rapid freezing and mounting on a cryostage may be the best approach for such samples. Although studies of large samples on an SEM may be useful, it is frequently necessary to examine thin sections. With thin sections of tissue, one can obtain much better spatial resolution than is possible with a thick specimen and it is possible to obtain element distribution patterns from selected proteinaceous or globoid crystal regions. Another major advantage of thin sections is that it is possible to locate the sample exactly. Since there are distinct organ-to-organ, tissue-to-tissue and cell-to-cell differences, this is an important factor.

With any fixation, dehydration and embedding procedure, one should be cautious, since elements could be extracted or redistributed. For globoid crystal studies the use of osmium tetroxide is of major concern since it can bring about differential extraction of elements (Lott *et al.*, 1978a). For example, most of the elements normally present in Brazil nut globoid crystals (Fig. 35) are missing following osmium tetroxide fixation (Fig. 36). Fortunately the natural electron density of globoid crystals allows them to be located in freeze-dried tissue powders. Lott and Vollmer (1979b) developed a procedure for obtaining freeze-dried powders from small pieces of seed tissues. Although the exact origin of a globoid crystal in a freeze-dried powder is not known since the tissue is disrupted, it is possible to determine the elemental content of unfixed globoid crystals from pieces of tissue the size of a radicle. Comparison of elements present in globoid crystals from freeze-dried powders (e.g., Fig. 32) with those present in fixed and embedded tissue (compare Fig. 29) allows one to rule out major extraction or redistribution artifacts.

Additional information on microanalysis procedures and methods can be obtained in Chandler (1977), Spurr (1980), Goldstein *et al.* (1981) and Lott (1983).

D. Phytic Acid Determinations

A number of procedures are available for the measurement of phytic acid in seeds. As most P in seeds occurs as phytin, for some purposes a total P measurement is adequate (Lolas et al., 1976). A number of chemical procedures are available, many of which are based on precipitation of ferric phytate. For discussions of procedures of phytic acid measurement, see Makower (1970), Oberleas (1971), Wheeler and Ferrell (1971) and Uppström and Svensson (1980).

A high-performance liquid chromatography (HPLC) procedure for measurement of phytic acid, inositol and inorganic phosphate has recently been developed (Tangendjaja et al., 1980). Use of HPLC allows the accurate measurement of phytic acid from relatively small samples.

VII. FUTURE RESEARCH

In his review of seed formation, Dure (1975) noted the paucity of literature concerned with the physiology of developing seeds. Although research into seed development has accelerated since then, there is still a great shortage of information.

There has been little careful research into where phytin is synthesized in the cell, how it is transported to the developing protein body, and how globoid crystals are assembled. The most detailed study is by Greenwood (1983). The Greenwood model, which was presented earlier (Fig. 8), is based upon studies of developing castor bean endosperm. It is important that similar studies should be carried out with embryo and endosperm systems from a number of species to see if this model can be generalized.

The Greenwood model suggests that phytin particles self-assemble into larger structures, the globoid crystals. It is not yet clear why in some species or tissues the phytin particles entering a protein body form into one or a few large globoid crystals, whereas in other species many small globoid crystals form. Since phytin is chemically similar in these different species, one would suspect that the other materials in the developing protein body must have some influence on the aggregation of small phytin particles during globoid crystal formation.

In most protein bodies, both phytin and storage proteins are present. Despite our knowledge that protein–phytin complexes can form, and that these complexes have altered solubility properties, a generally held view is that storage proteins and phytin are merely functionally unconnected occupants of the same vacuole. We should study whether these two materials do interact in a functionally significant way during protein body formation or degradation.

As detailed in Section III, some elements, such as Ca, Fe or Mn, are specifically sequestered in globoid crystals in certain cells. We have little knowledge as to how this occurs. Results from developmental studies of seeds would determine whether the mineral composition of these cells is unique from the start of phytin deposition, or whether the trace element differences are introduced at a later stage of development.

The process whereby minerals enter a developing seed is clearly complex and a great deal of research is needed to determine what factors control the amount of a given mineral entering a seed and the timing of that uptake. As seeds develop, the ovary wall or pericarp (exocarp, mesocarp, endocarp), integuments (testa, tegmen), endosperm and embryo may all influence mineral uptake. One way to study the regulation of mineral uptake involves the culture of seeds or embryos on defined media. In this way it may be possible to determine whether the embryo takes up the same types of elements in the presence and absence of the seed coats. Another approach to this question requires the breeding of interspecies hybrids. This method has recently been adopted in my own laboratory.

The development of reliable chemical procedures and adequate standards for elemental analysis is another area requiring considerable research. Given the range of composition of seeds from different species, the currently available standards are not sufficient. It is frequently difficult to ash or dissolve seed tissues and interferences may be more of a problem than the literature indicates.

Certain cells in seeds contain more phytin than others; for example, the aleurone cells of cereal grains contain numerous globoid crystals whereas the starchy endosperm cells nearby do not. Any information about the physiological control of these differences would be interesting and perhaps could be used to develop strains of crop plants with reduced phytin levels. Reduced phytin levels might be nutritionally advantageous to man.

ACKNOWLEDGEMENTS

Technical assistance for this manuscript was provided by Mr Douglas Holmyard. Mrs Irene Ockenden critically reviewed this manuscript. Mr E. Spitzer, Mr D. Holmyard and Dr J. Greenwood kindly provided illustrations. Much of the original research, which provided a base for writing this chapter, was supported by the Natural Science and Engineering Research Council of Canada.

REFERENCES

Ahluwalia, B., and Duffus, C. M. (1982a). *Ann. Bot.* **50**, 93-103.
Ahluwalia, B., and Duffus, C. M. (1982b). *Ann. Bot.* **50**, 105-109.
Amoa, B., and Muller, H. G. (1976). *Cereal Chem.* **53**, 365-375.

Asada, K., and Kasai, Z. (1962). *Plant, Cell Physiol.* **3**, 397-406.
Ashton, W. M., and Williams, P. C. (1958). *J. Sci. Food Agric.* **9**, 505-511.
Aulakh, M. S., and Pasricha, N. S (1978). *Indian J. Agric. Sci.* **48**, 143-148.
Barré, M. R. (1956). *Ann. Pharm. Fr.* **14**, 182-193.
Bassiri, A., and Nahapetian, A. (1977). *J. Agric. Food. Chem.* **25**, 1118-1122.
Beringer, H., and Forster, H. (1981). *Z. Pflanzenernaehr. Bodenkd.* **144**, 8-15.
Berlyne, G. M., Ben Ari, J., Nord, E., and Shainkin, R. (1973). *Amer. J. Clinical Nutr.* **26**, 910-911.
Bolley, D. S., and McCormack, R. H. (1952). *J. Am. Oil Chem. Soc.* **29**, 470-472.
Bourdillon, J. (1951). *J. Biol. Chem.* **189**, 65-72.
Bruce, H. M., and Callow, R. K. (1934). *Biochem. J.* **28**, 517-528.
Buttrose, M. S. (1978). *Aust. J. Plant Physiol.* **5**, 631-639.
Buttrose, M. S., and Lott, J. N. A. (1978a). *Can. J. Bot.* **56**, 2083-2091.
Buttrose, M. S., and Lott, J. N. A. (1978b). *Can. J. Bot.* **56**, 2062-2071.
Chandler, J. A. (1977). *In* 'Practical Methods in Electron Microscopy' (A. M. Glauert, ed.), North-Holland Publishing Company, Amsterdam.
Cosgrove, D. J. (1966). *Rev. Pure and Appl. Chem.* **16**, 209-224.
Davies, N. T., and Nightingdale, R. (1975). *Br. J. Nutr.* **34**, 243-258.
Deosthale, Y. G. (1979). *Curr. Sci.* **48**, 54-55.
Duffus, C. M., and Rosie, R. (1976). *J. Agric. Sci.* **87**, 75-79.
Dure, L. S. III. (1975). *Annu. Rev. Plant Physiol.* **26**, 259-278.
Dwarte, D., and Ashford, A. E. (1982). *Bot. Gaz.* **143**, 164-175.
Ergle, D. R., and Guinn, G. (1959). *Plant Physiol.* **34**, 476-481.
Ferguson, I. B., and Bollard, E. G. (1976). *Ann. Bot.* **40**, 1047-1055.
Fontaine, T. D., Pons, W. A., Jr., and Irving, G. W., Jr. (1946). *J. Biol. Chem.* **164**, 487-507.
Fulcher, R. G. (1982). *Food Microstructure* **1**, 167-175.
Fulcher, R. G., O'Brien, T. P., and Wong, S. I. (1981). *Cereal Chem.* **58**, 130-135.
Goldstein, J. I., Newbury, D. E., Echlin, P., Joy, D. C., Fiori, C., and Lifshin, E. (1981). 'Scanning Electron Microscopy and X-ray Microanalysis. A Text for Biologists, Materials Scientists and Geologists.' Plenum Press, New York.
Greenwood, J. S. (1983). 'Phytin Deposition Within the Protein Bodies of the Developing Endosperm of Castor Bean (*Ricinus communis* L. cv. Hale).' Ph.D. Thesis, University of Calgary, Alberta.
Griffiths, D. W., and Thomas, T. A. (1981). *J. Sci. Food Agric.* **32**, 187-192.
Guardiola, J. L., and Sutcliffe, J. F. (1971). *Ann. Bot.* **35**, 809-823.
Hall, J. R., and Hodges, T. K. (1966). *Plant Physiol.* **41**, 1459-1464.
Herman, E. M. Baumgartner, B., and Chrispeels, M. J. (1981). *European J. Cell Biol.* **24**, 226-235.
Hocking, P. J. (1980a). *Ann. Bot.* **45**, 383-396.
Hocking, P. J. (1980b). *Aust. J. Bot.* **28**, 633-644.
Hocking, P. J. (1982). *Ann. Bot.* **49**, 52-62.
Hofsten, A. v. (1973). *Physiol. Plantarum* **29**, 76-81.
Johnson, L. F., and Tate, M. E. (1969). *Can. J. Chem.* **47**, 63-73.
Kuo, T., Hocking, P. J., and Pate, J. S. (1982). *Aust. J. Bot.* **30**, 231-249.
Loewus, F. A., and Loewus, M. W. (1983). *Annu. Rev. Plant Physiol.* **34**, 137-161.
Lolas, G. M., and Markakis, P., (**75**). *J. Agric. Food Chem.* **23**, 13-15.
Lolas, G. M., Palamidis, N., and Markakis, P. (**76**). *Cereal Chem.* **53**, 867-871.
Lott, J. N. A. (1975). *Plant Physiol.* **55**, 913-916.
Lott, J. N. A. (1980). *In* 'The Biochemistry of Plants. Vol. 1. The Plant Cell' (N. E. Tolbert, ed.), pp. 589-623. Academic Press, New York.
Lott, J. N. A. (1981). *Nordic J. Bot.* **1**, 421-432.

Lott, J. N. A. (1983). *In* 'New Frontiers in Food Microstructure' (D. B. Bechtel, ed.), pp. 317-338. Am. Assoc. Cereal Chem., St Paul, Minnesota.
Lott, J. N. A., and Buttrose, M. S. (1978). *Can. J. Bot.* **56**, 2050-2061.
Lott, J. N. A., and Spitzer, E. (1980). *Plant Physiol.* **66**, 494-499.
Lott, J. N. A., and Vollmer, C. M. (1979a). *Can. J. Bot.* **57**, 987-992.
Lott, J. N. A., and Vollmer, C. M. (1979b). *Plant Physiol.* **63**, 307-311.
Lott, J. N. A., Greenwood, J. S., and Vollmer, C. M. (1978a). *Can. J. Bot.* **56**, 2408-2414.
Lott, J. N. A., Greenwood, J. S., Vollmer, C. M., and Buttrose, M. S. (1978b). *Plant Physiol.* **61**, 984-988.
Lott, J. N. A., Spitzer, E., and Vollmer, C. M. (1979). *Plant Physiol.* **63**, 847-851.
Lott, J. N. A., Greenwood, J. S., and Vollmer, C. M. (1982). *Plant Physiol.* **69**, 829-833.
Lui, N. S. T., and Altschul, A. M. (1967). *Arch. Biochem. Biophys.* **121**, 678-684.
McCance, R. A., and Widdowson, E. M. (1942). *J. Physiol.* **101**, 44-85.
Makower, R. U. (1969). *J. Sci. Food Agric.* **20**, 82-84.
Makower, R. U. (1970). *Cereal Chem.* **47**, 288-295.
Matile, Ph. (1975). 'The Lytic Compartment of Plant Cells'. Springer-Verlag, Heidelberg.
Michael, B., Zink, F., and Lantzsch, H. J. (1980). *Z. Pflanzenernaehr. Bodenkd.* **143**, 369-376.
Miller, G. A., Youngs, V. L., and Oplinger, E. S. (1980a). *Cereal Chem.* **57**, 189-191.
Miller, G. A., Youngs, V. L., and Oplinger, E. S. (1980b). *Cereal Chem.* **57**, 192-194.
Mix, G. P., and Marschner, H. (1976). *Z. Pflanzenphysiol.* **80**, 354-366.
Morris, E. R., and Ellis, R. (1976). *J. Nutr.* **106**, 753-760.
Morrison, I. N., Kuo, J., and O'Brien, T. P. (1975). *Planta* **123**, 105-116.
Mukhamedova, Kh. S., and Akramov, S. T. (1977). *Chem. Nat. Cmpd.* **13**, 422-424.
Murray, D. R. (1979). *Plant Physiol.* **64**, 763-769.
Murray, D. R. (1980). *Ann. Bot.* **45**, 273-281.
Nahapetian, A., and Bassiri, A. (1976). *J. Agric. Food. Chem.* **24**, 947-950.
Oberleas, D. (1971). *Methods Biochem. Anal.* **20**, 87-101.
Oberleas, D. (1973). *In* 'Toxicants Occurring Naturally in Foods' (D. Glick, ed.), pp. 363-371. Nat. Acad. Sci., U.S.A., Washington, D.C.
O'Dell, B. L., de Boland, A. R., and Koirtyohann, S. R. (1972). *J. Agr. Food Chem.* **20**, 718-721.
Ogawa, M., Tanaka, K., and Kasai, Z. (1975). *Agric. Biol. Chem.* **39**, 695-700.
Ogawa, M., Tanaka, K., and Kasai, Z. (1979a). *Plant, Cell Physiol.* **20**, 19-27.
Ogawa, M., Tanaka, K., and Kasai, Z. (1979b). *Soil Sci. Plant. Nutri.* **25**, 437-448.
Ogawa, M., Tanaka, K., and Kasai, Z. (1979c). *Agric. Biol. Chem.* **43**, 2211-2213.
Paul, A. A., and Southgate, D. A. T. (1978). *In* 'McCance and Widdowson's, The Composition of Foods' 4th edn. MRC Special Report No. 297. Her Majesty's Stationery Office, London.
Pernollet, J.-C. (1978). *Phytochemistry* **17**, 1473-1480.
Pfeffer, W. (1872). *Jahrbücher für Wissenschaftliche Botanik, Pringsheim's Jahrbuch* **8**, 429-574.
Poux, N. (1963). *J. Micros.* (Paris) **2**, 557-568.
Prattley, C. A., and Stanley, D. W. (1982). *J. Food. Biochem.* **6**, 243-253.
Ranhotra, G. S, Loewe, R. J., and Puyat, L. U. (1974). *Cereal Chem.* **51**, 323-329.
Redshaw, E. S., Martin, P. J., and Laverty, D. H. (1978). *Can. J. Anim. Sci.* **58**, 553-558.
Rozacky, E. E. (1968). 'The Isolation of a Protein Containing Manganese from the Aleurone Grains of the Seeds of *Arachis hypogaea* L.' *Diss. Abstr. Int. B.* **29**, Pt 1, 4519B.

Saio, K. (1964). *Plant, Cell Physiol.* **5**, 393-400.
Schwenke, K. D., Elkowicz, K., and Kozlowska, H. (1979). *Die Nährung* **23**, 967-970.
Shan, B. G., Giroux, A., Belonje, B., and Jones, J. D. (1979). *Nutr. Metab.* **23**, 275-285.
Sharma, C. B., and Dieckert, J. W. (1975). *Physiol. Plant.* **33**, 1-7.
Singh, B., and Reddy, N. R. (1977). *J. Food Sci.* **42**, 1077-1083.
Smith, A. K., and Rackis, J. J. (1957). *J. Am. Chem. Soc.* **79**, 633-637.
Sobolev, A. M. (1966). *Plant Physiol. U.S.S.R.* **13**, 177-183, translated from *Fiziologiya Rastenii* **13**, 193-200.
Spitzer, E., and Lott, J. N. A. (1980). *Can. J. Bot.* **58**, 699-711.
Spitzer, E., and Lott, J. N. A. (1982a). *Can. J. Bot.* **60**, 1381-1391.
Spitzer, E., and Lott, J. N. A. (1982b). *Can. J. Bot.* **60**, 1392-1398.
Spitzer, E., and Lott, J. N. A. (1982c). *Can. J. Bot.* **60**, 1399-1403.
Spitzer, E., Lott, J. N. A., and Vollmer, C. M. (1980). *Can. J. Bot.* **58**, 1244-1294.
Spitzer, E., Webber, M., and Lott, J. N. A. (1981). *Can. J. Bot.* **59**, 403-409.
Spurr, A. R. (1980). In 'Scanning Electron Microscopy 1980' (O. Johari, ed.), Vol. II, pp. 535-564, I.I.T. Research Institute, Chicago.
Srivastava, B. N., Biswas, T. D., and Das, N. B. (1955). *J. Indian Soc. Soil Sci.* **3**, 33-40.
Taira, H., Taira, H., and Saito, M. (1977). *Japanese J. Crop Sci.* **46**, 483-491.
Tanaka, K., Yoshida, T., Asada, K., and Kasai, Z. (1973). *Arch. Biochem. Biophys.* **155**, 136-143.
Tanaka, K., Yoshida, T., and Kasai, Z. (1974). *Plant, Cell Physiol.* **15**, 147-151.
Tanaka, K., Yoshida, T., and Kasai, Z. (1976). *Agr. Biol.Chem.* **40**, 1319-1325.
Tangendjaja, B., Buckle, K. A., and Wooton, M. (1980). *J. Chromatog.* **197**, 274-277.
Taylor, T. G. (1965). *Proc. Nutr. Soc.* **24**, 105-112.
Tluczkiewicz, J., and Berendt, W. (1977). *Acta Soc. Bot. Pol.* **46**, 3-14.
Toma, R. B., Tabekhia, M. M., and Williams, J. D. (1979). *Nutr. Reports Intern.* **20**, 25-31.
Tombs, M. P. (1967). *Plant Physiol.* **42**, 797-813.
Uppstrom, B., and Svensson, R. (1980). *J. Sci. Food Agric.* **31**, 651-656.
Van der Wilden, W., Herman, E. M., and Chrispeels, M. J. (1980). *Proc. Natl Acad. Sci. U.S.A.* **77**, 428-432.
Wada, T., and Maeda, E. (1980). *Japanese J. Crop Sci.* **49**, 475-481.
Webb, M. A., and Arnott, H. J. (1981). In 'Scanning Electron Microscopy, 1981' (O. Johari, ed.), Vol. III, pp. 285-292. I.I.T. Research Institute, Chicago.
Webb, M. A., and Arnott, H. J. (1982). In 'Scanning Electron Microscopy, 1982' (O. Johari, ed.), Vol. III, pp. 1109-1131. I.I.T. Research Institute, Chicago.
Weber, E., and Neumann, D. (1980). *Biochem. Physiol. Pflanzen* **175**, 279-306.
Welch, R. M., and Van Campen, P. R. (1975). *J. Nutr.* **105**, 253-256.
Wheeler, E. L., and Ferrell, R. E. (1971). *Cereal Chem.* **48**, 312-320.
Yiu, S. H., Poon, H., Fulcher, R. G., and Altosaar, I. (1982). *Food Microstructure* **1**, 135-143.
Zarei, J., Khyambashi, H., Emami, A., Farivar, H., Motamedi, G., Hedayat, H., Hekmatyar, F., and Barnett, R. (1972). *Acta Biochim. Iran.* **9**, 74-78.

CHAPTER 5

The Synthesis of Reserve Oligosaccharides and Polysaccharides in Seeds

N. K. MATHESON

I.   Introduction 167
II. Oligosaccharides 168
     A. Disaccharides and *Myo*-Inositol 168
     B. Trisaccharides 172
     C. Higher Oligosaccharides 173
     D. Accumulation and Location of Oligosaccharides 175
III. Polysaccharides 176
     A. α-Glucan 176
     B. Polysaccharides with a $(1\rightarrow 4)$-β-Glucan Core Structure 191
     C. Polysaccharides with a $(1\rightarrow 4)$-β-Mannan Core Structure .. 192
     D. Fructosylsucrose Oligosaccharides and Fructofuranan 200
     References 202

I. INTRODUCTION

Carbohydrates that accumulate in developing seeds can be oligosaccharides or polysaccharides. The major oligosaccharide is sucrose but a number of galactosylsucrose and fructofuranosylsucrose oligosaccharides are also found. Polysaccharides in seeds perform several roles and a single polymer may combine more than one function. The most common storage polysaccharide is starch which serves only as a glucose reserve. Many seeds contain polymers based on the $(1\rightarrow 4)$-β-mannan structure and one of these, galactomannan, probably maintains the germinating seedling in a moist environment as well as providing a carbohydrate reserve. Fructofuranans are found in some seeds and, in vegetative parts, these may contribute to frost hardiness.

A number of polysaccharides, whose role appears to be structural, are also depolymerized after imbibition and the released monosaccharides utilized by the seedling. Endosperm walls of barley (*Hordeum vulgare*) contain $(1\to 3)(1\to 4)$-β-glucan (Morall and Briggs, 1978) and, after imbibition, its hydrolysate is metabolized. On germination, the cotyledons of some species, for example, yellow lupin (*Lupinus luteus*) show selective partial depletion of polysaccharides which have the solubility and chemical composition of intercellular and cell wall polymers. The arabinose and galactose released are efficiently utilized by the seedling (Matheson and Saini, 1977; Saini and Matheson, 1981).

However, other seed polysaccharides are not hydrolysed on germination, for example, certain external mucilages. Linseed (*Linum usitatissimum*) has a surface mucilage which is a mixture of arabinoxylan and rhamnogalacturonan with galactosyl and fucosyl side chains (Hunt and Jones, 1962). Cress (*Lepidum sativum*) has a mixture of xylo-arabinofuranan, a complex polymer containing uronic acid and cellulose (Tyler, 1965a, b). These do not appear to be utilized endogenously after imbibition (N. K. Matheson, unpublished data).

II. OLIGOSACCHARIDES

A. Disaccharides and *Myo*-Inositol

1. Sucrose
Sucrose is a major, water-soluble component of developing seeds. The accumulation of sucrose in seeds includes its translocation from other parts of the plant, since carbohydrate is commonly transported as this disaccharide.

The first step in sucrose synthesis is the formation of sucrose phosphate from UDP-glucose and fructose-6-phosphate, catalysed by sucrose phosphate synthase [EC 2.4.1.14]. This is followed by hydrolysis with a specific sucrose phosphatase [EC 3.1.3.24] (Turner and Turner, 1975; Akazawa and Okamoto, 1980; Preiss, 1982a). The latter reaction is essentially irreversible. Control is effected at the first step in sucrose synthesis through sucrose phosphate synthase, for which sigmoidal kinetic curves have been reported, in some instances for both substrates. A kinetic study of the spinach (*Spinacia oleracea*) leaf enzyme (Amir and Preiss, 1982) showed sigmoidal and hyperbolic substrate saturation curves with fructose-6-phosphate and UDP-glucose respectively. Inorganic phosphate

(P_i) inhibited with a K_i of 1.75 mM and sucrose phosphate was also a competitive inhibitor towards UDP-glucose (K_i 0.4 mM). It was concluded that sucrose phosphate synthase is a regulatory enzyme, with its activity modulated by the concentrations of P_i, fructose-6-phosphate and UDP-glucose in the cytoplasm, where the enzyme is located. Hyperbolic kinetics have also been found for the spinach leaf enzyme. Glucose-6-phosphate activated and P_i inhibited (Doehlert and Huber, 1983). There are a number of schemes proposing mechanisms of transfer of sucrose to the vacuole (Giaquinta, 1980).

2. Trehalose

The compound α-α-trehalose, a disaccharide composed of two α-linked glucose units (Glcα1-1αGlc) has been detected in *Selaginella* spp., European beech (*Fagus sylvatica*), *Brassica oleracea*, *Botrychium lunaria* and ripening fruits of Apiaceae. In pteridophytes, much larger amounts are present than sucrose and it is translocated (Kandler and Hopf, 1980). The sequence of reactions for synthesis from glucose 6-phosphate (Glc-6-P) is

$$\text{Glc-6-P} \xrightleftharpoons{\text{UTP}}_{(a)} \text{Glc-1-P} \xrightleftharpoons{}_{(b)\ PP_i} \text{UDP Glc} \xrightleftharpoons{\text{Glc-6-P}}_{(c)} \alpha\text{-}\alpha\text{-trehalose-P}$$

$$\xrightarrow{(d)} \text{trehalose} + P_i \qquad (1)$$

The enzymes catalysing these reactions are (a) phosphoglucomutase [EC 2.7.5.1]; (b) glucose-1-phosphate uridylyltransferase [EC 2.7.7.9]; (c) α-α-trehalose phosphate synthase [EC 2.4.1.15] and (d) trehalose phosphatase [EC 3.1.3.12]. Information about the third and fourth enzymes has been mainly assembled from non-plant sources and trehalose phosphate synthase can use GDP-glucose as a substrate in some tissues (Nikaido and Hassid, 1971). Reaction (c) with UDP-glucose has been demonstrated in extracts of the pollen of lily (*Lilium longiflorum*) (Gussin and McCormack, 1970) and α-α-trehalose-6-phosphate has been identified chromatographically in extracts of *Selaginella kraussiana* after photosynthesis with $^{14}CO_2$ (Kandler and Hopf, 1980). The flow of ^{14}C from $^{14}CO_2$ was greater through α-α-trehalose than through sucrose. Phosphoglucomutase (a), which is also involved in other biosynthetic pathways, has been detected in pea (*Pisum sativum*) seeds (Small and Matheson, 1979), extracted from the seeds of jack bean (*Canavalia ensiformis*) (Cardini, 1956) and mung bean (*Vigna radiata*) Ramasarma *et al.*, 1954) and purified from *Cassia corymbosa* (Small and Matheson, 1979). The equilibrium for the reaction strongly favours the formation of glucose-6-phosphate (19:1); the pH optimum is near 7.5 and the magnesium ion (Mg^{2+}) is required at an optimal concentration of 1-5 mM, with higher concentrations inhibiting. Anions inhibit. Glucose-1-6-bis-phosphate is a co-factor and the molecular weight (MW) is about 60 000 daltons. Two forms have been separated from potato (*Solanum tuberosum*) tubers by chromatography (Takamiya and Fukui, 1978). The apparent Michaelis-Menten constant (K_m) for glucose-1-phosphate has been found to be in the range 55-120 μM with one value at 700 μM and for glucose-1,6-bis-phosphate, 0.5-2.2 μM. Fructose-1,6-bis-phosphate and glycerate-bis-phosphate inhibited the potato enzymes (K_i 20-40 μM).

Glucose-1-phosphate uridylyltransferase (b) (UDP-glucose pyrophosphorylase) has been detected in a range of plant tissues; for example, seeds of peas (Turner and Turner, 1958), mung beans (Ginsberg, 1958), wheat (*Triticum aestivum*) (Tovey and Roberts, 1970) and sorghum (*Sorghum bicolor*) seedlings (Gustafson and Gander, 1972), lily pollen (Hopper and Dickenson, 1972), Jerusalem artichoke (*Helianthus tuberosus*) tubers (Otozai *et al.*, 1973) and in cucumber (*Cucumis sativus*) fruit peduncles (Smart and Pharr, 1981). It catalyses a reaction in the pathway of sucrose synthesis and the product, UDP-glucose is also converted to cell wall polysaccharides. The pH optimum is broad, near 8, the equilibrium ratio [UTP][Glc-1-P]/[UDPGlc][PP$_i$] is 7.2 at pH 7.9 and Mg^{2+} activates with a maximal concentration of 1-5 mM. Two of the actual substrates are Mg UTP and Mg pyrophosphate. The K_m values for glucose-1-phosphate are in the range 0.1-1 mM and for UDP-glucose, pyrophosphate and UTP, about 0.1-0.3 mM. With UTP the lily enzyme is inhibited by UDP-glucose, K_i 0.1-0.2 mM, and other nucleoside diphospho sugars (e.g., UDP-glucuronic acid, 0.75 mM, UDP-xylose, 1.6 mM) and the effects are additive, but it is not affected by glycolytic intermediates. Activity is found in the cytoplasm.

3. Galactinol
Another disaccharide involved in oligosaccharide synthesis and found in seeds is galactinol (Galα-1-*myo*-inositol) (Kandler and Hopf, 1980). It is formed by reaction (2).

$$\text{UDP-Gal} + myo\text{-inositol} \rightleftharpoons \text{galactinol} + \text{UDP} \qquad (2)$$

UDP-galactose is derived from UDP-glucose via a reversible action (3) catalysed by UDP-glucose 4'-epimerase [EC 5.1.3.2].

$$\text{UDPGlc} \rightleftharpoons \text{UDPGal} \qquad (3)$$

This activity has been detected in extracts of mung beans (Neufeld et al., 1957), fenugreek (*Trigonella foenum-graecum*) seedlings (Clermont and Percheron, 1979) and wheat germ (Fan and Feingold, 1969), where the pH optimum was 9.0, the K_m for UDP-glucose 0.2 mM and for UDP-galactose 0.1 mM at this pH. NAD was required with 50% activation at 0.04 mM and NADH competitively inhibited with a K_i of 2 μM. The equilibrium [UDPGal]/[UDPGlc] was 3.1 and the MW of the enzyme about 100 000 daltons. The K_m for UDP-galactose with the pea seed enzyme was 0.08 mM (Rubery, 1973).

4. Myo-*Inositol*

Myo-inositol, important also as the precursor of phytin (Section 4.II,B), is formed after cyclization of glucose-6-phosphate catalysed by *myo*-inositol-1-phosphate synthase [EC 5.5.1.4]. This reaction requires NAD and produces 1-L-*myo*-inositol-1-phosphate. A Mg^{2+} dependent specific phosphatase [EC 3.1.3.25] then releases *myo*-inositol (Loewus and Loewus, 1971). The cyclase, which has been found in a number of plant sources, has a K_m for glucose-6-phosphate near 1 mM at pH 8 (Funkhouser and Loewus, 1975) and a MW of 150 000 daltons. Isomeric inositols and their methyl ethers, commonly found in plants are formed by further modification of *myo*-inositol (Kindl, 1969).

5. *Other Galactopinitols*

Galactopinitols (Galα1-(4-*o*-Me)-D-*chiro*-inositol) and galacto-*chiro*-D- have been detected in legume seeds (Beveridge et al., 1977; Schweizer et al., 1978; Schweizer and Horman, 1981) and may have a similar role to galactinol. Glcβ1-1-*myo*-inositol has also been found in some seeds (Hopf and Kandler, 1980).

The enzyme involved in reaction (2) is UDP-galactose: *myo*-inositol-1-galactosyl transferase and extracts of developing pea seeds catalyse this conversion (Frydman and Neufeld, 1963). Galactose was transferred less efficiently to *chiro*-inositol and even less so to *scyllo*-inositol. The enzyme required the manganese ion, Mn^{2+}, at an optimal concentration of 7 mM. Cobalt ions, Co^{2+}, and iron II ions, Fe^{2+}, could replace Mn^{2+}, but less effectively, and EDTA inhibited the reaction. Some other inositols reacted, hexitols were not acceptors and α-galactosidase did not hydrolyse the product formed with (−) *chiro*-inositol. The pH optimum was 5.6 and, at this pH in acetate buffer, the K_m for *myo*-inositol was 5 mM. Activity has also been found in broad bean (*Vicia faba*) seeds.

B. Trisaccharides

The trisaccharides that function as storage carbohydrate in seeds mainly have hexosyl–sucrose structures (Kandler and Hopf, 1980, 1982; Dey, 1980a). The smallest members of the fructofuranan series, the kestoses (fructosylsucroses) will be discussed with these polysaccharides in Section III.D. The major galactosyl–sucrose trisaccharides found are raffinose (Galα1-6Glcα1-2βFru$f$), planteose (Galα1-6Fru$f\beta$2-1αGlc), and umbelliferose (Galα1-2Glcα1-2βFru$f$).

The trisaccharide (Galα1-3Glcα1-2βFru$f$) has been found in representatives of *Lolium* and *Festuca*, including perennial ryegrass (*L. perenne*), Italian ryegrass (*L. italicum*), darnel (*L. temulentum*), meadow fescue (*F. pratensis*), tall fescue (*F. arundinacea*), red fescue (*F. rubra*) and sheep's fescue (*F. ovina*) (MacLeod and McCorquodale, 1958). Selaginose (Glcα1-2Glcα1-1αGlc), a glucosyltrehalose, has been detected in two groups of *Selaginella* spp. in the sub-genus Heterophyllum (Fischer and Kandler, 1975).

1. Raffinose
Raffinose occurs in a wide range of vascular plants but may not be present in all organs of a particular species. Concentrations are higher in seeds and in some of these with no clearly defined storage polysaccharide, for example lupins, the total content of sucrose and galactosyl–sucrose oligosaccharides may be very high (Matheson and Saini, 1977). As raffinose is translocated (Webb and Gorham, 1964; Kandler and Hopf, 1982) it can be synthesized in leaves and transported to the seed.

2. Umbelliferose
Umbelliferose is found in all parts of various species in the family Umbelliferae, such as *Aegopodium podagraria*, fennel (*Foeniculum vulgare*), and caraway (*Carum carvi*) (Hopf and Kandler, 1976). Umbelliferose accumulates in fruiting organs and can form up to 5% of dry weight. In ripe fruits the amount can exceed that of sucrose but turnover of umbelliferose is slower. It has also been detected in members of the Araliaceae and Pittosporaceae (Kandler and Hopf, 1980, 1982). Translocation has been shown in *A. podograria*.

3. Planteose
Planteose has been isolated from a more restricted range of families than raffinose (Plantaginaceae, Pedaliaceae, Solanaceae, Labiatae and Oleaceae) but exclusively from non-vegetative tissue. Both planteose and

umbelliferose are synthesized from sucrose and UDP-galactose by enzymes that transfer galactose to the 6-OH on the fructose moiety or the 2-OH on the glucose residue of sucrose respectively. An extract from sesame (*Sesamum indicum*) seeds synthesized labelled planteose on incubation with UDP-[^{14}C]-galactose and sucrose (Dey, 1980b) and an extract from leaves of *A. podagraria* synthesized umbelliferose (Hopf and Kandler, 1974). Whereas the biosynthesis of these two oligosaccharides conforms to the most general pattern of synthesis of glycosidic bonds, that is, via nucleoside diphospho sugars, the synthesis of raffinose takes place via transglycosylation of galactose from galactinol.

$$\text{Galactinol + sucrose} \rightleftharpoons \text{raffinose} + myo\text{-inositol} \qquad (4)$$

UDP-galactose has also been reported to function as a donor (Pridham and Hassid, 1965; Gomyo and Nakamura, 1966; Imhoff, 1973) and galactinol as a donor in planteose synthesis (Dey, 1980b).

Reaction (4) is catalysed by galactinol: sucrose galactosyl transferase [EC 2.4.1.82], which has been extracted from wheat (Lehle *et al.*, 1970), when it required protection with sulphydryl reagents. The enzyme from broad bean seeds (Lehle and Tanner, 1973) was purified by chromatography and was free of α-galactosidase activity. This enzyme also catalysed an exchange of a galactosyl residue between raffinose and sucrose, a reaction that had been previously reported for wheat germ extracts (Moreno and Cardini, 1966). In synthesis it was specific for sucrose, and UDP-galactose could not be substituted for galactinol. The MW estimated by gel chromatography was 80 000 daltons and by density gradient centrifugation, 100 000 daltons. The optimum pH was 7.0 and K_m values for galactinol, sucrose and raffinose were 7, 1 and 10 mM at pH 7.2. A trisaccharide, ciceritol, which is a digalactosyl pinitol (Galα1-6Galα1-2 (4-*o*-Me)-D-*chiro*-inositol) has been found in seeds of chick pea (*Cicer arietinum*), lentil (*Lens culinaris*) and lupin (*Lupinus albus*) (Quemener and Brillouet, 1983).

C. Higher Oligosaccharides

In addition to the trisaccharides raffinose and planteose, seeds of certain species have higher oligomers with extra galactosyl residues linked (1→6)α to the galactosyl moiety, for example, stachyose (Galα1-6Galα1-6Glcα1-2βFru*f*). Higher homologues are verbascose (a pentasaccharide) and ajugose (hexasaccharide) and unnamed oligosaccharides with a higher degree of polymerization (d.p.) have been detected in some plants.

Raffinose occurs most widely, followed by stachyose and then less commonly by verbascose and more rarely ajugose. The higher homologues are often present when the concentration of raffinose is high and they are usually associated with storage tissue. Stachyose is translocated (Gross and Pharr, 1982).

In the planteose series, sesamose (d.p.4) and the pentasaccharide are also found. The synthesis of these occurs by transfer of galactose from UDP-galactose to planteose and sesamose (Dey, 1980b). Certain members of the genus *Festuca* (*F. rubra* and *F. ovina*) contain a tetrasaccharide with two $(1 \rightarrow 4)\alpha$ linked galactosyl units linked $(1 \rightarrow 3)\alpha$ to the glucose moiety of sucrose (MacLeod and McCorquodale, 1958; Morganlie, 1970). Lychnose (Galα1-6Glcα1-2βFru$f$1-α1Gal) has been detected in sesame seeds (Hatanaka, 1959) and a higher homologue of umbelliferose has been found in *A. podagraria* (Kandler and Hopf, 1982). Higher homologues of galactinol, with one or two more galactosyl residues linked $(1 \rightarrow 6)\alpha$, also occur in seeds (Dey, 1980a).

The formation of stachyose is catalysed by galactinol: raffinose galactosyltransferase [EC 2.4.1.67]. An extract from bean (*Phaseolus vulgaris*) seeds (Tanner and Kandler, 1968) had K_m values for galactinol and raffinose of 7.3 mM and 0.84 mM at pH 7.0, the latter at a galactinol concentration of 0.3 mM. At a higher concentration (8.3 mM), the K_m value for raffinose increased to 9 mM. The extract also catalysed galactosylation of melibiose with a higher K_m value, 12 mM, at a galactinol concentration of 0.3 mM. Although the galactinol value is high for a synthetic enzyme it is believed that the seed content of galactinol, in conjunction with enzyme levels, is sufficient for this reaction to take place at a suitable rate. At equilibrium, the ratio [stachyose][*myo*-inositol]/[raffinose][galactinol] was 4. An extract from broad bean seeds (Tanner *et al.*, 1967), which also contain verbascose, had a K_m for galactinol of 11 mM and for raffinose of 0.85 mM. The extract also reacted with stachyose and melibiose, but less readily (K_m stachyose, 3.3 mM). The ratio of galactosyl transfer to raffinose relative to stachyose was 2.7. The enzyme in pumpkin (*Cucurbita pepo*) leaves (Gaudreault and Webb, 1981) has been purified by chromatography. It was optimally active at pH 6.9 and specific for galactinol and raffinose with K_m values of 7.7 and 4.6 mM respectively. There was no cation activation and Co^{2+} and Mn^{2+} were inhibitory. *myo*-Inositol inhibited competitively, (K_i 2 mM). Galactinol was hydrolysed in the absence of raffinose but there was no hydrolysis of either raffinose or stachyose. Exchange reactions occurred between substrates and products.

D. Accumulation and Location of Oligosaccharides

Sucrose is already present in the early stages of seed development (e.g., wheat, Abou-Guendia and D'Appolonia, 1972, 1973). The content varies during growth and the concentration may be lower at maturity than earlier.

On ripening of beans, stachyose and raffinose did not appear until seeds had reached 50% of maximum fresh weight (Kandler and Hopf, 1980), probably after the end of cell division. They then accumulated steadily. An increase in galactinol preceded raffinose and stachyose synthesis (Tanner, 1969). In a number of mature leguminous seeds the stachyose and verbascose contents are higher than that of raffinose (Schweizer et al., 1978; Quemener and Brillouet, 1983). The concentrations of stachyose and galactinol and the measured equilibrium constant for the galactosylation of raffinose would indicate that stachyose needs to be spatially separated within seed tissues for its synthesis to continue (Kandler and Hopf, 1980). The synthesis of stachyose appeared to be limited by the availability of raffinose. Galactinol: raffinose galactosyltransferase was present in the seeds during the early stages of maturation and increased several-fold, reaching maximal activity when stachyose content was increasing most rapidly. In pumpkin no activity was detected in young leaves (Gaudreault and Webb, 1981).

Raffinose series oligosaccharides occur in both the embryo and the endosperm, with the former attaining higher levels (Dey, 1980a). In beans they are present in the cotyledons and embryo axis at later stages of ripening, with a trace of raffinose in the testa, but they are absent from the pod (Gould and Greenshields, 1964). In honey locust (*Gleditsia triacanthos*) stachyose and raffinose were found in both the endosperm and the cotyledons (McCleary and Matheson, 1976), with more in the latter. Raffinose appeared late in the development of wheat grains (Abou-Guendia and D'Appolonia, 1972, 1973) and has been detected in the bran, embryo scutellum and aleurone.

Umbelliferose and trehalose were both first present in caraway when the fruits had reached about 80% of the full length. Umbelliferose then steadily accumulated, but the trehalose content increased rapidly and then fell slowly, until at ripeness it was about 60% of the maximum value, expressed as per gram fresh weight (Hopf and Kandler, 1976). In fennel fruits, umbelliferose was detected soon after the start of development and its content increased steadily to maturity.

Planteose is confined to the endosperm in European ash (*Fraxinus excelsior*) (Jukes and Lewis, 1974) and, in seeds of red fescue and sheep's fescue, $(1\rightarrow3)\alpha$ linked galactosyl–sucrose oligosaccharides co-exist with raffinose and stachyose (MacLeod and McCorquodale, 1958; Morganlie,

1970). Planteose appears soon after formation of sesame seeds and increases per unit weight of seed with time of development. The tetrasaccharides and pentasaccharides appear later but both also increase as the seeds mature (Dey, 1980b).

These water-soluble oligosaccharides appear to be rapidly utilized after imbibition, usually before degradation of any polysaccharide, thus providing an immediate carbon and energy source for germination.

III. POLYSACCHARIDES

A. α-Glucan

The most common form of polysaccharide reserve in the seeds of crop plants is starch, which is $(1\rightarrow 4)(1\rightarrow 6)$-α-glucan. There have been extensive chemical studies on starches from the seeds of cereals and legumes, and there are recent reviews of the chemistry (Banks and Greenwood, 1975; Banks and Muir, 1980), biochemistry (French, 1975; Duffus, 1979; Preiss and Levi, 1980; Preiss, 1982a, b) and physiology (Jenner, 1982) of accumulation.

Seeds consist of various organs and tissues in which the starch may have different characteristics, for example, in the embryo and endosperm of waxy cereal grains. Different parts of the seed may show distinct patterns of starch accumulation and utilization, for example, in the aleurone cells of endosperm, where starch is present soon after anthesis, but disappears before maturity (see Fig. 3 of Chapter 4). Some tissues are capable of starch synthesis because they possess chloroplasts. Developing cotyledons of dicotyledonous species commonly contain chloroplasts that may be converted to starch grains (Section 2.V). The pericarp surrounding the grains of barley and wheat initially contains chloroplasts, but that of maize (*Zea mays*) does not. Starch in both chloroplasts and amyloplasts disappears from the pericarp early in the development of barley and wheat grains (MacGregor et al., 1972).

In some dicotyledons there is no starch anywhere in the seed at maturity, but on germination, starch can be detected microscopically (Mlodzianowski and Weslowska, 1975) or chemically (Saini and Matheson, 1981). Other carbohydrate reserves, for example, galactomannan in fenugreek (Reid, 1971) or sucrose and galactosylsucrose oligosaccharides in yellow lupin are then utilized for starch synthesis (Saini and Matheson, 1981). This synthesis occurs in the dark and is also found in species that already contain reserve starch in their cotyledons: [^{14}C]glucose was incorporated into starch in pea cotyledons following imbibition and germination (Juliano and Varner, 1969).

1. Forms of Starch and Their Properties

Starch, which occurs as granules, is composed of an essentially unbranched component of $(1\rightarrow4)$-α-glucan (amylose) and a branched component (amylopectin), in which the branches are $(1\rightarrow6)$-α-linked. The amount of the unbranched component can be conveniently estimated from the isopotential absorption of iodine by potentiometric titration of the solubilized starch granules. The long $(1\rightarrow4)$-α-glucan chains of amylose bind iodine more strongly than the shorter chains of the amylopectin.

Various procedures have been developed to separate unbranched (amylose) from branched fractions. Solubilized starches from most mature seed sources have been separated into an insoluble fraction equated with amylose and a soluble fraction equated with amylopectin by complexing with compounds such as n-butanol and thymol. Some authors have noted small amounts of anomalous fractions, which may have intermediate structures, but it is always difficult to effect complete dispersion, and fractionation may be incomplete in a separation process that only involves several repetitions of the phase separation. Complex formation is less effective with immature starches and starches with a high amylose content, as well as with the smaller granules typical of leaf (chloroplast) starches.

Improved separation has resulted from the application of alternative procedures, such as ultracentrifugation and gel chromatography (Greenwood and Thomson, 1962; Matheson, 1971) and the affinity of concanavalin A (Section 3.II,A) for amylopectin (Chang, 1979). Some starches, such as those from *amylose-extender* maize have considerable amounts of material that is different from the conventional fractions.

Amylose has a lower average d.p. of 2000 to 5000 glucose units compared to amylopectin (c. 1 000 000). Treatment with debranching enzymes indicates some limited degree of branching in the whole amylose fraction. Other parameters that describe the structure of amylopectin are the average length of $(1\rightarrow4)$-α-glucan chains and the A:B chain ratio, which is the number of $(1\rightarrow4)$-α-chains linked only once to another chain (A), compared with the number of chains linked more than once (B).

The average chain length differs for starches from different sources, but is usually in the range of 20–25 glucose residues. The distribution of chain lengths determined after debranching (hydrolysis of the $(1\rightarrow6)$-α-linkages) is bimodal, different from the unimodal distribution found with glycogen. The A:B chain ratio for amylopectin has generally been estimated as just above one (Manners and Matheson, 1981). Physico–chemical properties, such as viscosity and ultracentrifugation behaviour, as well as chain length distribution, show a much less symmetrical structure than for glycogen, and there have been a number of proposals about the molecular architecture of these polymers.

Phytoglycogen is a branched form of α-glucan like amylopectin, but differs from amylopectin in having $(1\rightarrow4)$-α-chains of shorter average length. The distribution of these, determined by gel chromatography after debranching, is intermediate between the nearly symmetrical peak found for glycogen and the asymmetrical pattern of amylopectin. The molecular weight (MW) is also lower than that of amylopectin. The best known source of phytoglycogen is sweet corn (*Zea mays*), where it occurs in association with starch, but it has also been detected in the Müllerian bodies of *Cecropia peltata* (Marshall and Rickson, 1973) and the blue-green alga, *Anacystis nidulans* (Weber and Wöber, 1975).

As storage organs develop, the isopotential iodine absorption of the starch, that is the proportion of apparent amylose, usually increases. In normal genotypes, starch from immature seeds generally contains 10%-15% of amylose at the first suitable sampling time and this increases to about 20%-30% at maturity. The viscosity of starch also increases as seed storage organs develop, and where fractionation has been performed, it appears that the increase in viscosity is associated with the amylose fraction. Since each molecule of amylose is essentially unbranched, all of these observations indicate that the average d.p. increases. By comparison, amylopectin fractions have generally been shown to have little change in viscosity and average chain length during development, but at very early stages of development the MW is lower.

There are exceptions to the usual pattern of development, resulting in starch practically devoid of amylose, or at the other extreme, starch with a very high content of amylose (*c.* 70%). In the well known waxy varieties of cereals including maize, barley and rice (*Oryza sativa*), the endosperm starch consists entirely of amylopectin. The earliest formed starch of yellow lupin cotyledons (synthesized after imbibition and germination, as noted above) also appears to contain only amylopectin, as any amylose content is below the limit for detection (Saini and Matheson, 1981). The highest proportion of amylose occurs typically in a number of mutant maize genotypes, including *dull* and *amylose-extender* (genotype *ae ae*), and also in pea mutants. High amylose starches from the endosperm of maize kernels of the *amylose-extender* genotype contain an amylopectin fraction with a higher than usual average chain length. However, in the endosperm of kernels of the *dull* genotype, the amylopectin appears to have the usual structure.

2. Organization of Starch Granules
Starch granules are always contained within plastids which may be chloroplasts or amyloplasts. The amyloplasts of seed storage tissues are apparently derived either from chloroplasts or from previously undifferentiated plastids. Both simple and compound granules exist, and the

number of granules per amyloplast has been found to vary both with tissue and with species (Buttrose, 1960; 1963; Bain and Mercer, 1966; Duffus, 1979).

The microscopic morphology of starch granules is often characteristic of the botanical source. Their shapes have been described in various ways; for example, nearly spheroidal, ovoid, biconvex and lenticular. The average granule size generally increases during seed development, and the shape may change from oval to lenticular (Evers, 1971; Duffus, 1979). In a wrinkle-seeded variety of pea, cv. Victory Freezer, small starch granules were apparent in the stroma of some plastids 10 days after fertilization (Bain and Mercer, 1966) and most amyloplasts contained a single granule after 19 days. At 18 days the diameters of the starch granules had increased up to 10 μm and they continued to increase until the 45th day. The average diameter of isolated granules from smooth-seeded peas increased from 10 μm to 40 μm as starch content increased from 22% to 40%, and, in wrinkle-seeded peas, the increase in diameter was from 8 μm to 30 μm as starch content changed from 12% to 31% (Greenwood and Thomson, 1962). In normal barley the average diameter of isolated granules increased from 5 μm to 22 μm during the interval 9 to 46 days after anthesis, but there was little change in granule size in a high amylose variety of barley (Banks and Muir, 1980).

The size distribution of starch granules isolated from some mature cereal grains is clearly bimodal (May and Buttrose, 1959; Buttrose, 1960, 1963; Evers, 1971; Bathgate and Palmer, 1972; and see Fig. 1 of Chapter 4, Volume 2). In barley, the ratio of small to large granules changed from 3.9:1 at 25 days after anthesis to 9.2:1 at 60 days (Williams and Duffus, 1977). The fraction of granules with the smaller diameter had a lower blue value with iodine until late in development, when this value became higher than the blue value for the larger diameter fraction. A higher blue value indicates an increased proportion of unbranched component (amylose). May and Buttrose (1959) observed that the larger starch granules were initiated before the smaller granules during development of barley grains. Williams and Duffus (1978) have confirmed that the smaller granules are initiated only at a later stage of development by following the distribution of label in amyloplasts after photosynthetic assimilation of $^{14}CO_2$ by the flag leaf at different stages of development.

The gelatinization temperature of isolated starch granules has been shown to vary during development in some species. In pea, the gelatinization temperature increases with maturity, but in barley it is unchanged (Banks and Greenwood, 1975; Biliaderis, 1982). This property is related to granule size (Bathgate and Palmer, 1972).

What factors influence the shape and structure of starch granules? Microscopic examination of the endosperm starch granules from the *amylose-extender* type of maize showed that the granules are composed of a mixture of normal, nearly spherical and abnormal types and the latter range from spherical, with extensions of amorphous or optically isotropic starch, to very long, slender amorphous granules (Boyer et al., 1977). When high amylose peas (60%-70% amylose), which contain reduced starch levels, were compared with genotypes producing normal starch, the grains of the latter were found to remain oval and birefringent throughout development, but in the former, the small oval granules began to fragment two weeks after flowering, and ultimately appeared to be compound, with only a portion of the granules birefringent (Boyer, 1981).

The ability to form branches is clearly important for the normal development of a starch granule, and the average chain length appears to be critical. As noted above, phytoglycogen has a shorter average chain length than amylopectin. The phytoglycogen of sweet corn can be separated into two fractions, soluble and insoluble, the latter sedimenting below 28 000 g (Matheson, 1975). The precipitated material is associated with a small amount of amylose and has a slightly higher chain length than the soluble phytoglycogen. Taken in conjunction with the existence in *waxy* maize of starch granules composed entirely of amylopectin, these observations point to the chain length of the branched fraction as being one factor influencing granule formation.

Starch granules have been isolated from developing maize seeds by a non-aqueous extraction procedure (Liu and Shannon, 1981) and found to contain a number of metabolites required for starch synthesis. During development, apart from phytoglycogen, all $(1\rightarrow 4)(1\rightarrow 6)$-$\alpha$-glucan is found associated with granules and none as soluble polysaccharide. The properties of the granule suggest that it should not be viewed as an inert coacervate of polysaccharide but as a sub-organelle.

3. Enzymes Involved in Glucan Synthesis
In chloroplasts, starch synthesis proceeds from fructose 6-phosphate, a product of photosynthesis, by the sequence of reactions (e) to (g), reaction (5).

$$\text{Fru-6-P} \underset{(e)}{\rightleftharpoons} \text{Glc-6-P} \underset{(a)}{\rightleftharpoons} \text{Glc-1-P} \underset{(f)\ PP_i}{\overset{ATP}{\rightleftharpoons}} \text{ADP-Glc} \underset{(g)\ ADP}{\overset{(1\rightarrow 4)\text{-}\alpha\text{-Glc}_n}{\longrightarrow}} (1\rightarrow 4)\text{-Glc}_{n+1} \quad (5)$$

The enzymes catalysing these reactions are (e) glucose phosphate isomerase [EC 5.3.1.9], (a) phosphoglucomutase [EC 2.7.5.1], (f) glucose-1-phosphate adenylyltransferase (ADP-glucose pyrophosphorylase) [EC 2.7.7.27] and (g) starch synthase [EC 2.4.1.21]. Amylopectin formation is catalysed by $(1\rightarrow 4)$-α-glucan branching enzyme [EC 2.4.1.18].

All the enzymes of starch biosynthesis are found in chloroplasts. In spinach (*Spinacia oleracea*) leaves, ADP-glucose pyrophosphorylase, starch synthase and branching enzyme are located only in the chloroplasts (Okita *et al.*, 1979). Starch synthase and ADP-glucose pyrophosphorylase are located only in the amyloplasts in suspension cultures of soybean (*Glycine max*) (Macdonald and ap Rees, 1983).

Glucose phosphate isomerase (e) has been obtained from seeds of many species, including mung bean (Ramasarma and Giri, 1956), pea (Takeda *et al.*, 1967), lucerne (*Medicago sativa*) (McCleary and Matheson, 1976) and maize (Salamini *et al.*, 1972). Of the multiple forms separated from maize, one form was common to both endosperm and embryo, the second was confined to the developing endosperm, and the third to the embryo. Two forms have been isolated from both developing and mature seeds of *Cassia coluteoides* (Lee and Matheson, 1984). Chloroplast and cytosol forms have been detected in spinach leaves and indicated in a number of other C_3 plants. In leaves of C_4 plants, three forms were found: in maize two of these were in the bundle sheath strands and the other was found in mesophyll protoplasts (Herbert *et al.*, 1979). The pH optimum is high (8–9) and the K_m glucose-6-phosphate values range from 0.3 to 17 mM. Values for fructose-6-phosphate vary from 0.1 to 0.5 mM. The MW is near 120 000 daltons. The equilibrium value has been determined for [Glc-6-P]:[Fru-6-P] as 13:7 at 37°C. Erythrose-4-phosphate and 6-phosphogluconate inhibit competitively in the μM range, for example, the activities from pea (K_i 6-phosphogluconate, 13 μM; 10 μM erythrose-4-phosphate gave 84% inhibition) and sunflower cotyledons (10 μM erythrose-4-phosphate gave 82% inhibition) (Takeda *et al.*, 1967; Kelly and Latzko, 1980a). Citrate inhibits the enzyme from maize seeds.

Phosphoglucomutase (a) has been discussed in Section II,A.2. In spinach leaves and other C_3 plants, cytosolic and chloroplast isoenzymes were isolated and in C_4 plants the distribution was similar to that for phosphoglucoisomerase. Both of these enzymes have been detected in amyloplasts prepared from suspension cultures of soybean (Macdonald and ap Rees, 1983).

The reaction catalysed by ADP-glucose pyrophosphorylase (f) is the major point of control of starch synthesis in chloroplasts (Priess and Levi, 1980; Preiss, 1982a, b). In seeds this enzyme has been found in wheat (Espada, 1962; Turner, 1969a; Tovey and Roberts, 1970), maize (Dickenson and Preiss, 1969a, b; Amir and Cherry, 1972; Ozbun et al., 1973; Hannah and Nelson, 1975), ripening rice (Murata et al., 1964) and peas (Turner, 1969b). The leaf enzyme is subject to allosteric regulation, activated by glycerate-3-phosphate and inhibited by P_i. Other glycolytic intermediates, for example, fructose-1,6-bis-phosphate, activate at higher concentrations. Glycerate-3-phosphate decreases the K_m for all substrates and the substrate rate curves are hyberbolic. Stimulation of ADP-glucose synthesis is pH dependent. Concentrations of glycerate-3-phosphate required for 50% of maximal stimulation have been found to be in the range 20–400 μM with enzymes from different species and this stimulation can be up to 80 fold. P_i is inhibitory and glycerate-3-phosphate reverses inhibition. The glycerate-3-phosphate activation curve becomes sigmoidal when P_i is present and decreases of several hundred-fold in sensitivity to P_i inhibition have been noted. There is some inhibition by ADP. Spinach leaf ADP-glucose pyrophosphorylase is made up of non-identical subunits (44 000 and 48 000 daltons) and has an aggregate MW of 210 000 daltons (Copeland and Preiss, 1981).

With ADP-glucose pyrophosphorylase from maize endosperm, activation of glycerate-3-phosphate is much less pronounced and the stimulation at mM levels is only several fold. Also, a higher concentration of P_i is needed for 50% inhibition. A high concentration of glycerate-3-phosphate is required for half maximal stimulation of the enzyme from potato tubers (Sowokinos and Preiss, 1982) but both it and the activity from cassava (*Manihot esculenta*) tubers (Hawker and Smith, 1982) exhibit an almost complete dependence on glycerate-3-phosphate for activity. The ion Mg^{2+} is required, with optimal concentrations of 20 mM for wheat germ and 8 mM for maize enzymes. The equilibrium ratio [Glc-1-P][ATP]/[ADPGlc][PP_i] was 1.2 starting with ADP-glucose and 0.9 in the other direction at pH 7.9. The wheat enzyme had a K_m for ADP-glucose of 5 mM at pH 7.9 and for glucose-1-phosphate of 0.044 mM, at pH 8.0. Values of 0.8–0.3 mM were found for the $[A]_{0.5}$ (substrate concentration at which the rate is equal to half maximal velocity) for ADP-glucose in maize endosperm. In developing maize seeds two fractions were obtained by polyacrylamide gel electrophoresis, both of which had a K_m glucose-1-phosphate of near 0.1 mM, which changed to about 0.05 mM in the presence of 10 mM glycerate-3-phosphate at pH 8. The K_m for pyrophosphate was 0.05 mM and this was not altered by glycerate-3-phosphate. Without activator, P_i (3 mM) gave 50% inhibition but in the presence of 10 mM glycerate-3-phosphate, 10 mM P_i was required for the same inhibition. ADP (3 mM) caused 50% inhibition and fructose-6-phosphate activated. The subunit MW for the enzyme from normal maize endosperm was reported as 96 000 daltons (Fuchs and Smith, 1979), but from potato as 50 000 with four subunits in

5. *Carbohydrate Reserves* 183

the active enzyme. The K_m values for glucose-1-phosphate and ATP for the former preparation were 0.038 mM and 0.18 mM at pH 8.0. The enzyme from sweet corn, after purification by chromatography, showed K_m values at pH 7.4 of 0.10 mM for glucose-1-phosphate, 32 μM for ATP, 33 μM for pyrophosphate and 0.62 mM for ADP-glucose. An equilibrium value of 0.3 was determined for the synthesis of ADP-glucose at the same pH and pyrophosphate competitively inhibited, with a K_i of 0.38 μM. It was suggested that pyrophosphate could affect the production of nucleoside diphosphosugar *in vivo*.

In developing wheat and pea seeds, activity of this enzyme increases and then decreases, with a maximum that coincides with the most rapid rate of starch synthesis. Reduced levels of this enzyme were measured in endosperm from *shrunken-2* and *brittle-2* mutants of maize that produce low amounts of starch (Dickinson and Preiss, 1969b). At 22 days after pollination, activity in the *shrunken-4* mutant was one third that of normal (Ozbun *et al.*, 1973). Shrivelled triticale seeds developed only one third the activity of plump (Ching *et al.*, 1983).

Starch synthase (g) transfers a glucosyl moiety to a glucose residue at the non-reducing end of a $(1\rightarrow 4)$-α-glucan chain, which may or may not be a branch and granules can act as acceptors. This enzyme has most frequently been studied in maize (Schiefer *et al.*, 1978), both in the endosperm (Ozbun *et al.*, 1971, 1973; Hawker and Downton, 1974) and in the embryo (Tandecarz *et al.*, 1975). Other sources of the enzyme include rice grains (Tanaka and Akazawa, 1971; Perdon *et al.*, 1975; Pisigan and Del Rosario, 1976), peas and beans (Frydman and Cardini, 1967). Soluble and granule bound preparations have been studied and in some extracts the soluble activity has been separated by chromatography into different forms. Soluble fractions requiring and not requiring added primer, as well as forms in which citrate greatly increases the affinity for substrate, have been reported. In waxy rice, activity is mainly soluble, in contrast to normal lines (Murata and Akazawa, 1966). Activity in the latter can be solubilized with a dimethyl sulphoxide–water mixture (Perdon *et al.*, 1975).

The K_m values for $(1\rightarrow 4)$-α-glucan with maize enzymes are about 1 mM, calculated as non-reducing end groups, and for ADP-glucose from 0.1 mM to 3 mM. Monovalent cations activate and the potassium ion, K^+, is most effective.

In the *dull* maize mutant which accumulates diminished amounts of starch, a 60% reduction in the amount of one form of soluble starch synthase has been detected (Boyer and Preiss, 1981). In the *opaque-2* mutant, which also has a lower grain yield than normal lines, both primed and unprimed soluble and total activities paralleled those of normal maize up to about 25 days after pollination, but then decreased so that by 30 days total activity was about 20% of the normal value (Joshi et al., 1980). The shrunken-4 mutant also shows lower activities at 22 days after pollination than normal, waxy and amylose extender lines (Ozbun et al., 1973). Starch synthase activities in developing peas increase from 8 to 14 days after anthesis and then decrease (Matters and Boyer, 1981).

4. Branching Enzymes

Branching enzyme activity introduces $(1\rightarrow 6)$-α branch points, converting $(1\rightarrow 4)$-α-glucan to $(1\rightarrow 4)(1\rightarrow 6)$-$\alpha$-glucan. Most studies have been performed with mutant and normal maize. Chromatography has been used to separate three branching enzyme fractions from normal maize, designated types I, IIa and IIb (Boyer and Preiss, 1978). Two of these enzymes are immunologically similar (types IIa and IIb), but type I is distinct (Fisher and Boyer, 1983). Branching enzyme type I synthesizes phytoglycogen, and it also occurs in other mutants which do not make this polymer. An explanation is provided by the finding that starch granules isolated from *sugary* maize are susceptible to modification by the phytoglycogen-branching enzyme, whereas, those from normal maize are resistant (Boyer et al., 1982).

In *amylose-extender* maize, where the endosperm starch has elevated levels of amylose and an abnormal amylopectin fraction, there is an 80% reduction in the two branching enzyme fractions that do not make phytoglycogen, and one of these, type IIb, is reduced much more than the other. This and other evidence indicates that the *amylose-extender* gene (*ae*) is the structural gene for branching enzyme IIb (Boyer and Preiss, 1981; Fisher and Boyer, 1983). In the *dull* mutant, which also has a high amylose content, there is a reduction in branching enzyme activity type IIa (Boyer and Preiss, 1981). In another study of *amylose-extender* maize, the total branching enzyme activity was found to be one third of normal, and three fractions were present (Baba et al., 1982).

When amylopectin and phytoglycogen-branching activities from sugary maize kernels were separated by electrophoresis or gel chromatography (Manners et al., 1968) the pH optima were 6.6 and 7.0. When separated by diethylaminoethyl (DEAE-)cellulose chromatography (Hodges et al., 1969), the phytoglycogen-branching enzyme had a pH optimum of 7.4 and a K_m for soluble amylose of 0.2 mg/mL. The phytoglycogen-branching activity in normal maize had a K_m with amylose of 0.16 mg/mL and the amylopectin-branching fraction of 0.50 mg/mL at pH 7.0. Sodium dodecyl sulphate (SDS)-electrophoresis indicated that they consist of single polypeptides of estimated MW 70 000 to 90 000 daltons (Boyer and Preiss, 1978). In *amylose-extender* maize, the K_m value for amylose was 0.018 mg/mL (Baba et al., 1982). The pea seed enzymes (Matters and Boyer,

1981) differ in pH optima and the ability to branch amylose; total branching activity increased from 8 to 22 days after flowering and then decreased slightly at 26 days.

Branching activity from potato tubers (Borovsky et al., 1975, 1976) reacts with (1→4)-α-glucan chains with a degree of polmerization of greater than forty. The reaction probably occurs by interchain transfer. These authors proposed that the unbranched chains form a double helix, facilitating the transfer of a portion of one chain to the other. The minimum chain length requirement was related to the need for prior double helix formation.

5. Evidence that Phosphorylase Is Not Involved in Starch Synthesis
Before the discovery of starch synthase, the *in vitro* synthesis of (1→4)-α-glucan chains by phosphorylase [EC 2.4.1.1] was seen to indicate the method of synthesis *in vivo*, but its role is now believed to be degradative (Preiss and Levi, 1980). There have been several suggestions that it may catalyse the synthesis of (1→4)-α-glucan under certain conditions *in vivo* (de Fekete and Viewveg, 1974; de Fekete et al., 1980). However, the anabolic pathways of other polymers usually require different enzymes to the catabolic pathways, thereby allowing separate control of each process.

Incubation of ADP-glucose with starch synthase and branching enzyme allows the synthesis of a glucan with MW and iodine spectrum after debranching closely resembling those of native amylopectin (Doi, 1969), whereas the product from incubation of glucose-1-phosphate with phosphorylase and branching enzyme differs significantly in chain length distribution (Borovsky et al., 1975).

Estimates of glucose-1-phosphate concentrations in plant tissues are much lower than its K_m with phosphorylase and the concentrations of P_i are also more suitable for phosphorolysis. Glucose-1-phosphate is a substrate in an equilibrium reaction with glucose-6-phosphate catalysed by phosphoglucomutase, in which the equilibrium of glucose-1-phosphate: glucose-6-phosphate is 1:9. The latter phosphoric ester is also involved in an equilibrium reaction with fructose-6-phosphate. With conditions at or near equilibrium, the total concentration of these three compounds necessary to maintain glucose-1-phosphate levels near to the K_m value with phosphorylase is much higher than the levels measured in plants. Both glucose-6-phosphate and fructose-6-phosphate are substrates for enzymes with which they have low K_m values. Compartmentation, with large differences in concentration, could conceivably overcome these difficulties. However, estimation of the ratio of glucose-1-phosphate to P_i in starch granules from immature maize endosperm, indicated that a synthetic reaction could not occur via phosphorylase (Liu and Shannon, 1981).

The increase in phosphorylase activity associated with starch accumulation in developing seeds may reflect the need for some breakdown of starch to provide substrates for metabolism when translocated sugar is not available (Section 2.VII,A). In pea cotyledons, the increase in phosphorylase activity is much greater after germination (Matheson and Richardson, 1976) when starch is utilized. The kinetic behaviour of the phosphorylases in pea seeds (Matheson and Richardson, 1978), one of which corresponds in its physical and kinetic properties to the form in chloroplasts in spinach leaves (Steup and Schächtele, 1981; Shimomura et al., 1982), is consistent with a degradative role. The K_m for amylopectin underwent a seven-fold decrease when [P_i] increased from 10 mM to 100 mM and it was competitively inhibited by ADP-glucose with respect to P_i in phosphorolysis, with a K_i of 0.7 mM. ADP-glucose inhibited little in the direction of glucan synthesis. The non-chloroplastic phosphorylase from spinach leaves was found to degrade isolated starch granules (Steup et al., 1983).

6. Hexose Sources for Glucan Synthesis

In amyloplasts, hexose used for starch synthesis is derived from transported carbohydrate. Sucrose, the major sugar translocated, does not enter the chloroplast (Walker, 1976) and the enzymes of sucrose metabolism are not included in this organelle (Kelly and Latzko, 1980b). As transport across the amyloplast membrane probably resembles that of the chloroplast then the main sugar transported is triose phosphate (Liu and Shannon, 1981; Sowokinos and Preiss, 1982; Macdonald and ap Rees, 1983). One pathway of sucrose breakdown starts with sucrose synthase [EC 2.4.1.13] which catalyses reaction (6).

$$UDP + Suc \rightleftharpoons UDP\text{-}Glc + Fru \qquad (6)$$

Fructose is phosphorylated by a specific fructokinase [EC 2.7.1.4] (7), two forms of which have been purified from pea seeds with K_m values for fructose of 0.60 mM and for Mg ATP of about 0.1 mM (Copeland et al., 1978). Fructose inhibited fructokinase at concentrations of about four times the K_m value and excess Mg ATP was also inhibitory. In stem tissue the enzyme was located in the cytosol (Tanner et al., 1983). Fructose-6-phosphate may then enter glycolysis for conversion to triose-phosphate.

$$Fru \xrightarrow[ADP]{ATP} Fru\ 6\text{-}P \qquad (7)$$

5. Carbohydrate Reserves   187

The UDP-glucose can be utilized in cell wall polysaccharide synthesis or converted to glucose-1-phosphate by a reversal of reaction (6) catalysed by UDP-glucose pyrophosphorylase, and then incorporated into the glycolytic pathway. This pyrophosphorylase is located in the cytosol (Bird *et al.*, 1974). The alternative hydrolytic reaction catalysed by invertase would require that both products, glucose and fructose, be phosphorylated before taking part in further reaction. No significant control of invertase action is known.

In the pathway of resynthesis of fructose-6-phosphate from triose-phosphate in chloroplasts, one of the enzymes involved, fructose bisphosphatase [EC 3.1.3.11], is regulated by light. The activation of this enzyme in darkened chloroplasts by NADPH (Leegood and Walker, 1981) provides an alternative mechanism that could function in amyloplasts.

The equilibrium for the reaction catalysed by sucrose synthase, [sucrose][UDP]/[UDP Glc][Fru], varies from 1 to 8 with pH (Akazawa and Okamoto, 1980) and there are pH optima of 8 for synthesis and 6.5 for degradation. This enzyme is found at high levels in non-photosynthetic tissues and is associated with developing seeds, for example, rice (Murata, 1972) and maize (Su and Preiss, 1978), but levels at germination are low. The endosperm of the *shrunken-1* mutant of maize is deficient in starch relative to normal grain and contains less than 10% of the sucrose synthase activity of the latter (Chourey and Nelson, 1976). The rate saturation curves in the direction of sucrose breakdown are sigmoidal for the enzyme from sweet potato (*Ipomoea batatas*) (Murata, 1971; 1972). The $[A_{0.5}]$ values for sucrose are high, usually 20–40 mM, but for UDP are generally around the 0.1–1.0 mM range. The MW is about 400 000 daltons and the enzyme is made up of four identical subunits. A number of compounds affecting activity have been described, for example, NADP and pyrophosphate activated the mung bean enzyme in degradation but inhibited synthesis (Delmer, 1972). Evidence has also been obtained for a protein inhibitor (Pontis and Salerno, 1982).

In some plants galactosylsucrose oligosaccharides are transported and the utilization of the galactosyl moiety could proceed via sequence (8) (Gross and Pharr, 1982)

$$(Gal)_n\text{-Suc} \xrightarrow{\text{(h)}} Gal \xrightarrow{ATP} Gal\ 1\text{-P} \xrightarrow{UTP} UDP\ Gal \rightleftharpoons UDP\ Glc \qquad (8)$$
$$\qquad\qquad Suc\ (i) \quad ADP\ (j) \quad PP_i \quad (k)$$

These reactions are catalysed by (h), α-galactosidase [EC 3.2.1.22]; (i), galactokinase [EC 2.7.1.6]; (j), galactose-1-phosphate uridylyltransferase

(UDP-galactose-pyrophosphorylase) [EC 2.7.7.10] and (k), UDP-glucose-4'-epimerase.

α-D-Galactosidase has been detected and purified from seeds of a number of species (Dey and Pridham, 1972; Pridham and Dey, 1974). The pH optima are in the range 3–6 with most near 5, and the MW estimates range from 20 000 daltons to 80 000 daltons, with evidence for larger forms also (150 000–200 000). The K_m values for raffinose and stachyose range from 3 to 100 mM.

Galactokinase has been prepared from mung bean (Chan and Hassid, 1975), broad bean (Dey, 1983) and fenugreek seeds (Foglietti and Percheron, 1976). It required Mg^{2+}. In the last preparation the K_m values for galactose and Mg ATP at pH 7.6 were 0.54 and 5 mM. Galactose-1-phosphate inhibited competitively with galactose and non-competitively with Mg ATP. Mg ADP was a non-competitive inhibitor with both substrates. The K_m values for the broad bean enzyme were 0.5 mM galactose and 1.5 mM ATP. Activity increased in germinating seeds up to 24 hours and then decreased.

Conversion of the galactose to the glucose configuration is accomplished via UDP-galactose by epimerization, in a reversal of reaction (3), catalysed by UDP-glucose 4'-epimerase.

UDP-galactose pyrophosphorylase activity was detected in mung bean seedlings and required a bivalent cation, Mg^{2+}, Mn^{2+} or Co^{2+} (Neufeld et al., 1957). Two uridylyltransferases were resolved by chromatography of extracts from fruit peduncles of cucumber (Smart and Pharr, 1981), a species in which stachyose is the major carbohydrate translocated (Gross and Pharr, 1982). One form could utilize glucose-1-phosphate, whereas the other could utilize both glucose-1-phosphate and galactose-1-phosphate, with a preference for the latter. In the absence of Mg PP_i the K_m values for UTP and galactose-1-phosphate were 0.14 mM and 1.2 mM. Mg PP_i showed a mixed inhibition with respect to galactose-1-phosphate and non-competitive inhibition with UTP (K_i about 0.6 mM). This concentration stimulated UDP-glucose formation with either enzyme.

In another group of plants, carbohydrate is transported as hexitol (Ziegler, 1975). Mannitol has been detected in sieve tube sap of some Oleaceae, sorbitol in a number of sub-families of Rosaceae and galactitol in Celastraceae. Sorbitol has also been detected in maize kernels, mainly in the endosperm, although it is not transported in this species (Carey et al., 1982). The pathways of conversion of these into hexose phosphate are not clear but would require phosphorylation and oxidation to either an aldose or a ketose (Bieleski, 1982; Loescher et al., 1982). A number of relevant enzymic activities have been found. Loquat (*Eriobotrya piponica*) leaves (Hirai, 1981) and protoplasts of apple (*Malus sylvestris*) cotyledons (Yamaka, 1981) contain a NADP dependent dehydrogenase that intercoverts sorbitol-6-phosphate and glucose-6-phosphate. The K_m values for sorbitol-6-phosphate and glucose-6-phosphate were 2.2 mM and 11.6 mM. Apple leaves contain a sorbitol oxidase (Yamaki, 1980) that converts sorbitol to glucose, without involving NAD or NADP. The K_m for sorbitol is high (100 mM) and the pH optimum 4. Apple callus tissue contains a sorbitol dehydrogenase, which converts sorbitol to D-fructose and is NAD dependent, with K_m values for sorbitol of 86 mM and for fructose of 1.5 M (Negm and Loescher, 1979; Loescher et al., 1982). In celery (Apium graveolens), in which mannitol is both a major photosynthetic product and translocated, substantial activities of mannose-6-phosphate reductase and mannitol-1-phosphatase were detected (Rhumpho et al., 1983).

6. Integration of Starch Synthesis

Although sufficient enzymic activities have been obtained to account for the synthesis of amylose and amylopectin, the *in vitro* production of starch granules has not been performed. The characteristic morphology of granules from different botanical sources and the presence of lipid and protein suggests a more complex system of organization than that implied by simply the presence of the enzymes forming the starch molecule.

A major question in starch production concerns the synthesis of new glucan chains. The substrate for starch synthase is oligomeric, with maltose showing a very slow rate of reaction, which increases as the d.p. rises. Although some unprimed forms have been described, there is always uncertainty about low levels of substrate firmly attached to an isolated enzyme and whether added sugar substrates contain traces of higher oligomeric products (Hawker et al., 1974; Schiefer et al., 1978). There are two aspects of synthesis to consider, (i) the production of the first $(1\rightarrow 4)$-α-glucan chain in a newly formed organelle, and (ii) the production of new glucan chains for elongation and branching in organelles that already contain starch.

There have been several proposals for the production of a glucoprotein primer with different glucans, for example, for glycogen (Whelan, 1976; Butler et al., 1977) and paramylon (Deri Tomos and Northcote, 1978) and evidence presented for structures of this type. In the synthesis of starch in potato tubers, a proposal has been made for the prior synthesis of a glucoprotein precursor from a particulate fraction and UDP-glucose, which then acts as an acceptor for ADP-glucose (Lavintman et al., 1974). Experimentation on this topic is difficult, particularly because of problems of complete separation between protein and polysaccharide and of obtaining evidence of homogeneity. An enzyme has been reported from spinach leaves that converts two molecules of glucose-1-phosphate to maltose plus two P_i (Schilling, 1982). The K_m for glucose-1-phosphate was 1-2 mM, the pH optimum 6.8, and the MW 95 000 daltons. Gluconolactone, ATP and P_i were competitive inhibitors.

Once some soluble $(1\to 4)$-α-glucan chains have been synthesized, the production of new chains can occur by disproportionation. The D-enzyme [EC 2.4.1.25] 4-α-D-glucanotransferase, which transfers a segment of $(1\to 4)$-α-glucan to the 4 position of another $(1\to 4)$-α-glucan acceptor or glucose, has been found in potato tubers (Jones and Whelan, 1969). The pH optimum was 6.6 and the smallest donor substrate, maltotriose. Maltosyl groups were transferred most rapidly. In carrot roots (*Daucus carota*) and tomato fruits (*Lycopersicon esculentum*), this activity was separated by electrophoresis from α-glucosidase (Manners and Rowe, 1969) and had a pH optimum of 5.2–5.5. Other transferase activities have been described; for example, in mung bean seeds an extract converted maltose to maltotriose and maltotetraose by transglycosylation (Nigam and Giri, 1960). An intermolecular mechanism of branching (Borovsky et al., 1976) would give a short $(1\to 4)$-α-glucan chain as a branch formed. For α-amylolysis to provide new chains very effective control would be required to stop continuous hydrolysis of all newly formed $(1\to 4)$-α-glucan. Inhibitors of α-amylase have been described from wheat seeds and other plants (Buonocore et al., 1977; Kashlan and Richardson, 1981; Weselake et al., 1983): evidence has been obtained for endogenous reduction of hydrolytic activity but inhibition was not complete.

A further regulatory aspect of starch biosynthesis concerns the relationship between amylose and amylopectin and the pattern of synthesis of the latter. When wheat plants were injected with [^{14}C]glucose at the late milk to late dough stage and the starch from mature grains fractionated by complexing with *n*-butanol, the distribution of label indicated that deposition in the kernel was only partially reversible and that amylopectin was formed from amylose (McConnell et al., 1958). When wheat plants (Whistler and Young, 1960) were injected with labelled sucrose at the early dough stage and starch samples taken at times from one hour to 14 days and then fractionated by *n*-butanol complexing, amylose acquired more label more rapidly than amylopectin. It was concluded that amylose was not extensively converted to amylopectin once incorporated into the granule. Similar results were obtained with plants exposed to $^{14}CO_2$.

Maize cobs, 18 days after pollination, were treated with $^{14}CO_2$ for one hour (Shannon et al., 1970), the starch isolated up to 36 hours later fractionated by complexing with thymol and *n*-butanol and the distribution of label between amylose and amylopectin measured. Label was not incorporated into amylose earlier than amylopectin. The distribution of label along the amylose molecule was determined after partial β-amylolysis and was found to be scattered along the whole. Amylopectin was degraded to maltose and a β-limit dextrin, the latter debranched with pullulanase and chromatographed according to chain length. Label was distributed relatively uniformly among all fractions. These authors concluded that all interconversions were complete before deposition onto or into the starch granule and that the data were consistent with a model in which the polysaccharides are synthesized in the matrix of the amyloplast, followed by crystallization of the completed molecules in the starch granule.

Isolated bean leaf chloroplasts (Kovacs and Hill, 1974) were incubated with ADP-[^{14}C]glucose in pulse-chase experiments and the starch isolated and fractionated into amylose and amylopectin by gel chromatography. Label in amylose decreased during a chase of unlabelled substrate and a significant proportion of the decrease was found in the amylopectin fraction. The rate of incorporation into amylopectin was higher than into amylose in the first 30 minutes of incubation, indicating some degree of independence in synthesis. The results were interpreted as suggesting that amylopectin synthesis could involve direct reaction with ADP-glucose or synthesis from amylose.

The questions of why all unbranched $(1\rightarrow 4)$-α-glucan is not converted to amylopectin by branching enzymes and why the amylose:amylopectin ratio in a given genotype at equivalent stages of seed development remains constant under different environmental conditions have yet to be answered completely.

B. Polysaccharides with a $(1\rightarrow 4)$-β-Glucan Core Structure

Seeds of some species in certain families, for example, Leguminosae, Cruciferae and Annonaceae, contain 'amyloid' (Kooiman, 1960, 1967). In *Tamarindus indica* it occurs as a major component and is a $(1\rightarrow 4)$-β-glucan, with xylose linked $(1\rightarrow 6)$-α to some glucose residues, and galactose linked $(1\rightarrow 2)\beta$ to some of these xylose units. Rape (*Brassica napus*) seed contains a fuco-galacto-xyloglucan, in which side chains can be $(1\rightarrow 6)\alpha$ linked xylose or more extended structures with $(1\rightarrow 2)\beta$ linked galactose and $(1\rightarrow 2)\alpha$ linked L-fucose joined to xylose (Siddiqui and Wood, 1971; Aspinall *et al.*, 1977).

In white mustard (*Sinapis alba*) seeds (Gould *et al.*, 1971), the soluble xyloglucan fraction was utilized on germination and none could be detected after four days of imbibition in the dark. Similar polysaccharides have been found in cell walls of other tissues and an extract that hydrolysed the xyloglucan from etiolated soybean hypocotyls has been prepared from this source and etiolated mung beans (Koyama *et al.*, 1981).

A particulate fraction that transferred glucose from UDP-glucose and xylose from UDP-xylose into xyloglucan has been isolated from suspension-cultured soybean cells. Incorporation of [^{14}C]xylose was dependent on the presence of UDP-glucose and the incorporation of [^{14}C]glucose was dependent on the concentration of UDP-xylose (Hayashi and Matsuda, 1981). UDP-xylose is formed by an NAD dependent decarboxylation of UDP-glucuronic acid which in turn is synthesized from UDP-glucose by a dehydrogenase. The two enzymes are UDP-glucuronate decarboxylase [EC 4.1.1.35] and UDP-glucose dehydrogenase [EC 1.1.1.22] respectively. The K_m value for UDP-glucose with the latter enzyme was about 0.05 mM in pea tissues and 0.3 mM in lily pollen. The K_m values with NAD are 0.1–0.4 mM (Davies and Dickinson, 1972). UDP-xylose inhibits, with a K_i in the μM region (Neufeld and Hall, 1965). Two forms of the decarboxylase have been separated from wheat germ (John et al., 1977) with K_m values of 0.18 and 0.53 mM at the pH optimum of 7.0. The MW is 210 000 daltons. Both were activated by UDP-glucuronate and UDP-glucose and inhibited by UDP-xylose. Activity has also been detected in mung bean seedlings, bean leaves and bean cambial tissue.

C. Polysaccharides with a (1→4)-β-Mannan Core Structure

Mannan serves as a storage reserve in a number of plants (Meier and Reid, 1982). Palm seeds can contain large amounts of mannan; for example, the endosperms of ivory nuts (*Phytelephas macrocarpa*, Aspinall et al., 1958), date palm (*Phoenix dactylifera*, Keusch, 1968), doum-palm (*Hyphaene thebaica*, El Khadem and Sallam, 1967), *Erythea edulis* (Robic and Percheron, 1973), *Livistona australis* and *Archontophoenix cunninghamiana* (McCleary and Matheson, 1975; McCleary, 1978). These can be generally described as unbranched polymers of mannose, linked (1→4)β, although small amounts of other sugars such as galactose and glucose may also be detected. Galactose may be linked (1→6)α to mannose, and glucose may replace some mannose residues. These mannans are water-insoluble.

Two mannan fractions are obtained from palm seeds. If the endosperm is extracted with cuprammonium solution both are dissolved, but one is soluble in sodium hydroxide (mannan A) and the other insoluble (mannan B). Mannan occurs as thickened cell walls of endosperm cells. In date seeds and ivory nuts, mannan A appeared crystalline by X-ray diffraction, as both native and extracted material (Meier, 1958). Mannan B was microfibrillar, showing birefringence before solubilizing, but amorphous by X-ray diffraction, both before and after isolation. The d.p. of mannan A is low, about 15, and of mannan B between 40 and 80. From microscopic studies it was concluded that mannan A is built into the cell wall as small crystalline grains with a diameter of 10–20 nm, whereas mannan B lies in

5. Carbohydrate Reserves 193

the walls as microfibrils, in which cellulose may be admixed. The X-ray diagram of β-mannan from seaweed cell walls, which is almost identical with the mannan of ivory nuts and palm seeds, has a fibre repeat unit of 1.02–1.03 nm, which fits a two-fold screw axis (Frei and Preston, 1968). This is consistent with the calculated conformation of $(1\rightarrow 4)$-β linked mannose residues, leading to an extended, ribbon-like structure (Sundarajan and Rao, 1970).

Mannan also occurs in other seeds. It has been extracted by sodium hydroxide solution from coffee beans (*Coffea arabica*, Wolfrom *et al.*, 1961) and is present in caraway seeds (Hopf and Kandler, 1977). In this seed there is only a trace of mannan A, which is dominant in palm seeds. The embryo of lettuce (*Lactuca sativa*) is enveloped in an endosperm of only two to three cell layers (Halmer *et al.*, 1978) and these cells have thickened walls, composed largely of mannan, which is depleted between 15 and 25 hours after imbibition.

The $(1\rightarrow 4)$-β-glucomannans have been found in seeds of iris (*Iris ocroleuca* and *I. sibirica*) (Andrews *et al.*, 1953), bluebell (*Scylla non-scripta*) (Thompson and Jones, 1964), *Dracaena draco*, *Clivia miniata* (McCleary, 1978) and members of the Lilaceae (Jakimov-Barras, 1973). The glucomannan extracted with 10% aqueous sodium hydroxide from iris seeds and purified as the copper complex contained glucose and mannose in about equal proportions and a small amount of galactose was present. The d.p. was about 20–30. The polymer from bluebell seeds had a glucose:mannose ratio of 1:1.3, from *C. miniata* 43:57 and that from *D. draco* 49:51. Glucomannans are also found in tubers of Araceae, Iridaceae, Amaryllidaceae, Liliaceae and Orchidaceae. Some of these glucomannans have acetyl ester groups and can be dissolved in water. If the acetyl groups are removed they become water-insoluble. Both the glucose and the mannose residues are linked $(1\rightarrow 4)\beta$, with the glucose located non-regularly in the unbranched chain.

Galactomannans have been found in the endosperm of members of Caesalpiniaceae, Mimosaceae and Fabaceae (Dea and Morrison, 1975; Dey, 1978; Meier and Reid, 1982), and have also been reported in species of Annonaceae, Convolvulaceae, Ebenaceae, Loganiaceae and Palmae (Kooiman, 1971). The seed content of galactomannans at maturity can vary from low as in soybean up to nearly 40% in carob (*Ceratonia siliqua*). They are 6-α-galactosyl $(1\rightarrow 4)$-β-mannans, consisting of an unbranched $(1\rightarrow 4)$-β-mannan chain with mannose residues substituted by a single galactose. The galactose:mannose ratios of polymers from mature seeds vary according to the species. Although the values for individual molecules within a species are distributed about a mean, this mean is reasonably consistent for the same species from different sources (McCleary, 1981). The mean galactose content varies from about 15% to nearly 50% for the Leguminosae. In fenugreek seeds galactomannan content increases up to maturity (Reid and Meier, 1970) and the mannose: galactose ratio is unchanged during the various stages of development. This galactomannan contains nearly 50% galactose. In polymers with lower galactose contents a polysaccharide with a 1:1 ratio could first form and then some galactose residues be removed. However, no difference was found between young and mature seeds of *Gleditsia ferox* (Courtois and Le Dizet, 1963).

The spread of galactose content of individual molecules about the mean value probably varies between species. Presumably where the galactose: mannose ratio is nearly one as in some Trifolieae, very little variation between individual molecules can occur, but in carob and *Cassia* galactomannans, cold and hot water or alkali extractions have given fractions with very different galactose contents (Hui and Neukom, 1964; McCleary et al., 1976). The extent of galactose substitution affects water solubility and, when it is above about 20%, the polymer becomes soluble in cold water. Estimations of the MW of galactomannans have given a range of values by different methods but usually, using physico-chemical techniques, they have been found to be much higher than mannans or glucomannans, with d.p. values of at least 2000 being reported.

The pattern of distribution of galactose residues along the mannan chain is one aspect of the fine structure that may vary between species. From results with a degradative chemical method, sections of block distribution have been proposed for carob galactomannan and a pattern of alternate substituted and unsubstituted mannose residues proposed for the guar (*Cyamopsis tetragonolobus*) polysaccharide (Baker and Whistler, 1975). With another chemical procedure, it was suggested that in guar galactomannan, galactosyl substitution was in pairs or triplets (Hoffman and Svensson, 1978). The results of periodate oxidation indicated that a simple alternating structure of distribution or block substitution was excluded for both polymers (Hoffman et al., 1976). The action pattern of β-mannanase, considered in conjunction with the hetero-oligosaccharide products of carob galactomannan indicated a pattern that excluded both block substitution and a regular pattern (McCleary and Matheson, 1983). From the high degree of gel interaction and hydrolysis by β-mannanase in relation to the amount of galactose substitution, it was proposed that *Leucaena leucocephala* galactomannan has extensive regions with a regular pattern of unsubstituted mannose alternating with substituted mannose residues (McCleary, 1979).

Galactoglucomannan fractions have been extracted by alkali from seeds of Liliaceae and Iridaceae (Jakimow-Barras, 1973) and a water-soluble polysaccharide from the endosperm of asparagus (*Asparagus officinalis*) seeds contained galactose, glucose and mannose in a ratio of 7:43:49. In the endosperm of *Cercis siliquastrum* seeds, a galactoglucomannan is present as more than 10% of seed dry weight and it has a ratio of galactose to glucose to mannose of 2:1:11 (McCleary *et al.*, 1976). Hydrolysis of this galactomannan by β-mannanase has produced an oligosaccharide containing glucose, mannose and galactose (B. V. McCleary, unpublished data). This polysaccharide was not detected in seeds of another member of the tribe Cercidium, *Bauhinia purpurea*, but has since been detected in *B. galpinii* and *B. tomentosa*, indicating the possibility of another group of Leguminosae with a distinctive reserve polysaccharide.

Polymers with structures based on mannan are found in much lower concentrations as part of the plant cell wall and intercellular polysaccharide fraction; for example, galactoglucomannan in leaf and stem tissue of red clover (*Trifolium pratense*) and Townsville lucerne (*Stylosanthes humilis*) (Buchala and Meier, 1973) and glucomannan in softwoods (Timell, 1965).

The biosynthesis of $(1\rightarrow 4)$-β-mannan probably follows the sequence (9), although other possibilities such as epimerization at the nucleoside diphosphosugar level; for example, from GDP-glucose to GDP-mannose (Elbein and Hassid, 1966; Herold and Lewis, 1977) or the involvement of other nucleoside derivatives, have not been eliminated.

$$\text{Fru6-P} \rightleftharpoons \text{Man6-P} \rightleftharpoons \text{Man1-P} \overset{\text{GTP}}{\rightleftharpoons} \text{GDP-Man} \overset{(\text{Man})_n}{\rightarrow} (\text{Man})_{n+1} \quad (9)$$

$$\text{(l)} \quad\quad \text{(m)} \quad\quad \text{(n) PP}_i \quad\quad \text{(o)}$$

The enzymes catalysing these reactions are: (l), mannose phosphate isomerase [EC 5.3.1.8]; (m), mannose phosphate mutase [EC 2.7.5.—]; (n), GTP: α-D-mannose-1-phosphate guanylyltransferase (GDP-mannose pyrophosphorylase) [EC 2.7.7.13]; and (o), mannan: 4-β-mannosyl transferase (mannan synthase) [EC 2.4.1.32].

Mannose phosphate isomerase (l) catalyses a reaction in which the equilibrium Man6-P:Fru6-P is about 2:3. It also operates in the utilization of mannose formed from mannan on germination, when reaction proceeds in the other direction. Activity has been detected in a range of mature legume seeds and in lower amounts of other seeds (McCleary and Matheson, 1976). The enzyme from konjac (*Amorphophallus konjac*) tuber was partly purified by gel chromatography (Murata, 1975) and had a K_m for mannose 6-phosphate of 0.73 mM at pH 6.5 and a pH optimum in the range 6.5–7.0. It was inhibited by metal chelating agents and reactivated by Zn^{2+},

Co^{2+}, Fe^{2+}, Mn^{2+} and Cu^{2+}. The equilibrium ratio was 1.06 at pH 6.5 and the enzyme had a MW of 45 000 daltons. In germinated lucerne seeds (McCleary and Matheson, 1976) this activity was purified by chromatography on DEAE-cellulose in Tris HCl followed by re-chromatography with a phosphate gradient, which separated glucose phosphate isomerase, and then gel chromatography. The K_m for mannose-6-phosphate was 0.77 mM at pH 7.5, with a broad pH optimum and EDTA caused slow inactivation. In germinating honey locust seeds, activity was detected in cotyledons. When developing *Cassia coluteoides* pods were dissected, 95% of activity was found in the cotyledons and 5% in the mucilaginous matrix surrounding the seed (Lee and Matheson, 1984, in press). During development, total activity increased from 4 nmol of fructose-6-phosphate formed per minute per pod at 12 days after anthesis to 198 nmol at maturity (100 days). There was a maximum (48 nmol) at 42 days after anthesis, when the rate of dry weight increase was optimal, and again at maturity. The K_m for mannose-6-phosphate at the pH optimum of 7.0 was 1.6 mM. Activity was inhibited much less by erythrose-4-phosphate (K_i 0.98) than was glucose phosphate isomerase and 6-P-gluconate gave only 10% inhibition at 2 mM. The sensitive inhibition of glucose phosphate isomerase by 6-P-gluconate and erythrose-4-phosphate, in conjunction with the much less sensitive inhibition of mannose phosphate isomerase, would allow an increase in the concentration of these two metabolites from pentose phosphate metabolism to selectively decrease the activity of glucose phosphate isomerase in *C. coluteoides* cytoplasm. This could direct the flow of fructose-6-phosphate, derived from translocated sucrose or from triose-phosphate transported out of chloroplasts and surplus to the needs of glycolysis, towards the production of galactomannan. The MW was measured as 75 000 daltons by density gradient centrifugation and 68 000 daltons by gel chromatography. The MW values estimated by gel chromatography for the enzymes from other developing seeds are soybean, 59 000 daltons; guar, 54 000 daltons; and fenugreek, 63 000 daltons.

Mannose phosphate mutase from konjac tubers has been studied using a preparation partly separated from phosphoglucomutase (Murata, 1976). It had a pH optimum of 6.5–7.0, a K_m for mannose-1-phosphate of 0.2 mM and for the cofactor glucose-1,6 bis-phosphate of 1.8 µM at pH 7.0. The ion Mg^{2+} was required and a high concentration of mannose-1-phosphate was inhibitory. The equilibrium for [Man 6-P]/[Man 1-P] was determined as 8.5 at pH 7.0, strongly in favour of the reverse reaction. The MW was estimated as 62 000 daltons by gel chromatography.

In developing *Cassia corymbosa* seeds (Small and Matheson, 1979) this enzyme was separated from glucose phosphate mutase and mannose phosphate isomerase on DEAE-cellulose, followed by chromatography on phosphocellulose with elution by fructose-1,6 bis-phosphate. It required Mg^{2+} at an optimal concentration of 2.5 mM and histidine at 10 mM as a metal ion chelator. The K_m value at pH 6.8 for mannose-1-phosphate was 0.15 mM, for mannose-6-phosphate 0.3–0.5 mM, for glucose-1,6 bis-phosphate 0.87 μM and mannose-1,6 bis-phosphate could replace this, with a K_m value of 2.5 μM. The pH optimum was 6.5–7.0 and the equilibrium in favour of the 6-phosphate 81:19 at pH 7.5 and 30°C. Activity was also detected in pea seeds and *Cymbidium* tubers.

Enzyme (n), GDP-mannose pyrophosphorylase, has been found in extracts of *Gleditsia macracantha* seeds (Jimenez de Asua et al., 1966) and required divalent cations, with a maximal activity for Mg^{2+} at 10 mM. In the alga *Acetabularia mediterranea* (Bachmann and Zetsche, 1979), activity was high while cells grew and synthesized cell wall mannan but decreased to a low value when growth ceased. Also, there was high activity in the apical region of the cell, where mannan synthesis mainly occurs and low activity in the basal region, where growth and mannan synthesis are low.

Mannan synthase (o) has been detected in a particulate fraction from mung bean seedlings (Elbein and Hassid, 1966; Elbein, 1969; Heller and Villemez, 1972; Villemez, 1974). The pH optimum was 7.5, Mg^{2+} was required at an optimal concentration of 10 mM and activity was partly solubilized. The K_m for GDP-mannose was 0.1 mM. It has also been found in orchid (*Orchis morio*) tubers (Franz, 1973), where the pH optimum was 6.5 and added Co^{2+} stimulated activity, peas (Hinman and Villemez, 1975) developing fenugreek seeds (Campbell and Reid, 1982) and seedlings (Clermont et al., 1982). In fenugreek seeds transfer was to galactomannan and activity was associated with particulate fractions. During seed development an initial increase in activity per seed, with a maximum at the start of galactomannan deposition, was followed by a decrease and then a further increase when galactomannan was deposited most rapidly. At seed maturity the level of activity was negligible. Activity in *A. mediterranea* (Bachmann and Zetsche, 1979) followed the rate of mannan synthesis and a low level of this activity led to the suggestion that mannan synthase is the limiting factor in mannan production.

The mechanism of biosynthesis of heteropolymers, like glucomannan and galactomannan, where there is apparently no regularly repeating structure, has not been clarified. This is in contrast to polysaccharides that are synthesized via oligosaccharide–lipid intermediates, (e.g., bacterial capsular polysaccharides), or those that have a regularly repeating, unbranched heterodisaccharide unit, (e.g., the animal connective tissue proteoglycans), where one of the sugars attached to the growing polymer is the substrate for the second sugar derivative. Although some polysaccharides, for example, alginic acid and heparin, undergo post-polymerization modification, this process appears to be limited.

In the biosynthesis of glucomannan in mung bean, an enzyme preparation solubilized by detergent contains mannosyl and glucosyl transferases (Elbein, 1969; Heller and Villemez, 1972; Villemez, 1974). On incubation with GDP-mannose, a β-mannan forms, and with GDP-glucose, a β-glucan, but if both substrates are present a glucomannan results. The MW of the product with GDP-mannose is lower than the polymer formed from both sugars and it was proposed that the presence of GDP-glucose is required for rapid production of a polymer of similar molecular size to native glucomannan. The glucosyl transferase requires the continual production of acceptor molecules with mannose at the non-reducing end, but the mannosyl transferase does not require glucose containing acceptors. The latter is, however, inhibited by GDP-glucose. These properties evidently allow the synthesis of glucomannan with irregular replacement of mannose by glucose residues, instead of two homopolymers.

Synthesis of galactomannan requires the activities of mannan β-4-mannosyl transferase and mannan α-6-galactosyl transferase. It has been found with another plant polysaccharide with some structural similarities, a $(1\rightarrow 6)$-α-xylo-$(1\rightarrow 4)$β-glucan from suspension-cultured soybean cells (Hayashi and Matsuda, 1981), that incorporation of [^{14}C]xylose into polysaccharide is dependent on the presence of UDP-glucose and incorporation of [^{14}C]glucose is dependent on the concentration of UDP-xylose. This shows that the heteropolymeric chain is not formed by the addition of xylose to a pre-formed completed glucan chain but that xylose substitution occurs as the glucan extends. A model can be proposed for the synthesis of galactomannans where the galactose to mannose ratio is species specific. In lucerne seed for instance, which synthesizes a polymer with a ratio of nearly one, the presence of a galactosyl substituent on the penultimate mannose residue of the growing chain and the next mannose residue towards the reducing end, would not sterically hinder placement of a galactose on the end mannose (9) and Figure 1. The mannosyl transferase would also be associated with the growing polymer. The favoured conformation of $(1\rightarrow 4)$-β-mannan is a ribbon-like structure with a two-fold screw axis, which places neighbouring galactosyl substituents on opposite sides of this chain (Sundarajan and Rao, 1970; McCleary *et al.*, 1976; McCleary and Matheson, 1983).

In a species with less substitution, it can be proposed that a galactose substituent on the penultimate and further mannose residues sterically hinders the approach of the galactosyl transferase substrate complex. Species differences in the amount of galactose substitution would then be produced by differences in the degree of steric hindrance. The element of randomness in this hindrance would lead to the distribution of ratio values found for a single species. A galactose substituent on the third mannose residue from the non-reducing end of the mannan chain would inhibit reaction, because it and the new galactose unit would lie on the same side of the ribbon-like conformation. A sequence of three or more substituted

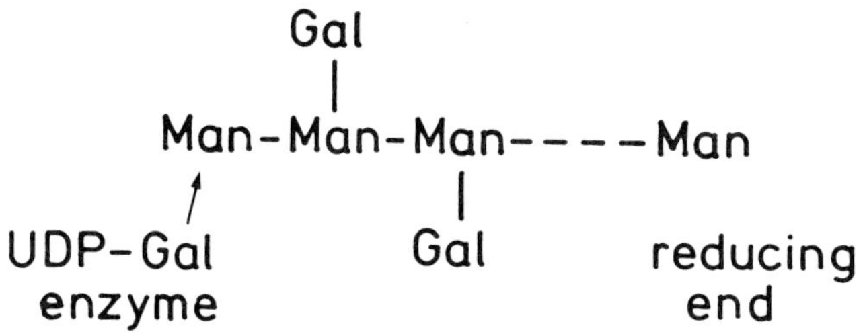

```
              Gal
               |
      Man-Man-Man----Man
       ↗       |
    UDP-Gal   Gal      reducing
    enzyme                end
```

Fig. 1. Insertion of a galactosyl substituent at the non-reducing end of a (1→4)-β-mannan chain.

mannose residues would be very unlikely in lightly substituted galactomannans. If hindrance by the penultimate galactose substituent were greater than that by the galactose on the third mannose residue, the galactomannan would contain a preponderance of substituents separated by one unsubstituted mannose (that is on the same side of the chain). If the converse held, the polymer would mainly contain pairs of neighbouring substituents. Block substitution would be most unlikely where steric restriction occurred.

A number of plant cell polysaccharides are secreted to the outside of the plasmalemma. The mode of deposition (secretion) of galactomannan in cells of fenugreek seed endosperm has been studied by microscopy (Meier and Reed, 1977). Stacks of rough endoplasmic reticulum were detected in cells that had just begun to secrete galactomannan. The intracisternal space swelled, became vacuolated and formed a voluminous network, with pockets of cytoplasm entrapped, and material was included that reacted with periodate, thiocarbazide-silver proteinate in a similar manner to the galactomannan already deposited in the plant cell wall and intercellular space. It was proposed that galactomannan is formed in the intracisternal space of the rough endoplasmic reticulum and then expelled to the outside of the plasmalemma. This mechanism is different from that for many plant cell wall polysaccharides which involves the Golgi apparatus.

Galactomannan deposition starts next to the embryo and proceeds out towards the seed coat as a cell by cell process. All cells, except those of the aleurone layer, become almost completely filled with polysaccharide. Sometimes very small, irregularly distributed protoplasmic residues persist. In aleurone cells, some galactomannan is deposited but a large cell lumen, filled with protoplasm, remains at the end of maturation. Scanning electron microscopy of the seeds of guar indicates that galactomannan deposition is a thickening of the secondary cell wall (McClendon et al., 1976).

D. Fructosylsucrose Oligosaccharides and Fructofuranan

The trisaccharides 6^F-kestose (Fru$f\beta$2-6Fru$f\beta$2-1αGlc) (kestose), 1^F-kestose (Fru$f\beta$2-1Fru$f\beta$2-1αGlc) (isokestose) and 6^G-kestose (Fru$f\beta$2-1αGlc6-2βFru$f$) (neokestose) have been isolated from some seeds and, more commonly, other parts of plants (Bacon, 1960; Kandler and Hopf, 1980; Meier and Reid, 1982).

A number of oligosaccharides of higher d.p. have been found, some of which have mixed linkages. The higher members are called fructofuranans, which if linked (2→1)β may have a d.p. of up to about 40. Higher values (up to 250) have been found for the (2→6)-β-fructofuranans from plants. Kestoses have been detected in cereal seeds (Bacon, 1960). Fructofuranan has been extracted from ears of wheat (Schlubach and Müller, 1952) and extraction of mature endosperm gave a fructofuranan fraction in about 1% yield (Montgomery and Smith, 1957). The glucose content and methylation analysis of this indicated an average d.p. of 10 and the presence of both 2→1 and 2→6 linkages. Chromatographic evidence has been obtained for two separate linkage series in wheat (White and Secor, 1953; Medcalf and Cheung, 1971). The seeds of *Aster tripolium* contain fructan (Binet *et al.*, 1974). Wheat bran has a 2→1 linked fructosyl raffinose (Fru$f$2β-1Fru$f\beta$2-1αGlc6-1αGal) (Saunders, 1971). Seeds of the horse chestnut (*Aesculus hippocastanum*) contain 6^F-kestose and 1^F-kestose and a tetrasaccharide homologue (Kahl *et al.*, 1969).

Many fructofuranans are characteristically isolated in a continuous range of d.p. values, often from oligosaccharides upwards, for example, from *Arnica montana* roots (Lombard *et al.*, 1981). In some polymers the linkage is exclusively 2→1^F, for example, from Jerusalem artichoke tubers (Edelman and Jefford, 1968) and in others 2→6^F, for example, from perennial rye grass (Tomasić *et al.*, 1978). A tetrasaccharide with 2→1 and 2→6 linkages (**1**) has been isolated from rye (*Secale cereale*) stems (Schlubach and Koehn, 1958). Both the oligosaccharides and polysaccharides are very water-soluble.

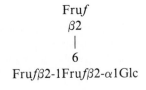

(1)

In wheat grains the fructan content is high, about 18% dry weight, before rapid starch synthesis occurs, and then it decreases to several percent at maturity (Escalada and Moss, 1976).

The biosynthesis of the $2 \rightarrow 1^F$ linked polymers has been studied in extracts of Jerusalem artichoke tubers (Edelman and Jefford, 1968). It is a two stage process, with the initial reaction catalysed by sucrose:sucrose 1^Ffructosyl-transferase in reaction (10) to give 1^F-kestose and glucose. This was essentially irreversible, with the enzyme showing no hydrolytic activity, and the reaction could not occur effectively with 6^F-kestose or with glucose.

$$\text{Fru}f\beta 2\text{-}1\alpha\text{Glc} + \text{Fru}f\beta 2\text{-}1\alpha\text{Glc} \rightarrow \text{Fru}f\beta 2\text{-Suc} + \text{Glc} \quad (10)$$

The second reaction is catalysed by 2,1-β-fructofuranan: 2,1-β-fructofuranan 1β-fructofuranosyl transferase, which transfers a single terminal $(2 \rightarrow 1)\beta$ linked fructosyl residue from one molecule to the same position in another molecule, as in reaction (11). Sucrose cannot donate but it can act as an acceptor; that is,

$$(\text{Fru}f\beta 2)_n\text{-}1^F\text{-Suc} + (\text{Fru}f\beta 2)_m\text{-}1^F\text{-Suc} \rightleftharpoons (\text{Fru}f\beta 2)_{m-1}\text{-}1^F\text{-Suc} + (\text{Fru}f\beta 2)_{m+1}\text{-}1^F\text{-Suc} \quad (11)$$

where n can be ≥ 1 and $m \geq 0$, but the rate is affected by the values of m and n. Polymers with a d.p. of about 20 have a very high affinity as an acceptor so that in a mixture of 1^F-kestose and polymers, the preferred reaction is to polymers with a higher d.p., not to oligomers.

In leaves of wheat and barley, induced accumulation of fructofuranan was associated with a several fold increase in sucrose: sucrose 1^F-fructosyltransferase. Fructofuranan and enzyme activity were found exclusively in the vacuoles of isolated protoplasts (Wagner et al., 1983).

In asparagus roots, a series of oligosaccharides has been identified that includes 1^F-kestose, 6^G-kestose and higher oligosaccharides with both $2 \rightarrow 1^F$ and $2 \rightarrow 6^G$ linkages which may be further substituted $2 \rightarrow 1^F$. Sucrose: sucrose 1^F-β-fructofuranosyl transferase has been purified (Shiomi and Izawa, 1980) from this source and has a MW of 65 000 daltons. Reaction with sucrose is reversible and it also transfers fructose from sucrose to the end fructosyl residue in the $2 \rightarrow 1$ linked portion of 6^G-kestose and homologues, such as Fru$f\beta$2-1αGlc6-(2β Fru$f)_n$. The K_m for sucrose was 0.11 M with a pH optimum of 5 and it appeared to be similar to, but not identical with, the enzyme from Jerusalem artichokes. The enzyme 6^G-fructosyl transferase has also been purified from asparagus roots (Shiomi,

1981) and catalyses reaction (12) where $m > 0$ and $n \geq 0$. Reaction was more rapid with lower members of the series but sucrose was not a donor substrate. The terminal fructosyl residue of 1^F-kestose was transferred and reaction proceeded most rapidly when n was 1 in the acceptor.

$$(\text{Fru}f\beta 2)_m\text{-}1\text{Fru}f 2\beta\text{-}1\alpha\text{Glc} + (\text{Fru}f\beta 2)_n\text{-}1\text{Fru}f 2\beta\text{-}1\alpha\text{Glc} \rightleftharpoons$$
$$(\text{Fru}f\beta 2)_{m-1}\text{-}1\text{Fru}f 2\beta\text{-}1\alpha\text{Glc} + (\text{Fru}f\beta 2)_n\text{-}1\text{Fru}f 2\beta\text{-}1\alpha\text{Glc}6\text{-}2\beta\text{Fru}f \quad (12)$$

The enzyme 1^F-fructosyl transferase (Shiomi, 1982) has been purified from asparagus roots. Although it catalysed the transfer of a fructofuranosyl moiety from oligosaccharides of the 1^F-kestose series to the 1 position of fructose in another fructosylsucrose oligomer it differed from the activity in artichoke tubers in that transfer occurred more readily to the 6^G-kestose series than the 1^F-kestose series and it did not transfer fructosyl residues to inulin. In leaves of cocksfoot grass (*Dactylis glomerata*), no low MW fructosyloligosaccharides besides sucrose were detected and it was proposed that fructofuranan synthesis occurred by direct transfer from sucrose to polymer (Pollack, 1979). In contrast, in leaves of darnel trisaccharide intermediates were detected and two distinct series of oligofructosides were isolated, one of which had chromatographic properties of the (1→2) linked series (Pollack, 1982).

Fructosylsucrose oligosaccharides are found in some seeds that contain starch and raffinose series oligosaccharides, but they are not found in those containing galactomannan. Fructofuranan in vegetative parts is accumulated for use over a short time and is consumed during shoot elongation and flowering (Staesche, 1966; Meier and Reid, 1982). Unlike starch and galactomannan it is very water-soluble. High levels of sucrose are associated with fructofuranan and it may serve to reduce osmotic effects, acting as a concentration buffer. In cereal grains it may still have this role when the grain hydrates, since on germination of seeds of wheat, barley, rye and oats (*Avena sativa*), changes in fructosylsucrose trisaccharide follow sucrose levels, which decrease initially and then increase considerably (Täufel et al., 1959).

REFERENCES

Abou-Guendia, M., and D'Appolonia, B. L. (1972). *Cereal Chem.* **49**, 664–676.
Abou-Guendia, M., and D'Appolonia, B. L. (1973). *Cereal Chem.* **50**, 723–734.
Akazawa, T., and Okamoto, K. (1980). *In* 'The Biochemistry of Plants, Carbohydrates: Structure and Function' (J. Preiss, ed.), pp. 199–220. Academic Press, New York.
Amir, J., and Cherry, J. H. (1972). *Plant Physiol.* **49**, 893–897.
Amir, J., and Preiss, J. (1982). *Plant Physiol.* **69**, 1027–1030.
Andrews, P., Hough, L., and Jones, J. K. N. (1953). *J. Chem. Soc.* 1186–1192.

5. Carbohydrate Reserves 203

Aspinall, G. O., Rashbrook, R. B., and Kessler, G. (1958). *J. Chem. Soc.* 215-221.
Aspinall, G. O., Krishnamurthy, T. N., and Rosell, K.-G. (1977). *Carbohydr. Res.* **55**, 11-19.
Baba, T., Arai, Y., Ono, T., Munakata, A., Yamaguchi, H., and Itoh, T. (1982). *Carbohydr. Res.* **107**, 215-230.
Bachmann, P., and Zetsche, K. (1979). *Planta* **145**, 331-337.
Bacon, J. S. D. (1960). *Bull. Soc. Chim. Biol.* **42**, 1441-1449.
Bain, J. M., and Mercer, F. V. (1966). *Aust. J. Biol. Sci.* **19**, 49-67.
Baker, C. W., and Whistler, R. L. (1975). *Carbohydr. Res.* **45**, 237-242.
Banks, W., and Greenwood, C. T. (1975). 'Starch and Its Components.' Edinburgh University Press, Edinburgh.
Banks, W., and Muir, D. D. (1980). *In* 'The Biochemistry of Plants, Vol. 3, Carbohydrates: Structure and Function' (J. Preiss, ed.), pp. 321-369. Academic Press, New York.
Bathgate, G. N., and Palmer, G. H. (1972). *Staerke* **24**, 336-341.
Beveridge, R. J., Ford, C. W, and Richards, G. N. (1977). *Aust. J. Chem.* **30**, 1583-1590.
Bieleski, R. L. (1982). *In* 'Encyclopedia of Plant Physiology, Vol. 13A, Plant Carbohydrates I, Intracellular Carbohydrates' (F. A. Loewus and W. Tanner, eds), pp. 158-192. Springer Verlag, New York.
Biliaderis, C. G. (1982). *Phytochemistry* **21**, 37-39.
Binet, P., Collin, A., and Duyme, M. (1974). *Bull. Soc. Bot. Fr.* **121**, 323-328.
Bird, I. F., Cornelius, M. J., Keys, A. J., and Wittingham, C. P. (1974). *Phytochemistry* **13**, 59-64.
Borovsky, D., Smith, E. E., and Whelan, W. J. (1975). *Eur. J. Biochem.* **59**, 615-625.
Borovsky, D., Smith, E. E., and Whelan, W. J. (1976). *Eur. J. Biochem.* **62**, 307-312.
Boyer, C. D. (1981). *Am. J. Bot.* **68**, 659-665.
Boyer, C. D., and Preiss, J. (1978). *Carbohydr. Res.* **61**, 321-324.
Boyer, C. D., and Preiss, J. (1981). *Plant Physiol.* **67**, 1141-1145.
Boyer, C. D., Daniels, R. R., and Shannon, J. C. (1977). *Am. J. Bot.* **64**, 50-56.
Boyer, C. D., Simpson, E. K. G., and Damewood, P. A. (1982). *Staerke* **34**, 81-85.
Buchala, A. J., and Meier, H. (1973). *Carbohydr. Res.* **31**, 87-92.
Buonocore, V., Petrucci, T., and Silano, V. (1977). *Phytochemistry* **16**, 811-820.
Butler, N. A., Lee, E. Y. C., and Whelan, W. J. (1977). *Carbohydr. Res.* **55**, 73-82.
Buttrose, M. S. (1960). *J. Ultrastruct. Res.* **4**, 231-257.
Buttrose, M. S. (1963). *Aust. J. Biol. Sci.* **16**, 305-317.
Campbell, J. McA., and Reid, J. S. G. (1982). *Planta* **155**, 105-111.
Cardini, C. E. (1952). *Enzymologia* **15**, 44-48.
Carey, E. E., Dickinson, D. B., Wei, L. Y., and Rhodes, A. M. (1982). *Phytochemistry* **21**, 1909-1911.
Chan, P. H., and Hassid, W. Z. (1975). *Anal. Biochem.* **64**, 372-379.
Chang, C. W. (1979). *Plant Physiol.* **64**, 833-836.
Ching, T. M., Poklemba, C. J., and Metzger, R. J. (1983). *Plant Physiol.* **73**, 652-657.
Chourey, P. S., and Nelson, O. E. (1976). *Biochem. Genet.* **14**, 1041-1055.
Clermont, S., and Percheron, F. (1979). *Phytochemistry* **18**, 1963-1965.
Clermont, S., Saïd, R., and Percheron, F. (1982). *Phytochemistry* **21**, 1951-1954.
Copeland, L., and Preiss, J. (1981). *Plant Physiol.* **68**, 996-1001.
Copeland, L., Harrison, D. D., and Turner, J. F. (1978). *Plant Physiol.* **62**, 291-294.
Courtois, J. E., and Le Dizet, P. (1963). *Bull. Soc. Chim. Biol.* **45**, 731-741.
Davies, M. D., and Dickinson, D. B. (1972). *Arch. Biochem. Biophys.* **152**, 53-61.
Dea, I. C. M., and Morrison, A. (1975). *Adv. Carbohydr. Chem. Biochem.* **31**, 241-312.
Delmer, D. P. (1972). *Plant Physiol.* **50**, 469-472.
Deri Tomas, A., and Northcote, D. H. (1978). *Biochem. J.* **174**, 283-290.

Dey, P. M. (1978). *Adv. Carbohydr. Chem. Biochem.* **35**, 341-376.
Dey, P. M. (1980a). *Adv. Carbohydr. Chem. Biochem.* **37**, 283-372.
Dey, P. M. (1980b). *FEBS Lett.* **114**, 153-156.
Dey, P. M. (1983). *Eur. J. Biochem.* **136**, 155-159.
Dey, P. M., and Pridham, J. G. (1972). *Adv. Enzymol. Relat. Areas Mol. Biol.* **36**, 91-130.
Dickinson, D. B., and Preiss, J. (1969a). *Arch. Biochem. Biophys.* **130**, 119-128.
Dickinson, D. B., and Preiss, J. (1969b). *Plant Physiol.* **44**, 1058-1062.
Doelhert, D. C., and Huber, S. C. (1983). *Plant Physiol.* **73**, 989-994.
Doi, A. (1969). *Biochim. Biophys. Acta* **184**, 477-485.
Duffus, C. M. (1979). *In* 'Recent Advances in the Biochemistry of Cereals' (D. L. Laidman and R. G. Wyn Jones, eds), pp. 209-238. Academic Press, New York.
Edelman, J., and Jefford, T. G. (1968). *New Phytol.* **67**, 517-531.
El Khadem, H., and Sallam, M. A. E. (1967). *Carbohydr. Res.* **4**, 387-391.
Elbein, A. D. (1969). *J. Biol. Chem.* **244**, 1608-1616.
Elbein, A. D., and Hassid, W. Z. (1966). *Biochem. Biophys. Res. Commun.* **23**, 311-318.
Escalada, J. A., and Moss, D. N. (1976). *Crop Sci.* **16**, 627-631.
Espada, J. (1962). *J. Biol. Chem.* **237**, 3577-3581.
Evers, A. D. (1971). *Staerke* **23**, 157-192.
Fan, D.-F., and Feingold, D. S. (1969). *Plant Physiol.* **44**, 599-604.
de Fekete, M. A. R., and Vieweg, G. H. (1974). *In* 'Plant Carbohydrate Biochemistry' (J. B. Pridham, ed.), pp. 127-144. Academic Press, New York.
de Fekete, M. A. R., Vieweg, G. H., and Thomas, W. (1980). *In* 'Mechanisms of Saccharide Polymerization and Depolymerization' (J. J. Marshall, ed.), pp. 145-159. Academic Press, New York.
Fischer, M., and Kandler, O. (1975). *Phytochemistry* **14**, 2629-2633.
Fisher, M. B., and Boyer, C. D. (1983). *Plant Physiol.* **72**, 813-816.
Foglietti, M.-J., and Percheron, F. (1976). *Biochimie* **58**, 499-504.
Franz, G. (1973). *Phytochemistry* **12**, 2369-2373.
Frei, E., and Preston, R. D. (1968). *Proc. Roy. Soc. London Ser. B* **169**, 127-145.
French, D. (1975). *In* 'Biochemistry of Carbohydrates' (W. J. Whelan, ed.), pp. 267-335. Butterworths, London.
Frydman, R. B., and Cardini, C. E. (1967). *J. Biol. Chem.* **242**, 312-317.
Frydman, R. B., and Neufeld, E. F. (1963). *Biochem. Biophys. Res. Commun.* **12**, 121-125.
Fuchs, R. L., and Smith, J. D. (1979). *Biochim. Biophys. Acta* **566**, 40-48.
Funkhouser, E. A., and Loewus, F. A. (1975). *Plant Physiol.* **56**, 786-790.
Gaudreault, P.-R., and Webb, J. A. (1981). *Phytochemistry* **20**, 2629-2633.
Giaquinta, R. T. (1980). *In* 'Biochemistry of Plants, Vol. 3, Carbohydrates: Structure and Function' (J. Preiss, ed.), pp. 271-320. Academic Press, New York.
Ginsburg, V. (1958). *J. Biol. Chem.* **232**, 55-61.
Gomyo, T., and Nakamura, M. (1966). *Agric. Biol. Chem.* **30**, 425-427.
Gould, M. F., and Greenshields, R. N. (1964). *Nature* **202**, 108-109.
Gould, S. E. B., Rees, D. A., and Wight, N. J. (1971). *Biochem. J.* **124**, 47-53.
Greenwood, C. T., and Thomson, J. (1962). *Biochem. J.* **82**, 156-164.
Gross, K. C., and Pharr, D. M. (1982). *Plant Physiol.* **69**, 117-121.
Gussin, A. E. S., and McCormack, J. H. (1970). *Phytochemistry* **9**, 1915-1920.
Gustafson, G. L., and Gander, J. E. (1972). *J. Biol. Chem.* **247**, 1387-1397.
Halmer, P., Bewley, J. D., and Thorpe, T. A. (1978). *Planta* **139**, 1-8.
Hannah, L. C., and Nelson, O. E. (1975). *Plant Physiol.* **55**, 297-302.
Hatanaka, S. (1959). *Arch. Biochem. Biophys.* **82**, 188-194.
Hawker, J. S., and Downton, W. J. S. (1974). *Phytochemistry* **13**, 893-900.
Hawker, J. S., and Smith, G. M. (1982). *Aust. J. Plant Physiol.* **9**, 509-518.

Hawker, J. S., Ozbun, J. L., Ozaki, H., Greenberg, E., and Preiss, J. (1974). *Arch. Biochem. Biophys.* **160**, 530-551.
Hayashi, T., and Matsuda, K. (1981). *J. Biol. Chem.* **256**, 11 117-11 122.
Heller, J. S., and Villemez, C. L. (1972). *Biochem. J.* **129**, 645-655.
Herbert, M., Burkhard, Ch., and Schnarrenberger, C. (1979). *Planta* **145**, 95-104.
Herold, A., and Lewis, D. H. (1977). *New Phytol.* **79**, 1-40.
Hinman, M. B., and Villemez, C. L. (1975). *Plant Physiol.* **56**, 608-612.
Hirai, M. (1981). *Plant Physiol.* **67**, 221-224.
Hodges, H. F., Creech, R. G., and Loerch, J. D. (1969). *Biochim. Biophys. Acta* **185**, 70-79.
Hoffman, J., and Svensson, S. (1978). *Carbohydr. Res.* **65**, 65-71.
Hoffman, J., Lindberg, B., and Painter, T. (1976). *Acta. Chem. Scand. B.* **30**, 365-367.
Hopf, H., and Kandler, O. (1974). *Plant Physiol.* **54**, 13-14.
Hopf, H., and Kandler, O. (1976). *Biochem. Physiol. Pflanz.* **169**, 5-36.
Hopf, H., and Kandler, O. (1977). *Phytochemistry* **16**, 1715-1717.
Hopf, H., and Kandler, O. (1980). *Z. Planzenphysiol.* **100**, 189-195.
Hopper, J. E., and Dickinson, D. B. (1972). *Arch. Biochem. Biophys.* **148**, 523-535.
Hui, P. A., and Neukom, H. (1964). *Tappi* **47**, 39-42.
Hunt, K., and Jones, J. K. N. (1962). *Can. J. Chem.* **40**, 1266-1279.
Imhoff, V. (1973). *Z. Physiol. Chem.* **354**, 1550-1554.
Jakimow-Barras, N. (1973). *Phytochemistry* **12**, 1331-1339.
Jenner, C. F. (1982). *In* 'Encyclopedia of Plant Physiology, Vol. 13A, Plant Carbohydrates I, Intracellular Carbohydrates' (F. A. Loewus and W. Tanner, eds), pp. 700-747. Springer Verlag, New York.
Jimenez de Asua, L., Carminatti, H., and Passeron, S. (1966). *Biochim. Biophys. Acta* **128**, 582-585.
John, K. V., Schutzback, J. S., and Ankel, H. (1977). *J. Biol. Chem.* **252**, 8013-8017.
Jones, G., and Whelan, W. J. (1969). *Carbohydr. Res.* **9**, 483-490.
Joshi, S., Lodha, M. L., and Mehta, S. L. (1980). *Phytochemistry* **19**, 2305-2309.
Jukes, C., and Lewis, D. H. (1974). *Phytochemistry* **13**, 1519-1521.
Juliano, B. O., and Varner, J. E. (1969). *Plant Physiol.* **44**, 886-892.
Kahl, W., Roszkowski, A., and Zurowska, A. (1969). *Carbohydr. Res.* **10**, 586-588.
Kandler, O., and Hopf, H. (1980). *In* 'The Biochemistry of Plants, Vol. 3, Carbohydrates: Structure and Function' (J. Preiss, ed.), pp. 221-270. Academic Press, New York.
Kandler, O., and Hopf, H. (1982). *In* 'Encyclopedia of Plant Physiology', Vol. 13A, 'Plant Carbohydrates I, Intracellular Carbohydrates' (F. A. Loewus and W. Tanner, eds), pp. 348-383. Springer Verlag, New York.
Kashlan, N., and Richardson, M. (1981). *Phytochemistry* **20**, 1781-1784.
Kelly, G. J., and Latzko, E. (1980a). *Photosynth. Res.* **1**, 181-187.
Kelly, G. J., and Latzko, E. (1980b). *In* 'The Biochemistry of Plants, Vol. 1, The Plant Cell' (N. E. Tolbert, ed.), pp. 183-208. Academic Press, New York.
Keusch, L. (1968). *Planta* **78**, 321-350.
Kindl, H. (1969). *Ann. N.Y. Acad. Sci.* **165**, 615-623.
Kooiman, P. (1960). *Acta Bot. Neerl.* **9**, 208-219.
Kooiman, P. (1967). *Phytochemistry* **6**, 1665-1673.
Kooiman, P. (1971). *Carbohydr. Res.* **20**, 329-337.
Kovacs, M. I. P., and Hill, R. D. (1974). *Phytochemistry* **13**, 1335-1339.
Koyama, T., Hayashi, T., Kato, Y., and Matsuda, K. (1981). *Plant Cell Physiol.* **22**, 1191-1198.
Lavintman, N., Tandecarz, J., Carceller, M., Mendiara, S., and Cardini, C. E. (1974). *Eur. J. Biochem.* **50**, 145-155.
Lee, B. T., and Matheson, N. K. (1984). *Phytochemistry* **23**, 983-987.

Leegood, R. C., and Walker, D. A. (1981). *Arch. Biochem. Biophys.* **212**, 644-650.
Lehle, L. and Tanner, W. (1973). *Eur. J. Biochem.* **38**, 103-110.
Lehle, L., Tanner, W., and Kandler, O. (1970). *Z. Physiol. Chem.* **351**, 1494-1498.
Liu, T.-T. Y., and Shannon, J. C. (1981). *Plant Physiol.* **67**, 525-529.
Loescher, W. H., Marlow, G. C., and Kennedy, R. A. (1982). *Plant Physiol.* **70**, 335-339.
Loewus, M. W., and Loewus, F. (1971). *Plant Physiol.* **48**, 255-260.
Lombard, A., Rossetti, V., Buffa, M., and Congiu, G. (1981). *Carbohydr. Res.* **96**, 131-133.
McCleary, B. V. (1978). *Phytochemistry* **17**, 651-653.
McCleary, B. V. (1979). *Carbohydr. Res.* **71**, 205-230.
McCleary, B. V. (1981). *Lebensm. Wiss. Technol.* **14**, 188-191.
McCleary, B. V., and Matheson, N. K. (1975). *Phytochemistry* **14**, 1187-1194.
McCleary, B. V., and Matheson, N. K. (1976). *Phytochemistry* **15**, 43-47.
McCleary, B. V., and Matheson, N. K. (1983). *Carbohydr. Res.* **119**, 191-219.
McCleary, B. V., Matheson, N. K., and Small, D. M. (1976). *Phytochemistry* **15**, 1111-1117.
McClendon, J. H., Nolan, W. G., and Wenzler, H. F. (1976). *Am. J. Bot.* **63**, 790-797.
McConnell, W. B., Mitra, A. K., and Perlin, A. S. (1958). *Can. J. Biochem. Physiol.* **36**, 985-991.
Macdonald, F. D., and ap Rees, T. (1983). *Biochim. Biophys. Acta* **755**, 81-89.
MacGregor, A. W., Gordon, A. G., Meredith, W. O. S., and Lacroix, L. (1972). *J. Inst Brew.* **78**, 174-179.
MacLeod, A. M., and McCorquodale, H. (1958). *Nature* **182**, 815-816.
Manners, D. J., and Matheson, N. K. (1981). *Carbohydr. Res.* **90**, 99-110.
Manners, D. J., and Rowe, K. L. (1969). *Carbohydr. Res.* **9**, 441-450.
Manners, D. J., Rowe, J. J. M., and Rowe, K. L. (1968). *Carbohydr. Res.* **8**, 72-81.
Marshall, J. J., and Rickson, F. R. (1973). *Carbohydr. Res.* **28**, 31-37.
Matheson, N. K. (1971). *Phytochemistry* **10**, 3213-3219.
Matheson, N. K. (1975). *Phytochemistry* **14**, 2017-2021.
Matheson, N. K., and Richardson, R. H. (1976). *Phytochemistry* **15**, 887-892.
Matheson, N. K., and Richardson, R. G. (1978). *Phytochemistry* **17**, 195-200.
Matheson, N. K., and Saini, H. S. (1977). *Phytochemistry* **16**, 59-66.
Matters, G. L., and Boyer, C. D. (1981). *Phytochemistry* **20**, 1805-1809.
May, L. H., and Buttrose, M. S. (1959). *Aust. J. Biol. Sci.* **12**, 146-159.
Medcalf, D. G., and Cheung, P. W. (1971). *Cereal Chem.* **48**, 1-8.
Meier, H. (1958). *Biochim. Biophys. Acta* **28**, 229-240.
Meier, H., and Reid, J. S. G. (1977). *Planta* **133**, 243-248.
Meier, H., and Reid, H. S. G. (1982). In 'Encyclopedia of Plant Physiology, Vol. 13A Plant Carbohydrates I. Intracellular Carbohydrates' (F. A. Loewus and W. Tanner, eds), pp. 418-471. Springer Verlag,· New York.
Mlodzianowski, F., and Wesolowska, M. (1975). *Acta Soc. Bot. Pol.* **44**, 529-536.
Montgomery, R., and Smith, F. (1957). *J. Am. Chem. Soc.* **79**, 446-450.
Moreno, A., and Cardini, C. E. (1966). *Plant Physiol.* **41**, 909-910.
Morgenlie, S. (1970). *Acta Chem. Scand.* **24**, 2149-2155.
Morrall, P., and Briggs, D. E. (1978). *Phytochemistry* **17**, 1495-1502.
Murata, T. (1971). *Agric. Biol. Chem.* **35**, 1441-1448.
Murata, T. (1972). *Agric. Biol. Chem.* **36**, 1815-1818.
Murata, T. (1975). *Plant Cell Physiol.* **16**, 953-961.
Murata, T. (1976). *Plant Cell Physiol.* **17**, 1099-1109.
Murata, T., and Akazawa, T. (1966). *Arch. Biochem. Biophys.* **114**, 76-87.
Murata, T., Minamikawa, T., Akazawa, T., and Sugiyama, T. (1964). *Arch. Biochem. Biophys.* **106**, 371-378.

Negm, F. B., and Loescher, W. H. (1979). *Plant Physiol.* **64**, 69-73.
Neufeld, E. F., and Hall, C. W. (1965). *Biochem. Biophys. Res. Commun.* **19**, 456-461.
Neufeld, E. F., Ginsberg, V., Putman, E. W., Fanshier, D., and Hassid, W. Z. (1957). *Arch. Biochem. Biophys.* **69**, 602-616.
Nigam, V. N., and Giri, K. V. (1960). *J. Biol. Chem.* **235**, 947-950.
Nikaido, H., and Hassid, E. Z. (1971). *Adv. Carbohydr. Chem. Biochem.* **26**, 351-483.
Okita, T. W., Greenberg, E., Kuhn, D. N., and Preiss, J. (1979). *Plant Physiol.* **64**, 187-192.
Otozai, K., Taniguchi, H., and Nakamura, M. (1973). *Agric. Biol. Chem.* **37**, 531-537.
Ozbun, J. L., Hawker, J. S., and Preiss, J. (1971). *Plant Physiol.* **48**, 765-769.
Ozbun, J. L., Hawker, J. S., Greenberg, E., Lammel, C., Preiss, J., and Lee, E. Y. C. (1973). *Plant Physiol.* **51**, 1-5.
Perdon, A. A., Del Rosario, E. J., and Juliano, B. O. (1975). *Phytochemistry* **14**, 949-951.
Pisigan, R. A., and Del Rosario, E. J. (1976). *Phytochemistry* **15**, 71-73.
Pollock, C. J. (1979). *Phytochemistry* **18**, 777-779.
Pollock, C. J. (1982). *Phytochemistry* **21**, 2461-2465.
Pontis, H. G., and Salerno, G. L. (1982). *FEBS Lett.* **141**, 120-123.
Preiss, J. (1982a). *Annu. Rev. Plant Physiol.* **33**, 431-454.
Preiss, J. (1982b). *In* 'Encyclopedia of Plant Physiology, Vol. 13A, Plant Carbohydrates I, Intracellular Carbohydrates' (F. A. Loewus and W. Tanner, eds), pp. 397-417. Springer Verlag, New York.
Preiss, J., and Levi, C. (1980). *In* 'The Biochemistry of Plants, Vol. 3, Carbohydrates: Structure and Function' (J. Preiss, ed.), pp. 371-423. Academic Press, New York.
Pridham, J. B., and Dey, P. M. (1974). *In* 'Plant Carbohydrate Biochemistry' (J. B. Pridham, ed.), pp. 83-96. Academic Press, New York.
Pridham, J. B., and Hassid, W. Z. (1965). *Plant Physiol.* **40**, 984-986.
Quemener, B., and Brillouet, J.-M. (1983). *Phytochemistry* **22**, 1745-1751.
Ramasarma, T., and Giri, K. V. (1956). *Arch. Biochem. Biophys.* **62**, 91-96.
Ramasarma, T., Sri Ram, J., and Giri, K. V. (1954). *Arch. Biochem. Biophys.* **53**, 167-173.
Reid, J. S. G. (1971). *Planta* **100**, 131-142.
Reid, J. S. G., and Meier, H. (1970). *Phytochemistry* **9**, 513-520.
Robic, D., and Percheron, F. (1973). *Phytochemistry* **12**, 1369-1372.
Rubery, P. H. (1973). *Planta* **111**, 267-269.
Rumpho, M. E., Edwards, G. E., and Loesher, W. H. (1983). *Plant Physiol.* **73**, 869-873.
Saini, H. S., and Matheson, N. K. (1981). *Phytochemistry* **20**, 641-645.
Salamini, F., Tsai, C. Y., and Nelson, O. E. (1972). *Plant Physiol.* **50**, 256-261.
Saunders, R. M. (1971). *Phytochemistry* **10**, 491-493.
Schiefer, S., Lee, E. Y. C., and Whelan, W. J. (1978). *Carbohydr. Res.* **61**, 239-252.
Schilling, N. (1982). *Planta* **154**, 87-93.
Schlubach, H. H., and Koehn, H. O. A. (1958). *Justus Leibigs Ann. Chem.* **614**, 126-136.
Schlubach, H. H., and Müller, H. (1952). *Justus Liebigs Ann. Chem.* **578**, 194-198.
Schweizer, T. F., and Horman, I. (1981). *Carbohydr. Res.* **95**, 61-71.
Schweizer, T. F., Horman, I., and Würsch, P. (1978). *J. Sci. Food Agric.* **29**, 148-154.
Shannon, J. C., Creech, R. G., and Loerch, J. D. (1970). *Plant Physiol.* **45**, 163-168.
Shimomura, S., Nagai, M., and Fukui, T. (1982). *J. Biochem.* (Tokyo) **91**, 703-717.
Shiomi, N. (1981). *Carbohydr. Res.* **96**, 281-292.
Shiomi, N. (1982). *Carbohydr. Res.* **99**, 157-169.
Shiomi, N., and Izawa, M. (1980). *Agric. Biol. Chem.* **44**, 603-614.
Siddiqui, I. R., and Wood, P. J. (1971). *Carbohydr. Res.* **17**, 97-108.
Small, D. M., and Matheson, N. K. (1979). *Phytochemistry* **18**, 1147-1150.

Smart, E. L., and Pharr, D. M. (1981). *Planta* **153**, 370-375.
Sowokinos, J. R., and Preiss, J. (1982). *Plant Physiol.* **69**, 1459-1466.
Staesche, K. (1966). *Planta* **71**, 268-282.
Steup, M., and Schächtele, C. (1981). *Planta* **153**, 351-361.
Steup, M., Robenek, H., and Melkonian, M. (1983). *Planta* **158**, 428-436.
Su, J.-C., and Preiss, J. (1978). *Plant Physiol.* **61**, 389-393.
Sundararajan, P. R., and Rao, V. S. R. (1970). *Biopolymers* **9**, 1239-1247.
Takamiya, S., and Fukui, T. (1978). *J. Biochem.* (Tokyo) **84**, 569-574.
Takeda, Y., Hizukuri, S., and Nikuni, Z. (1967). *Biochem. Biophys. Acta* **146**, 568-575.
Tanaka, Y., and Akazawa, T. (1971). *Plant Cell Physiol.* **12**, 493-505.
Tandecarz, J., Lavintman, N., and Cardini, C. E. (1975). *Phytochemistry* **14**, 103-106.
Tanner, G. J., Copeland, L., and Turner, J. F. (1983). *Plant Physiol.* **72**, 659-663.
Tanner, W. (1969). *Ann. N.Y. Acad. Sci.* **165**, 726-742.
Tanner, W., and Kandler, O. (1968). *Eur. J. Biochem.* **4**, 233-239.
Tanner, W., Lehle, L., and Kandler, O. (1967). *Biochem. Biophys. Res. Commun.* **29**, 166-171.
Täufel, K., Romminger, K., and Hirschfeld, W. (1959). *Z. Lebensm. Unter. Forsch.* **109**, 1-12.
Thompson, J. L., and Jones, J. K. N. (1964). *Can. J. Chem.* **42**, 1088-1091.
Timell, T. E. (1965). *Adv. Carbohydr. Chem.* **20**, 409-483.
Tomasić, J., Jennings, H. J., and Glaudemans, C. P. J. (1978). *Carbohydr. Res.* **62**, 127-133.
Tovey, K. C., and Roberts, R. M. (1970). *Plant Physiol.* **46**, 406-411.
Turner, D. H., and Turner, J. F. (1958). *Biochem. J.* **69**, 448-452.
Turner, J. F. (1969a). *Aust. J. Biol. Sci.* **22**, 1321-1327.
Turner, J. F. (1969b). *Aust. J. Biol. Sci.* **22**, 1145-1151.
Turner, J. F., and Turner, D. H. (1975). *Annu. Rev. Plant Physiol.* **26**, 159-186.
Tyler, J. M. (1965a). *J. Chem. Soc.* 5288-5300.
Tyler, J. M. (1965b). *J. Chem. Soc.* 5300-5310.
Villemez, C. L. (1974). *Arch. Biochem. Biophys.* **165**, 407-412.
Wagner, W., Keller, F., and Wiemken, A. (1983). *Z. Pflanzenphysiol.* **112**, 359-372.
Walker, D. A. (1976). *In* 'Encyclopedia of Plant Physiology, Vol. 3, Intracellular Interactions and Transport Processes' (C. R. Stocking and U. Heber, eds), pp. 85-136. Springer-Verlag, New York.
Webb, J. A., and Gorham, P. R. (1964). *Plant Physiol.* **39**, 663-672.
Weber, M., and Wöber, G. (1975). *Carbohydr. Res.* **39**, 295-302.
Weselake, R. J., MacGregor, A. W., and Hill, R. D. (1983). *Plant Physiol.* **72**, 809-812.
Whelan, W. J. (1976). *Trends Biochem. Sci.* **1**, 13-15.
Whistler, R. L., and Young, J. R. (1960). *Cereal Chem.* **37**, 204-211.
White, L. M., and Secor, G. E. (1953). *Arch. Biochem. Biophys.* **43**, 60-66.
Williams, J. M., and Duffus, C. M. (1977a). *J. Inst. Brewing* **84**, 47-50.
Williams, J. M., and Duffus, C. M. (1977b). *Plant Physiol.* **59**, 189-192.
Wolfrom, M. L., Laver, M. L., and Patin, D. L. (1961). *J. Org. Chem.* **26**, 4533-4535.
Yamaki, S. (1980). *Plant Cell Physiol.* **21**, 591-599.
Yamaki, S. (1981). *Plant Cell Physiol.* **22**, 359-367.
Ziegler, H. (1975). *In* 'Encyclopedia of Plant Physiology, Vol. 1, Transport in Plants I. Phloem Transport' (M. H. Zimmermann and J. A. Milburn, eds), pp. 59-100. Springer-Verlag, New York.

CHAPTER 6

Synthesis of Storage Lipids in Developing Seeds

C. R. SLACK and J. A. BROWSE

I.	Introduction	209
II.	Occurrence and Composition of Storage Lipids	210
III.	Formation of Fatty Acids from Sucrose	214
	A. Conversion of Sucrose to Acetyl-Coenzyme A	214
	B. Production of Palmitate (16:0) and Oleate (18:1) from Acetyl-Coenzyme A	217
	C. Relationship of Fatty Acid Synthesis to Glycolysis and the Pentose Phosphate Pathway	221
IV.	Changes in Lipid Composition and Content during Seed Development	222
V.	Triacylglycerol Synthesis in Developing Seeds	223
	A. The Glycerol-3-Phosphate Pathway	223
	B. Formation of Triacylglycerols and Wax Esters Containing Long-chain Fatty Acids	225
	C. Synthesis of Some 'Unusual' Fatty Acids	229
	D. Selective Incorporation of 'Unusual' Fatty Acids into Storage Lipids	230
	E. Synthesis of Polyunsaturated Triacylglycerols	231
VI.	Composition and Origin of Oil Bodies	236
VII.	Environmental Effects on the Fatty Acid Composition of Oil Seed Lipids	237
VIII.	Conclusions	239
	References	240

I. INTRODUCTION

In plants the reserves of reduced carbon derived from photosynthesis are mainly stored as carbohydrate, either as soluble or insoluble forms

(Chapter 5). This is true for leaves, stems, roots, tubers and bulbs. However, in many species the seed stores some and often most of its reserves of energy and reduced carbon needed for germination as fatty acids. The storage of fatty acid as a seed reserve rather than carbohydrate is a way of maximizing the quantity of stored energy in a small volume of tissue. The carbons in fatty acids are more highly reduced than in carbohydrate and as a consequence the amounts of energy released during oxidation of the two types of compound are about 38 kJ/g and 17 kJ/g, respectively. Energy must be expended in the conversion of sucrose to the more highly reduced fatty acids during seed development and hence the seed yields of oil-rich crops are invariably less than those of crops that produce seed containing mainly carbohydrate. De Wit (1978) and McDermitt and Loomis (1981) have shown that about 0.86 g of starch compared with 0.36 g of fatty acid can be formed from 1 g of glucose. Sinclair and De Wit (1975) have calculated that from each g of photosynthate, rice can produce 0.75 g of starchy seed, whereas peanut can form only about 0.45 g of seed material rich in fatty acid.

Because of the commercial importance of seed oils both as foods and for industrial use a vast amount of information is available about the chemical composition of lipid-rich seeds (Section 6.II,A of Volume 2). By comparison, the present knowledge about how storage lipids are produced in seeds is limited. Appelqvist (1975) and Gurr (1980) have discussed why this should be so. Both agreed that the difficulties involved in obtaining sufficient quantities of seed material may be an important reason. Apart from the organization needed to ensure a supply of plants with seed at the correct stage of development, the task of obtaining adequate amounts of tissue is extremely time consuming. Despite these difficulties, sufficient progress has been made in recent years for us to describe in general terms the metabolic route by which sucrose is converted into storage lipid.

II. OCCURRENCE AND COMPOSITION OF STORAGE LIPIDS

The fatty acids that accumulate during seed development are nearly always present in triacylglycerol. In this molecule a fatty acid is esterified to each of the primary hydroxyl groups at positions 1 and 3 and to the secondary hydroxyl group at position 2 of glycerol (Fig. 1). A stereochemical numbering system denoted by *sn* is used to describe the stereochemistry of glycerolipids. If, for example, R_1 = palmitate, R_2 = oleate and R_3 = linoleate, the resulting triacylglycerol would be 1-palmitoyl, 2-oleoyl, 3-linoleoyl-*sn*-glycerol. This generalization applies to both Gymnosperms and Angiosperms. The only exception is the perennial desert shrub, jojoba

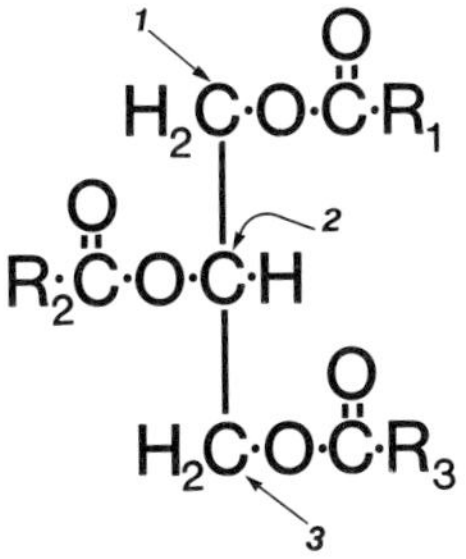

$$\begin{array}{c}
\overset{1}{H_2}\overset{}{C}\cdot O\cdot\overset{O}{\overset{\|}{C}}\cdot R_1 \\
\overset{O}{\overset{\|}{R_2}}\overset{}{C}\cdot O\cdot\overset{2}{C}H \\
\overset{}{H_2}\underset{3}{C}\cdot O\cdot\overset{O}{\overset{\|}{C}}\cdot R_3
\end{array}$$

Fig. 1. Triacylglycerol — a neutral lipid in which a fatty acid is esterified to each of the hydroxyl groups of a glycerol molecule.

(*Simmondsia chinensis*). The storage lipid in the seed of this plant is a wax in which fatty acids are esterified to long-chain alcohols (Section 6.II,C of Volume 2). Both triacylglycerols and wax esters contain no ionizable group, are completely hydrophobic and occur in the seed as water-free lipid droplets (Section VI). These uncharged lipids probably represent the two most innocuous forms in which fatty acids can be stored in large amounts within a cell. Detailed accounts of the quantities of triacylglycerol that occur in the seeds of different species can be found in Eckey (1954), Hilditch and Williams (1964) and Section 6.II of Volume 2. The amounts vary considerably, from values commonly greater than 50% by weight, to as little as 1%–2% by weight of the cotyledons of most legumes. In cereal grains, where endosperm is rich in carbohydrate, the embryo axis and scutellum can contain up to 50% by weight of triacylglycerol (Weber, 1973).

Palmitate, stearate, oleate, linoleate and linolenate (Table I) are the most widely occurring fatty acids in seeds. These are found as the major constituents of the seed triacylglycerols of most Gymnosperms and Angiosperms, including economically important species (Table II). The long-chain monoenoic acids, erucate and eicosenoate, are the major fatty acids in seeds of the *Cruciferae* and jojoba wax consists of these two acids esterified to alcohols derived from them (Yermanos, 1975). In addition to the above fatty acids, a great many others have been found in seeds (Hilditch and Williams, 1964; Smith, 1970; Hitchcock and Nichols, 1971).

The fatty acid composition of seed triacylglycerols is species and often variety specific. Within many of the agriculturally important species, a considerable degree of heritable diversity in composition has been found (Table II). In rape (*Brassica napus*), variation occurs mainly in the quantity of oleate relative to that of long-chain fatty acids. In other species, the relative amounts of oleate and polyunsaturated fatty acids vary. Oleate is the precursor of both long-chain and polyunsaturated fatty acids (Sections

Table I. Structures of some fatty acids commonly occurring in seed triacylglycerols.

Common name	Structure	'Shorthand' formulae[a]
Palmitate	$CH_3.(CH_2)_{14}.COOH$	16:0
Stearate	$CH_3.(CH_2)_{16}.COOH$	18:0
Oleate	$CH_3.(CH_2)_7.CH:CH(CH_2)_7.COOH$	18:1, 18:1 (9c)
Linoleate	$CH_3.(CH_2)_3.(CH_2.CH:CH)_2.(CH_2)_7.COOH$	18:2, 18:2 (9c, 12c)
Linolenate	$CH_3.(CH_2.CH:CH)_3.(CH_2)_7.COOH$	18:3, 18:3 (9c, 12c, 15c)
Eicosenoate	$CH_3.(CH_2)_7.CH:CH.(CH_2)_9.COOH$	20:1, 20:1 (11c)
Erucate	$CH_3.(CH_2)_7.CH:CH.(CH_2)_{11}.COOH$	22:1, 22:1 (13c)

[a] The number before the colon denotes the number of carbons in the molecule and the number after the colon the number of double-bonds. The fatty acid can be described more completely by denoting the position, counted from the carboxyl end of the molecule, of the double-bonds. All the double-bonds in these commonly occurring fatty acids have a cis-configuration, indicated c.

Table II. Within species variability in the fatty acid composition of the triacylglycerols of major oil-seed crops.

Species	Range of fatty acid composition mol/100 mol						
	16:0	18:0	18:1	18:2	18:3	20:1	22:1
Soybean[a]	9.3–17.4	2.2–7.2	15.6–29.6	33.8–59.6	4.3–15.0		
Sunflower[b]	7.3	4.2–7.0	16.5–30.5	55.2–72.0			
Peanut[a]	6.7–12.0	4.3–5.2	36.0–71.4	11.1–41.0			
Rape[a,c]	4.0– 4.8	1.0–1.2	8.0–63.0	11.0–20.0	6.0– 8.0	1.0–14.0	0.0–54.0
Maize[d]	6.0–22.0	0.6–5.0	14.0–64.0	19.0–71.0	0.0– 2.0		
Safflower[e]	6.4– 6.7	1.4–5.5	14.0–71.6	21.4–78.2			
Linseed[a]	6.0– 7.0	—	14.0–39.0	15.0–17.0	35.0–58.0		

[a] Downey and McGregor (1975).
[b] Putt et al. (1969).
[c] Downey and Craig (1964).
[d] Jellum (1970).
[e] Knowles (1972).

V,B and V,E) and hence presumably the variation in the amounts of these fatty acids must result from differences in the relative rates of synthesis and metabolism of oleate.

Some information is available about the genetic control of fatty acid composition in most oil seed crops (Downey and McGregor, 1975), but it is probably best understood in rape and safflower (*Carthamus tinctorius*). In both, the genetic constitution of the embryo determines the proportions of the different fatty acids synthesized, and there is no maternal effect. Two genes with no dominance and acting additively govern the amount of erucate produced in rape (Harvey and Downey, 1964). In safflower, the relative amounts of oleate and linoleate are determined by major genes at one locus. High linoleate is dominant, but two different alleles give either 75% or 45% oleate in homozygous recessives. The content of stearate is controlled by a single gene at a separate locus, high stearate being recessive (Knowles, 1972).

During the biosynthesis of triacyglycerols each of the hydroxyl groups on the glycerol molecule is biochemically distinct and fatty acids are not esterified at random to the different hydroxyl groups. The first measurements of the proportions of the different fatty acids esterified at each of the three positions established that saturated fatty acids were essentially absent from position 2, whereas the fatty acid compositions at positions 1 and 3, though not identical, were very similar (Brockeroff and Yurkowski, 1966). More recent analyses have shown considerably greater differences between the fatty acid composition at positions 1 and 3, with more saturated fatty acids at position 1 (Table III).

Since four or more fatty acids are present in the triacylglycerols of a particular species (Tables II and III), it is apparent that the reserve lipid can consist of many molecular species that differ in the combinations of the various fatty acids they contain, and also in the position that each occupies within the molecule. Molecular species can be separated into different classes on the basis of the sum of the double bonds present in the constituent fatty acids. Classes containing from one to eight double bonds have been separated from the seed triacylglycerols of maize (*Zea mays*) (de la Roche *et al.*, 1971) and soybean (*Glycine max*) (Fatemi and Hammond,

Table III. Variation in the fatty acid composition at each of three different positions of seed triacylglycerols.

Species	Position	Fatty acid composition mol/100 mol						
		16:0	18:0	18:1	18:2	18:3	20:1	22:1
Maize[a]	1	15.6	3.9	21.4	57.8	1.3		
	2	0.7	0.2	21.6	76.6	0.8		
	3	7.0	1.6	19.2	70.6	1.6		
Soybean[b]	1	21.1	6.6	20.7	42.4	10.2		
	2	0.6	0.1	21.2	69.2	9.0		
	3	12.2	4.2	29.3	46.1	8.1		
Crambe[c]	1	9.9	2.4	18.6	5.1	0	7.8	54.0
	2	0	0	51.3	24.0	20.4	1.5	3.0
	3	1.8	2.4	7.2	2.4	1.2	7.1	73.2

[a] Weber *et al.* (1971).
[b] Fatemi and Hammond (1977a, b).
[c] Gurr *et al.* (1972).

1977b). In these studies the positions occupied by the different fatty acids in each molecular class were not determined. To our knowledge the study of Gurr et al. (1972) with crambe (*Crambe abyssinica*) seed triacylglycerol is the only complete stereochemical analysis. They found that this lipid consisted of five major species.

III. FORMATION OF FATTY ACIDS FROM SUCROSE

The main reactions involved in the production of acetyl-CoA from carbohydrates and the subsequent synthesis of long-chain fatty acids have been known for some time, but until recently there was less information about the subcellular compartmentation of these reactions in seeds and other plant tissues. In contrast to some reports (Yang and Stumpf, 1965; Harwood et al., 1971; Harwood and Stumpf, 1972; Mazliak et al., 1972; Harwood, 1979) it is now clearly established that the sole site of *de novo* fatty acid synthesis in plant cells is the plastid, both in seeds (Zilkey and Canvin, 1969, 1971; Yamada and Usami, 1975) and other plant tissues (Weaire and Kekwick, 1975; Vick and Beevers, 1978; Ohlrogge et al., 1979). This finding emphasizes the importance of using rigorous methods for isolating and characterizing organelles when attempting to determine the subcellular location of an enzyme or pathway. As detailed below, most of the steps in the conversion of sucrose to acetyl-CoA in oil seeds may also occur in the plastids, as does the production of oleate (18:1) from palmite (16:0). Further chain elongation, desaturation and hydroxylation reactions, as well as the final assembly of the triacylglycerol or wax ester storage product, involves sites outside the plastid as described in Section V.

A. Conversion of Sucrose to Acetyl-Coenzyme A

The ultimate carbon source for the formation of storage lipids by developing seeds is usually sucrose and is translocated via the phloem from elsewhere in the plant. Yamada and coworkers used a 10 000 g pellet (Yamada et al., 1974) and a purified plastid fraction (Yamada and Usami, 1975) from developing castor bean (*Ricinus communis*) endosperm to study the incorporation of labelled carbohydrates into long-chain fatty acids. These studies showed that [2-^{14}C]pyruvate, [^{14}C]glucose-6-phosphate and [^{14}C]glucose-1-phosphate supported fatty acid synthesis at similar rates, suggesting the presence of a complete glycolytic sequence as well as a pyruvate dehydrogenase [EC 1.2.4.1] complex. More recent studies with plastids purified on continuous sucrose density gradients have demon-

6. Lipid Reserves   215

strated that all the glycolysis enzymes show activities sufficient to explain the observed rates of fatty acid accumulation by developing castor bean seeds (Simcox et al., 1977; Dennis and Miernyck, 1982). Several key glycolysis enzymes have also been reported in plastids of safflower, soybean and sunflower (*Helianthus annuus*) (Ireland and Dennis, 1980). In most cases the plastid enzymes have been shown to be isoenzymes distinct from those operating in cytoplasmic glycolysis (De Luca and Dennis, 1978; Dennis and Miernyck, 1982).

The pentose phosphate pathway also operates at high activity in castor bean endosperm (Agrawal and Canvin, 1971a, b) and plastids isolated from this tissue (Yamada and Usami, 1975). Again, distinct plastid isoenzymes are involved (Simcox and Dennis, 1978a, b; Ireland and Dennis, 1980). However, Simcox et al. (1977) were unable to demonstrate glucose-6-phosphate dehydrogenase [EC 1.1.1.49] activity in isolated plastids, which implies that the first step in the pentose phosphate pathway may occur in the cytoplasm (cf. Yamada and Usami, 1975). The pentose phosphate pathway initially involves conversion of glucose-6 phosphate via 6-phosphogluconate to ribulose-5-phosphate (Fig. 2). The pentose phosphate molecules are then involved in a series of rearrangement reactions to yield fructose-6-phosphate and glyceraldehyde phosphate with the overall stoichiometry

$$6 \text{ hexose-P} + 12 \text{ NADP}^+ \rightarrow 4\text{-hexose-P} + 2 \text{ glyceraldehyde-3-P} + 6 \text{ CO}_2 + 12 \text{ NADPH} \qquad (1)$$

In the presence of triosephosphate isomerase [EC 5.3.1.1] and aldolase [EC 4.2.13] the glyceraldehyde-3-phosphate may be reconverted to hexose phosphate and the pathway can thus operate as a closed cycle (Beevers, 1961). However, in oil seed plastids the two molecules of triose phosphate produced for each six molecules of glucose-6-phosphate entering the cycle might equally well enter the glycolytic sequence. The interrelationship between glycolysis and the pentose phosphate pathway in supplying reduced pyridine nucleotides and ATP for fatty acid synthesis is discussed in Section III,C.

The final conversion of pyruvate to acetyl-CoA probably involves a plastid pyruvate dehydrogenase complex (Reid et al., 1975, 1977). However, the fact that many oil seed tissues and isolated plastids rapidly incorporate exogenous [1-^{14}C]acetate into fatty acids indicates the presence of acetyl-CoA synthase in the organelle. *In vivo* this enzyme could enable free acetate diffusing from other sites in the cell to be activated for use by the plastid fatty acid synthetase. Although some evidence that free acetate from outside the plastid is a potential substrate for fatty acid synthesis by

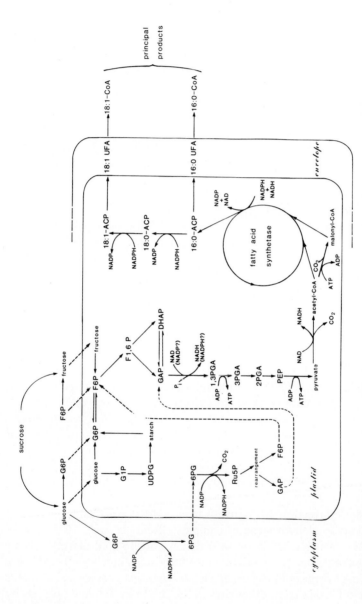

Fig. 2. Pathways involved in conversion of sucrose to palmitate and oleate by developing oil seed tissues. Abbreviations: ACP, acyl carrier protein; DHAP, dihydroxyacetone phosphate; F6P, fructose-6-phosphate; F1,6P, fructose-1,6-bisphosphate; GAP, glyceraldehyde-3-phosphate; G1P, glucose-1-phosphate; G6P, glucose-6-phosphate; PEP, phosphoenolpyruvate; PGA, phosphoglyceric acid; 6PG, 6-phosphogluconate; UDPG, UDP-glucose; UFA, unesterified fatty acid; Ru5P, ribulose-5-phosphate.

soybean (Nelson and Rinne, 1977) and in leaves (Kuhn et al., 1981; Murphy and Stumpf, 1981), the evidence reviewed above indicates that this is not the normal situation in oil seeds. In some oil seeds exogenously supplied [1-^{14}C]acetate appears to be used almost entirely for elongation reactions outside the plastid; for example, rape (Downey and Craig, 1964; Wright, 1980), crambe (Gurr et al., 1974), jojoba (Ohlrogge et al., 1978a), nasturtium (*Tropaeolum majus*) (Pollard and Stumpf, 1980a) and meadowfoam (*Limnanthes alba*) (Pollard and Stumpf, 1980b). These results suggest that free acetate produced in the mitochondrion or cytoplasm would be unavailable for *de novo* fatty acid synthesis in the plastid (see Section V.B).

The pathway by which hexose phosphates are synthesized from the imported sucrose is less clear. Yamada et al. (1974) argued for the operation of sucrose synthetase [EC 2.4.1.13], but this has been found at only low levels in castor bean seeds (Simcox et al., 1977). Instead, evidence suggests that sucrose is converted to glucose and fructose by a cytoplasmic invertase [EC 3.2.1.26] and that these sugars are phosphorylated in the cytoplasm or plastid by hexokinase [EC 2.7.1.1] (Simcox et al., 1977; Dennis and Miernyck, 1982). Starch grains have been observed in plastids from a variety of oil seeds (Appleqvist, 1975; Gurr, 1980), so it is assumed that the enzymes for the synthesis and breakdown of this storage carbohydrate are also present. A summary of these findings is shown in Figure 2.

B. Production of Palmitate (16:0) and Oleate (18:1) from Acetyl-Coenzyme A

Both acetyl-CoA and malonyl-CoA are required as substrates for the fatty acid synthetase, the second being formed from acetyl-CoA by the action of acetyl-CoA carboxylase. This carboxylation is the first committed step in fatty acid synthesis in animal systems and the enzyme has been shown to be highly regulated and rate-limiting to the overall process (Lane et al., 1974). In leaves and oil seeds it may be more correct to view acetyl-CoA synthesis as the first committed step, since this compound once produced in the plastid is apparently not available for the tricarboxylic acid (TCA) cycle or other syntheses.

Acetyl-CoA carboxylase [EC 6.4.1.2] from *Escherichia coli* dissociates during purification into three protein components: (i) biotin carboxylase; (ii) carboxyl carrier protein (which contains a biotin prosthetic group); and (iii) carboxyl transferase; whereas animal cells contain a multifunctional enzyme which exists, in its active form, as a high molecular weight polymer (Lane et al., 1974). Most of the studies of acetyl-CoA carboxylases from plant sources suggest it is a soluble multifunctional enzyme with a molecular weight of 6.0–6.5 × 10^5 daltons (Hatch and Stumpf, 1961; Heinstein and Stumpf, 1969; Reitzel and Nielsen, 1976; Brock and Kannangara, 1976; Mohan and Kekwick, 1980; Nikolau et al., 1981). Some breakdown during purification often occurs, but the subunit sizes appear to vary and the enzyme is clearly not of the prokaryotic type isolated from *E. coli*. In contrast to these reports, Kannangara and Stumpf (1972) suggested that the spinach chloroplast enzyme is of the prokaryotic type and that the biotin carboxyl carrier protein is covalently bound to the thylakoid membranes in chloroplasts of spinach and other plant species (Kannangara and Stumpf, 1973; Kannangara and Jensen, 1975). No studies of acetyl-CoA carboxylase from oil seeds have been reported.

Fatty acid synthetase complex performs a cycle of reactions during which malonate from malonyl-CoA is transferred to acyl carrier protein (ACP), then added to an ACP-bound acyl chain and the product converted by the sequential action of β-ketoacyl-ACP reductase, β-hydroxyacyl-ACP dehydrase and enoyl-ACP reductase to form a new acyl chain which is two carbon atoms longer than at the start of the cycle (Fig. 3). The plant fatty acid synthetase is like the easily dissociated prokaryotic type rather than the multienzyme complex described from yeast and animal sources (Hitchcock and Nichols, 1971). Despite their prokaryotic nature and location in the plastids, it seems likely that the enzymes of the complex are assembled outside the plastid (and presumably coded for in the nuclear genome), since a barley (*Hordeum vulgare*) mutant that lacked plastid ribosomes could synthesize fatty acids (Dorne et al., 1982). Shimikata and Stumpf (1982a) have distinguished all the component enzyme activities of the fatty acid synthetase from developing safflower seeds and determined the pyridine nucleotide requirements of the reductases. The β-ketoacyl-ACP reductase requires NADPH for activity. However, two enoyl-ACP reductases are present. Enoyl-ACP reductase I shows absolute specificity for NADH, whereas enoyl-ACP reductase II shows maximum rates with NADPH, although NADH also supports some activity (Table IV).

Previously, Saito et al. (1980) showed that in developing castor bean seeds the β-ketoacyl-ACP reductase is stereospecific for the B-side hydrogen of NADPH, while the enoyl-ACP reductase uses the A-side hydrogen of NADPH. It is not clear from this study whether castor bean seeds also contain a NADH-dependent enoyl-ACP reductase as is the case in safflower, but several other laboratories have shown that fatty acid synthesis by castor bean seed preparations requires both NADPH and NADH (Drennan and Canvin, 1969; Nakamura and Yamada, 1974).

Plastid fatty acid synthetases are typically regulated so that 16:0–ACP is the sole product. In all cases described longer chain fatty acids are formed

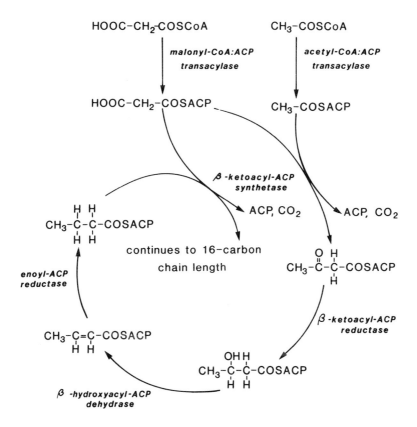

Fig. 3. Reactions and enzymes involved in *de novo* fatty acid synthesis by plant plastid enzymes. The enzyme activities are separated during purification, but the degree of association *in vivo* is not known.

by distinct chain elongation systems (see Section V.B.I). It is not clear whether the appearance of short-chain fatty acids, which are major constituents of seed triacylglycerol of some species, is related to the premature release of fatty acids from a normal fatty acid synthetase, or to some other mechanism (Section V,C).

In most seeds a large proportion of the 16:0-ACP formed is converted to 18:1-ACP by the action of an elongation system and an 18:0-ACP desaturase [EC 1.14.99.5]. The 16:0-elongation system from safflower uses malonyl-CoA and requires NADPH; NADH is less effective (Joworski *et al.*, 1974). A number of lines of evidence distinguishing the elongation system from the fatty acid synthetase have been reviewed by Stumpf (1980).

Table IV. Pyridine nucleotide-linked reactions during the conversion of sucrose to palmitate and oleate by oil seed plastids.

Enzymes	Pyridine nucleotide specificity
Glycolysis	
Glyceraldehyde-3-phosphate dehydrogenase	
D-Glyceraldehyde-3-phosphate: NAD^+ oxidoreductase (phosphorylating) [EC 1.2.1.12]	NAD (usual cytoplasmic enzyme)
or D-Glyceraldehyde-3-phosphate: $NADP^+$ oxidoreductase (phosphorylating) [EC 1.2.1.13]	NADP/NAD (usual chloroplast enzyme)
Pyruvate dehydrogenase complex	
Pyruvate: lipoamide oxidoreductase (decarboxylating and acceptor-acetylating) [EC 1.2.4.1]	
+ Acetyl-CoA: dihydrolipoamide 5-acetyltransferase [EC 2.3.1.12]	
+ NADH: lipoamide oxidoreductase [EC 1.6.4.3]	NAD
Pentose phosphate pathway	
Glucose-6-phosphate dehydrogenase	
D-Glucose-6-phosphate: $NADP^+$ 1-oxidoreductase [EC 1.1.1.49]	NADP (cytoplasm only?)
Phosphogluconate dehydrogenase	
6-Phospho-D-gluconate: $NADP^+$ 2-oxidoreductase (decarboxylating) [EC 1.1.1.44]	NADP
Fatty acid synthetase and 16:0-ACP elongation system	
β-Ketoacyl-ACP reductase	
D-3 Hydroxyacyl [acyl-carrier-protein]: $NADP^+$ oxidoreductase [EC 1.1.1.100]	NADP
Enoyl-ACP reductase	
Acyl-[acyl-carrier-protein]: NAD^+ oxidoreductase [EC 1.3.1.9]	NAD
and/or Acyl-[acyl-carrier-protein]: $NADP^+$ oxidoreductase [EC 1.3.1.10]	NADP (some NAD activity)
18:0-ACP desaturation	
18:0-ACP desaturase	
Acyl-[acyl-carrier-protein], ferredoxin: oxygen oxidoreductase [EC 1.14.99.6]	
+ Ferredoxin: $NADP^+$ oxidoreductase [EC 1.18.1.2]	NADP

In addition, it now appears that the two processes may involve distinct β-ketoacyl-ACP synthetases (Shimikata and Stumpf, 1982b).

An 18:0-ACP desaturase from plant sources was first described by Nagai and Bloch (1965, 1966, 1968). The enzyme is specific for the ACP-thioester and requires ferredoxin, NADPH, an NADPH oxidase and oxygen for activity. The same enzyme has been found in safflower preparations (Jaworski and Stumpf, 1974). Exogenous ^{14}C-labelled palmitate and oleate do not serve as substrates for elongation and desaturation reactions apparently because they are not converted to ACP-thioesters.

The immediate steps in the metabolism of acyl-ACPs in oil seed plastids are not well known. By analogy with leaf chloroplasts it is likely that a small proportion are used for the acylation of glycerol-3-phosphate and the synthesis of plastid lipids (Roughan et al., 1980; Roughan and Slack, 1982; Gardiner et al., 1982; Frentzen et al., 1983), whereas most are hydrolysed to free fatty acids (Ohlrogge et al., 1978b), which are then converted to acyl-CoA thioesters during or immediately after transport through the plastid envelope (Roughan and Slack, 1977; Joyard and Douce, 1977). Certainly 16:0 and 18:1 acyl-CoAs appear to be the substrates for almost all subsequent reactions in the cytoplasm and other subcellular compartments (Fig. 2; Section V).

C. Relationship of Fatty Acid Synthesis to Glycolysis and the Pentose Phosphate Pathway

The quantities of both reduced pyridine nucleotides and ATP required for fatty acid synthesis by plastids are approximately equal to the quantities produced by glycolysis (Figs 2, 3). Although the precise stoichiometry varies with the carbohydrate source and the ratio of 16:0/18:1 produced, it is tempting to view metabolism in oil seed plastids as including the self-sustaining conversion of carbohydrate to long-chain fatty acids (Yamada et al., 1974). However, NADH is the co-factor produced during conventional (cytoplasmic) glycolysis (Beevers, 1961), whereas fatty acid synthesis in plastids requires at least a portion (and probably more than 50%) of its reducing equivalents as NADPH (Section III,B). The plastid isoenzyme of glyceraldehyde phosphate dehydrogenase [EC 1.1.1.59] from castor bean was assayed as a NAD-requiring enzyme by Simcox et al. (1977), but the enzyme from chloroplasts has been shown to be a single protein capable of both NAD-linked and NADP-linked activity (Muller et al., 1969, Yonuschot et al., 1970; Preiss and Kosuge, 1977). Preliminary observations of NADP-linked activity in partially purified plastids of linseed (*Linum usitatissimum*) (J. A. Browse, unpublished data) suggest that a similar bifunctional enzyme may be present in oil seeds. Thus it is possible that the plastid glycolytic pathway and pyruvate dehydrogenase complex do provide both NADH and NADPH for 16:0 and 18:1 synthesis. Table IV shows the pyridine nucleotide-requiring enzymes involved in the conversion of sucrose to palmitate and oleate, together with their specificity for NAD or NADP.

Despite the possible balancing of glycolysis and fatty acid synthesis in energy and reducing equivalents, operation of the pentose phosphate pathway in oil seed plastids is well established (Section III,A). From studies using [3-^{3}H]glucose, Agrawal and Canvin (1971b) concluded that the pentose phosphate pathway contributed at least 20%-27% of the reducing equivalents used during fat synthesis (more precisely 10%-13% derived from the 6-phosphogluconate dehydrogenase [EC 1.1.1.44] step) in slices of developing castor bean seeds. On this basis (whether triose phosphate from the pentose phosphate pathway is recycled or converted to pyruvate and acetyl-CoA, Fig. 2), the minimum stoichiometry for conversion of carbohydrate to palmitate and oleate via glycolysis and the pentose phosphate pathway involves excess production of reduced pyridine nucleotides. The excess will be equivalent to 30% of the total reducing equivalents required for fatty acid synthesis. Rather less than half of this excess will be generated by glucose-6-phosphate dehydrogenase which may be located outside the plastid (Section III,A).

IV. CHANGES IN LIPID COMPOSITION AND CONTENT DURING SEED DEVELOPMENT

The accumulation of total lipids has been studied in a range of oil seeds, including castor bean (Canvin, 1963), crambe (Gurr et al., 1972), soybean (Privett et al., 1973), rape (Norton and Harris, 1975), almond (*Prunus dulcis*) (Hawker and Buttrose, 1980), flax (*Linum usitatissimum*) and safflower (Sims et al., 1961). Invariably a sigmoidal pattern is followed, which may be divided approximately into three periods. During the initial stage, before the onset of rapid lipid accumulation, structural lipids (phospholipids and glycolipids) are present as a small proportion of total seed weight, while triacylglycerols are essentially absent (e.g., Privett et al., 1973; Norton and Harris, 1975). The predominant fatty acids present at this stage, 16:0, 18:1, 18:2 and 18:3, are those that are also typical of leaves and other plant organs.

A switch to more rapid seed growth then takes place and this coincides with the start of triacylglycerol accumulation. If the oil is characterized by an unusual fatty acid, then this fatty acid also appears in increasing amounts from this time, although it may have been absent or nearly absent during the first stage of development (Canvin, 1963; Gurr et al., 1972; Norton and Harris, 1975). The rapid increase in total seed dry weight continues for at least two weeks, during which time triacylglycerol becomes the predominant lipid component. Usually the quantities of phospholipids and glycolipids also increase during this period, but they soon constitute only a small proportion (< 10%) of the total oil (Privett et al., 1973). The data of Norton and Harris (1975) for rape seed shows that phospholipids declined from 64% to 8% of the total lipid between three and five weeks after petal fall, but during this time the actual weight of phospholipids increased from 4.6 to 8.7 mg per 100 seeds. Similarly, glycolipids fell from 14.4% to 3.2% of the total during the same period while increasing from 1.0 to 3.5 mg per 100 seeds. Thus it is possible to show that the mass of any particular fatty acid does not decrease during this period of rapid oil accumulation, even though it may decline sharply as a proportion of the total fatty acid complement (Canvin, 1963; Appelqvist, 1975). By the end of this period, triacylglycerols constitute about 90% of the total lipids and the overall fatty acid composition is essentially that of the triacylglycerols.

During the final period leading to full seed maturity, dry weight and weight of oil per seed increase very little, and there is a progressive decrease in moisture content. There may also be some net loss of phospholipids and glycolipids, especially in seeds that are green during development. This is undoubtedly related to the breakdown of chloroplasts and other organelles that are no longer required following the completion of oil synthesis.

It has been possible to show that fatty acid synthesis from [^{14}C]acetate and the activities of enzymes involved in lipid synthesis reach a peak at the time of rapid oil accumulation (McMahon and Stumpf, 1966; Simcox et al., 1979). Some enzymes of the glycolytic and pentose phosphate pathways have also been shown to exhibit maximum activity at this stage of seed development (Simcox et al., 1979) and it has been demonstrated in the case of pyruvate kinase, hexose phosphate isomerase [EC 5.3.1.9] and 6-phosphogluconate dehydrogenase that the plastid isoenzyme activity increases proportionally more than the activity of the isoenzyme localized in the cytoplasm (Ireland and Dennis, 1981).

V. TRIACYLGLYCEROL SYNTHESIS IN DEVELOPING SEEDS

A. The Glycerol-3-Phosphate Pathway

Considerable indirect evidence shows that the glycerol-3-phosphate pathway of triacylglycerol synthesis (Fig. 4), found in animal tissues (Kennedy, 1961), also operates in developing seeds. Fatty acids from acyl-CoA esters are transferred sequentially to the free hydroxyls of glycerol-3-phosphate by two acyltransferases to form phosphatidate. The phosphate group is removed from this lipid by a specific phosphatase and the resulting

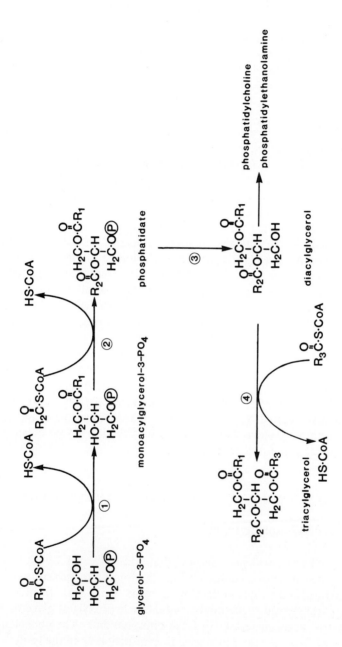

Fig. 4. The glycerol-3-phosphate pathway. By this pathway glycerol, from glycerol-3-phosphate, and fatty acids, from acyl-CoA esters, are incorporated into triacylglycerols and certain other lipids. Enzymes involved are: 1, glycerol-3-phosphate acyltransferase; 2, monoacylglycerol-3-phosphate acyltransferase; 3, phosphatidate phosphatase; 4, diacylglycerol acyltransferase.

6. Lipid Reserves

diacylglycerol finally acylated to produce triacylglycerol. Diacylglycerols are also intermediates in the formation of certain phospholipids such as phosphatidylcholine (PC) and phosphatidylethanolamine.

Evidence for the operation of this pathway in developing seeds is derived largely from the labelling of intermediates in detached cotyledons (Gurr et al., 1974) or in cell-free preparations with labelled substrates (Barron and Stumpf, 1962; Hitchcock and Nichols, 1971; Gurr, 1980). In seed tissues, enzymes of the pathway are located outside the plastid and are bound to membranes, but their precise intracellular locations have not been determined.

Nothing is known about the specificity of the enzymes for particular acyl-CoA esters or about the specificity of the diacylglycerol acyltransferase for particular diacylglycerol species. Whether the sequential acylation of the glycerol-3-phosphate involves two separate acyltransferases has not been established (Macher et al., 1975; Vick and Beevers, 1977). However, since it is a general rule that different fatty acids are esterified to positions 1 and 2 of seed triacylglycerols (Section II), it seems reasonable to assume that two separate acyltransferases are involved in phosphatidate formation and that each must have a different specificity for particular acyl-CoA esters. Diacylglycerol acyltransferase from avocado (*Persea americana*) mesocarp tissue appears rather non-specific with regard to the acyl donor, since acyl-ACP as well as acyl-CoA esters of varying chain length were utilized as substrates (Shine et al., 1976). It should be noted that this extra-plastid enzyme would not encounter acyl-ACP esters *in vivo* since ACP is localized in the plastid (Section III,B). When removed from the testa cotyledons retain the ability to synthesize lipids and readily incorporate label from [^{14}C]acetate and [2-^{3}H]glycerol. The labels are incorporated specifically into the fatty acid and glycerol moieties, respectively, of triacylglycerols and other lipids. It is now clear from the results of such labelling experiments and also from enzyme studies that the pathway of synthesis of long-chain monoenoic acids and their incorporation into triacylglycerols differs in some important respects from the synthesis and then incorporation of C_{18} polyunsaturated fatty acids into triacylglycerols. In the following sections we have discussed the formation of the two kinds of triacylglycerol separately.

B. Formation of Triacylglycerols and Wax Esters Containing Long-chain Fatty Acids

1. Synthesis of Erucate and Eicoseonoate

The C_{20} and C_{22} monoenoic acids eicoenenoate and erucate (Table III) are formed in developing cotyledons of rape, crambe, nasturtium and jojoba from oleoyl-CoA, exported from the plastid, by the sequential addition of acetate units to the carboxyl end of the fatty acid (Fig. 5). Downey and Craig (1964) first proposed that oleate was the precursor of these fatty acids because the level of erucate in rape seed could be varied (Table II) without changing the quantity of triacylglycerol, a reduction in erucate being compensated for by an increase in oleate. Analysis of label along the carbon chain of these long-chain acids from cotyledons of rape (Downey and Craig, 1964; Wright, 1980), crambe (Gurr et al., 1974), jojoba (Ohlrogge et al., 1978a) and nasturtium (Pollard and Stumpf, 1980a) after provision of ^{14}C-acetate showed that the carbons derived from oleate were uniformly labelled, whereas carbons at the carboxyl end of the chain had a considerably higher specific radioactivity than those in the rest of the molecule. It was suggested (Ohlrogge et al., 1978a) that these differences in specific radioactivities must mean that oleate synthesis and chain elongation occur in different cell compartments from different acetate pools. This idea is supported by cell fractionation studies of crambe (Gurr et al., 1974), jojoba (Pollard et al., 1979) and rape (Wright, 1980), showing that chain elongation occurs after oleate has been exported from the plastid. Why the specific radioactivity of the acetate pool inside the plastid should be low compared with that outside is unclear. However, plastids in developing cotyledons often contain starch (Section III,A) and perhaps, therefore, the ^{14}C-acetate that enters the plastid is diluted by unlabelled acetyl-CoA derived from starch. It should be noted that this argument holds only if the plastid envelope is impermeable to acetyl-CoA, preventing equilibration of the acetyl-CoA pool inside the plastid with that outside.

Considerable success has been achieved in studying chain elongation *in vitro*. Cross-contamination between different cell fractions has not been eliminated, but it would appear probable that the oil bodies in crambe (Gurr et al., 1974; Appleby et al., 1974) and rape (Wright, 1980) and the wax bodies in jojoba seeds (Pollard et al., 1979) are the sites of chain elongation. Acyl-CoAs and malonyl-CoA are the substrates for the enzyme(s) catalysing chain elongation and acyl-CoAs are the products. The preferred reductant is NADPH. Neither acetyl-CoA nor acetate are effective as acetate donors (Appleby et al., 1974; Pollard et al., 1979), indicating that oil bodies probably do not possess either acetyl-CoA synthetase [EC 6.2.1.1] or acetyl-CoA carboxylase [EC 6.4.1.2]. These enzymes are presumably present both in an extra-plastid compartment as well as in the plastid in order to generate the discrete pools of malonyl-CoA for fatty acid synthesis and for chain elongation. We have placed the extra-plastid enzymes in the cytoplasm (Fig. 5), but have no evidence for this apart from the solubility in extraction medium of these enzymes from chloroplasts (Roughan and Slack, 1977; Nikolou et al., 1981), potato (*Solanum*

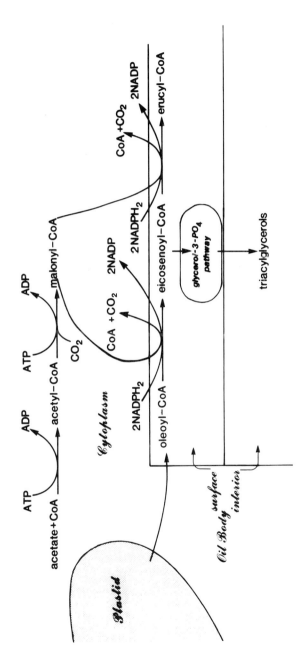

Fig. 5. The possible pathway of synthesis of triacylglycerols containing long-chain fatty acids from oleoyl-CoA exported from the plastid. Malonyl-CoA sequentially donates two acetate units to oleoyl-CoA forming eicosenoyl-CoA then erucyl-CoA. These long-chain fatty acids are then incorporated into triacylglycerol via the glycerol-3-phosphate pathway.

tuberosum) tubers (Huang and Stumpf, 1970) and wheat (*Triticum aestivum*) embryos (Hatch and Stumpf, 1961).

The chain length of the fatty acids generated by chain elongation are probably limited to 22 carbons because of the inability of the enzyme system to elongate erucyl-CoA (Pollard *et al.*, 1979). It is not known whether the two elongation steps converting oleoyl-CoA to erucyl-CoA are catalysed by the same enzyme system, or whether each step requires a separate one. However, the relative amounts of erucate and eicosenoate can be varied independently by selection (Downey and Craig, 1964) and hence separate elongating enzymes may be involved. Plant breeding has lowered the levels of long-chain fatty acids in rape seed to essentially zero (Downey and McGregor, 1975). The biochemical changes responsible for the absence of long-chain fatty acid formation are not known. However, it is apparent from Figure 5 that the synthesis of these acids could be blocked by eliminating either the chain-elongating enzymes or the enzymes situated outside the plastid that generate malonyl-CoA.

2. Incorporation of Long-Chain Fatty Acids into Triacylglycerols and Waxes

A large proportion of labelled long-chain fatty acids produced by preparations of oil bodies in assays containing glycerol-3-phosphate was esterified in triacylglycerols (e.g. Appleby *et al.*, 1974; Pollard and Stumpf, 1980a). In similar assays containing labelled glycerol-3-phosphate, radioactivity was incorporated into diacylglycerols and into the phospholipid fraction as well as triacylglycerols (Gurr *et al.*, 1974). Phosphatidate was the only phospholipid labelled during the synthesis of acyl-labelled triacylglycerols (Wright, 1980). These lipids are all intermediates of the glycerol-3-phosphate pathway (Fig. 4), hence we presume that the entire enzyme complement of this pathway is associated with the oil bodies (Fig. 5).

The wax-body fraction from homogenates of jojoba seed contains both the enzyme that reduces long-chain acyl-CoAs to long-chain alcohols and the enzyme that catalyses wax formation (Pollard *et al.*, 1979). The four electron reduction step converting the acyl-CoA to the free alcohol has an absolute requirement for NADPH (Fig. 6A), but an aldehyde has not been detected as an intermediate (Pollard *et al.*, 1979). Wax formation is catalysed by an acyl-CoA alcohol transacylase (Wu *et al.*, 1981) as shown in Figure 6B. A combination of the specificities of the chain elongating enzymes and the acyl-CoA reductase for acyl-CoAs of different chain length appears to determine the relative amounts of different acyl-CoAs and alcohols available to the acyl-CoA alcohol transacylase and hence the types of wax esters that are formed (Section 6.II,B of Volume 2).

A $R\overset{O}{\overset{\|}{C}}\text{-CoA} + 2\text{NADPH}_2 \longrightarrow R\text{-}H_2\text{C-OH} + \text{CoA} + 2\text{NADP}$

acyl-CoA reduction

B $R_1\overset{O}{\overset{\|}{C}}\text{-CoA} + \text{HO-CH}_2R_2 \longrightarrow R_1\overset{O}{\overset{\|}{C}}\text{-O-CH}_2R_2 + \text{CoA}$

acyl-CoA:alcohol acyltransferase

Fig. 6(A) Reduction of long-chain fatty acids esterified to CoA to fatty alcohols. This four electron reduction step is catalysed by enzyme(s) associated with wax bodies. A similar reduction of acyl-CoAs to fatty alcohols occurs in epidermal cells. Here two soluble enzymes are involved, an acyl-CoA reductase and an aldehyde reductase (Kolattukudy, 1980). (B) Wax ester formation catalysed by acyl-CoA alcohol acyltransferase.

C. Synthesis of Some 'Unusual' Fatty Acids

Little is known about the synthesis of many of the unusual fatty acids that are found in seeds, but recently some interesting information has been obtained about the synthesis of some of them. Meadowfoam triacylglycerols contain the 20:1(11*c*) and 22:1(13*c*) fatty acids that occur in seeds of jojoba and the *Cruciferae* and, in addition, isomeric forms of these fatty acids in which the double bond is located between carbons 5 and 6, that is, 20:1(5*c*) and 22:1(5*c*). The former are synthesized by chain elongation from oleoyl-CoA as described above (Section V.B), whereas palmitoyl-CoA is the precursor of the latter. Chain elongation from palmitoyl-CoA produces a series of saturated fatty acid-CoA esters that are then desaturated by a specific acyl-CoA desaturase (Pollard and Stumpf, 1980b).

A start has been made in the study of the synthesis of medium-chain-length fatty acids that occur in the triacylglycerols of certain species. Oo and Stumpf (1979) have used coconut (*Cocos nucifera*) endosperm in which C_{12} and C_{14} fatty acids are major components and Slabas *et al.* (1982) have studied the cotyledons of *Cuphea pubescens* in which a C_{10} fatty acid is predominant. In both studies [^{14}C] acetate was incorporated into these medium-chain fatty acids, but as yet the mechanism by which these fatty acids are released from the fatty acid synthetase system is unknown. An analogous situation occurs in the lactating mammary gland where medium-chain length fatty acids are produced. Here chain elongation by the fatty acid synthetase can be terminated at C_8 and C_{14} by a tissue-specific thioesterase that preferentially hydrolyses the thioester bonds that link these medium-chain length fatty acids to the multienzyme synthetase (Smith, 1980). Ohlrogge et al. (1978b) have concluded that this mechanism does not operate in coconut endosperm because C_{12}-ACP was hydrolysed only very slowly by the acyl-ACP hydrolase from this tissue.

Ricinoleate is the predominant fatty acid of the triacylglycerols in castor bean endosperm. This C_{18} fatty acid contains one double bond between carbons 9 and 10, as in oleate, and a hydroxyl group at carbon 12. Labelling studies with slices of developing endosperm established that ricinoleate is formed by hydroxylation from oleate (Canvin, 1965a), without any alteration of the stereochemical configuration at carbon 12 (Morris, 1967). The hydroxylation is catalysed by a microsomal-bound enzyme and the reaction requires both O_2 and reduced pyridine nucleotide (Galliard and Stumpf, 1966). The form in which oleate is hydroxylated is still in doubt. In the initial *in vitro* studies oleoyl-CoA was used as the substrate (Galliard and Stumpf, 1966), but since developing seed tissues invariably contain a highly active acyl transferase that incorporates oleate from the CoA ester into PC, the possibility that oleoylphosphatidylcholine could be the real substrate for hydroxylation has recently been studied (Moreau and Stumpf, 1981). This investigation demonstrated that oleate esterified to PC was converted to ricinoleate, but whether or not the oleate was hydroxylated while esterified to the phospholipid remains uncertain since ricinoleoylphosphatidylcholine did not accumulate during the assays.

D. Selective Incorporation of 'Unusual' Fatty Acids into Storage Lipids

It is a characteristic of the seeds of all species discussed above that the unusual fatty acids are almost exclusively found in storage lipids. The membrane phospholipids from these seeds contain the normal range of saturated and C_{18} unsaturated fatty acids that are found; for example, in the phospholipids of leaves. Röbbelen (1975) has put forward the intriguing suggestion that these 'unusual' fatty acids are relics of evolution which have been retained because there is no selective disadvantage in their use as energy reserves. This idea seems to be as good an explanation as any for their occurrence in storage lipids, but we would stress that the plant must possess some mechanism for excluding them from the important membrane phospholipids of the seed. Appleby et al. (1974) have suggested that phospholipids and triacylglycerols are probably synthesized from metabolically discrete pools of diacylglycerols. This seems to be so, at least in species that synthesize storage lipid with long-chain monoenoic acids, since the formation of these appears to occur inside oil bodies, whereas phospholipid synthesis from diacylglycerol is localized in the endoplasmic reticulum (Mudd, 1980).

E. Synthesis of Polyunsaturated Triacylglycerols

Species such as soybean, linseed, safflower and sunflower, in which the triacylglycerols contain polyunsaturated C_{18} fatty acids, differ from those discussed above because these same fatty acids are also the predominant fatty acids of the phospholipids and galactolipids of the seed. It has long been accepted that in plants linoleate and linolenate are produced from oleate by the sequential introduction of additional double bonds into the carbon chain, but the precise form in which oleate and lineolate are desaturated has been, until recently, the subject of much debate (Roughan and Slack, 1982). Now there is considerable support for the belief that the metabolism of phospholipids, in particular of PC, is intimately involved in the formation of polyunsaturated fatty acids. As a consequence, the fatty acids that finally accumulate in the triacylglycerols are derived from PC.

The idea that a membrane phospholipid can serve a metabolic as well as a structural role originated from labelling studies with developing cotyledons (Dybing and Craig, 1970; Slack et al., 1978; Wilson et al., 1980). Oleate synthesized de novo from [^{14}C]acetate and also [^{14}C]oleate supplied to cotyledons initially accumulates mainly in PC and diacylglycerols. With increasing time, radioactivity appears in the linoleate then the linolenate of this phospholipid. There are longer delays before polyunsaturated fatty acids in diacylglycerols become labelled and a pronounced lag before labelled fatty acids appear in triacylglycerols. After pulse-labelling and while unlabelled oleate is being supplied, radioactivity originally in the oleate of PC accumulates in the polyunsaturated fatty acids of the triacylglycerol.

The idea that the oleoyl-PC and linoleoyl-PC are the substrate and product, respectively, of oleate desaturation in cotyledons of developing seeds has been supported by in vitro studies (Stymne and Appelqvist, 1978; Slack et al., 1979). The possibility that oleate is transferred from the phospholipid to CoA and desaturated as the acyl-CoA ester, then the linoleate inserted back into the phospholipid, cannot be entirely discounted, but appears unlikely. The oleoyl-PC desaturase is membrane-bound, probably localized in the endoplasmic reticulum (Slack et al., 1977; Stymne, 1980). This enzyme desaturates oleate at both positions 1 and 2 of the phospholipid (Slack et al.,1979). Less success has been achieved, so far, in obtaining a highly active linoleate desaturase from cotyledons, but the available evidence suggests that this enzyme probably desaturates linoleoyl-

PC to linolenyl-PC and is also located in the endoplasmic reticulum (Stymne and Appelqvist, 1980; Browse and Slack, 1981). Both of these desaturases require O_2 and reduced pyridine nucleotide for activity.

At present the steps that must be involved before the C_{18} carbon chains, exported as oleoyl-CoA from the plastid, finally become esterified as polyunsaturated fatty acids in triacylglycerols are only incompletely understood. One mechanism for the incorporation of oleate into PC and the release of polyunsaturated fatty acids from this phospholipid is the reaction catalysed by an acyl-CoA, phospholipid acyltransferase (Fig. 7). This enzyme was identified in microsomal preparations from developing soybean cotyledons (Stymne and Glad, 1980) and catalysed acyl exchange between oleoyl-CoA and fatty acids at position 2 of PC (Stymne et al., 1983) and probably other phospholipids. Stymne and Glad (1981) proposed that this reaction provides polyunsaturated acyl-CoAs from phospholipids which can, by de novo lipid synthesis, be incorporated into triacylglycerols and phospholipids as shown in Figure 8.

Labelling studies are consistent with certain molecular species of diacylglycerol and PC being synthesized by this route notably those containing either palmitate or oleate at position 1 and linolenate at position 2 (Slack et al., 1978; Roughan and Slack, 1982; Slack et al., 1983). The kinetics of labelling of lipids with [2-³H]glycerol have indicated, however, that this is not the sole route for the transfer of polyunsaturated fatty acids from PC into triacylglycerols. If this storage lipid were synthesized directly by the glycerol-3-phosphate pathway from a pool of acyl-CoAs, one would expect labelled glycerol to enter the lipid at a linear rate. In fact, the glycerol moiety of the triacylglycerols was labelled at a linear rate only after a pronounced lag, whereas glycerol entered PC and diacylglycerol at linear rates from zero time (Slack et al., 1978). Initially, much of the labelled glycerol was present in molecular species of PC and diacylglycerol that contained oleate. This suggests that there must be a rapid rate of de novo synthesis of both these lipids from an acyl-CoA pool that contains a high proportion of oleoyl-CoA. After pulse-labelling with [2-³H]glycerol there was a flow of label from oleate-containing PC species into polyunsaturated triacylglycerols that was almost as fast as the movement of labelled fatty acids between the two lipids. Based on this observation, Slack et al. (1978) suggested that PC somehow donates diacylglycerol moieties from which triacylglycerols are formed. The possibility that diacylglycerols could be generated from phospholipids and then incorporated into triacylglycerols has also been proposed on the basis of other evidence (Wilson et al., 1980; Wilson, 1981).

In the developing cotyledons of seeds that accumulate polyunsaturated triacylglycerols, there is a marked similarity between the fatty acid

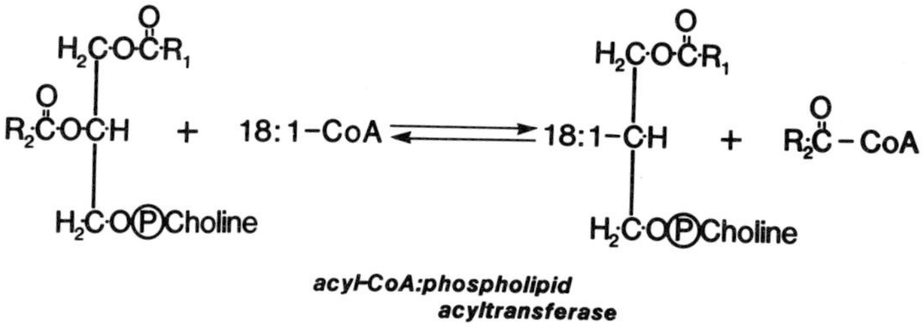

6. Lipid Reserves 233

$$\text{H}_2\text{COCR}_1 \qquad\qquad \text{H}_2\text{COCR}_1$$
$$\text{R}_2\text{COCH} + 18{:}1\text{-CoA} \rightleftharpoons 18{:}1\text{-CH} + \text{R}_2\text{C-CoA}$$
$$\text{H}_2\text{CO}\textcircled{P}\text{Choline} \qquad\qquad \text{H}_2\text{CO}\textcircled{P}\text{Choline}$$

acyl-CoA:phospholipid acyltransferase

Fig. 7. Acyl-CoA, phospholipid acyltransferase catalyses the exchange of fatty acids between CoA esters and phospholipids. The reaction has been studied by the exchange of [^{14}C]oleoyl-CoA with polyunsaturated fatty acids in the phospholipid.

composition of PC and that of diacylglycerol (Slack et al., 1978). Furthermore, the specific radioactivities of the equivalent molecular species of these two lipids rapidly attain similar values in cotyledons labelled with [^{3}H]glycerol (Slack et al., 1983). These two observations have led to the suggestion that there could be a rapid, reversible exchange of the diacylglycerol moiety of PC with the diacylglycerol pool (Slack et al., 1978; 1983). Choline phosphotransferase [EC 2.7.8.2] has been implicated as the catalyst of this diacylglycerol exchange. Numerous reports state that this enzyme catalyses the reversible exchange of phosphorylcholine between PC and CMP in microsomal preparations from mammalian tissues (Fig. 9) (see Roughan and Slack, 1982) and the microsomal enzyme from developing seed cotyledons also catalyses this reversible reaction (C. R. Slack, unpublished data). It is difficult to obtain conclusive proof that this reaction is also reversible in vivo, but markedly different rates of incorporation of ^{32}P compared with [^{3}H]glycerol into certain species of PC and a low specific radioactivity of CDP-choline compared with phosphorylcholine in cotyledons labelled with [^{14}C]choline suggest that the reaction must be freely reversible in developing cotyledons (Slack et al., 1983). A scheme for polyunsaturated triacylglycerol synthesis that incorporates choline phosphotransferase-catalysed diacylglycerol exchange is shown in Figure 10. This reaction, together with the incorporation of oleate into PC via the glycerol-3-phosphate pathway and by acyl exchange (Fig. 7), then the desaturation of oleate esterified to the phospholipid, would result in a flow of oleate and glycerol into polyunsaturated triacylglycerols. All the enzymes implicated in the scheme have been isolated with membranes that pellet during centrifugation at high speed ($\geq$ 20 000 g) and are presumably derived from the endoplasmic reticulum. At present there is no clear understanding of how polyunsaturated diacylglycerols and acyl-CoA esters derived from

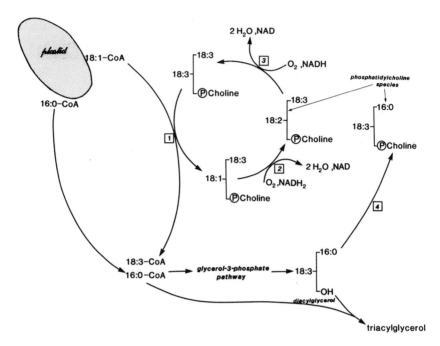

Fig. 8. Scheme showing the possible involvement of acyl-CoA, phospholipid acyltransferase and phosphatidylcholine fatty acid desaturases in the production of polyunsaturated acyl-CoA esters. Acyl-CoA, phospholipid acyltransferase (1) catalyses the exchange of oleate (18:1) from the CoA ester with linoleate (18:2) or linolenate (18:3) at position 2 of phosphatidylcholine, producing polyunsaturated acyl-CoAs (only the reaction with 18:3-phosphatidylcholine is shown). Oleate introduced into the phospholipid is desaturated sequentially by oleoylphosphatidycholine desaturase (2), then linoleoylphospatidyl-choline desaturase (3). Polyunsaturated acyl-CoAs, together with acyl-CoAs derived directly from the plastid are incorporated via the glycerol-3-phosphate pathway either into triacylglycerols or phospholipids.

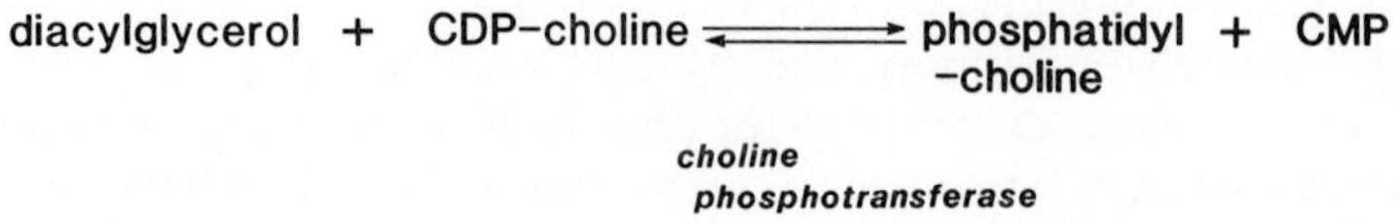

diacylglycerol + CDP-choline ⇌ phosphatidyl + CMP
-choline

choline
phosphotransferase

Fig. 9. Choline phosphotransferase catalyses the reversible exchange of phosphorylcholine between CDP-choline and diacylglycerol.

6. Lipid Reserves 235

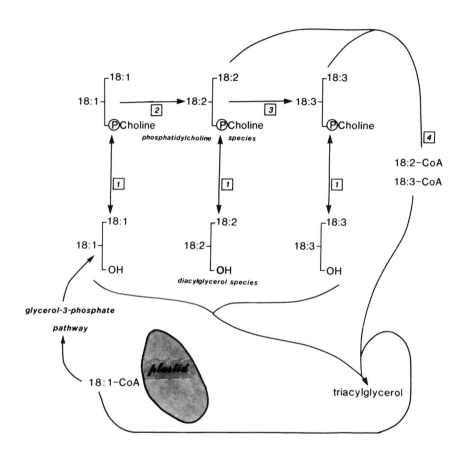

Fig. 10. Scheme showing the possible flow of oleate and glycerol through phosphatidylcholine and diacylglycerols into polyunsaturated triacylglycerols. Oleate-containing diacylglycerols, synthesized *de novo* by the glycerol-3-phosphate pathway are incorporated into phosphatidylcholine by choline phosphotransferase (*1*), then desaturated by the sequential action of the oleoyl- and linoleoylphosphatidylcholine desaturases (*2*) and (*3*). The diacylglycerol moiety of each phosphatidylcholine species is maintained in equilibrium with equivalent diacylglycerol species by the reversible reaction catalysed by choline phosphotransferase (Fig. 9). Diacylglycerols are incorporated into triacylglycerols by the addition of either oleate or palmitate from acyl-CoAs derived directly from the plastid or polyunsaturated fatty acids from acyl-CoAs derived from phosphatidylcholine by the action of acyl-CoA, phospholipid acyltransferase (Fig. 7).

PC in the endoplasmic reticulum could finally become the constituents of triacylglycerols stored in oil bodies.

VI. COMPOSITION AND ORIGIN OF OIL BODIES

Triacylglycerols accumulate in the cell cytoplasm as lipid droplets. These are spherically shaped during the early and mid-stages of seed development, but become tightly packed and compressed in the oil-rich cells of the mature seed. We have called the lipid-containing organelles oil bodies, but they have also been named spherosomes, oleosomes, reserve lipid droplets and lipid-containing organelles.

To date, studies of the ultrastructure of developing seeds have not provided any unifying concept about how oil bodies originate. However, there is general agreement on some points. Oil bodies in the seed of all the species so far examined, rape (Frey-Wyssling et al., 1963; Wright, 1980), mustard (*Sinapis alba*) (Rest and Vaughan, 1972; Bergfeld et al., 1978), crambe (Smith, 1974; Gurr et al., 1974), cotton (*Gossypium hirsutum*) (Yatsu, 1965; Yatsu et al., 1971), peanut (*Arachis hypogaea*) (Yatsu and Jacks, 1972) and linseed (Slack et al., 1980) are all about the same size with a mean diameter of about 1 μm. During seed development the size of the oil bodies remains almost constant and triacylglycerol accumulation is accompanied by an increase in the number of oil bodies within the cell (Rest and Vaughan, 1972; Smith, 1974). Oil bodies do not coalesce either *in situ*, or after isolation, and because of this it is believed that they must be bounded by some form of membrane. Often a thin electron-dense layer, about 3 nm wide, has been observed at the surface of each oil body and it is now generally agreed that this layer represents one half of a normal, tripartate unit membrane with the lipophilic side oriented inwards to the lipid matrix (Yatsu and Jacks, 1972; Bergfeld et al., 1978; Wanner et al., 1981). The limited information that is available about wax-containing organelles in jojoba seeds indicates that these are similar in size to oil bodies and are also probably membrane-bound (Muller et al., 1975; Fig. 1B of Chapter 6, Volume 2).

Phospholipids and protein remain associated with oil bodies after isolation and repeated washing (e.g. Yatsu et al., 1971), which supports the idea that the oil bodies are bounded by a phospholipid and protein-containing membrane. However, the amounts of these components found in isolated oil bodies vary considerably. For instance, crambe oil bodies were reported to contain 19% protein (Gurr et al., 1974) compared with only 0.2% protein in peanut oil bodies. This variation is far too large to be ascribed to differences in analytical technique, but may have resulted from severe cross-contamination of the crambe oil bodies with other cell components, such as protein bodies. In contrast to the above variations, oil bodies have been prepared with essentially constant composition of 2.5% protein, 0.7% phospholipid and 97% neutral lipid from developing and mature cotyledons of both linseed and safflower (Slack *et al.*, 1980). In the presence of sodium dodecylsulphate most of the protein from these oil body preparations dissociates into four small polypeptides, of molecular weight 14 000–18 000 daltons, which are not present in other cell fractions (Slack *et al.*, 1980; Slack and Roughan, 1980). Polypeptides of very similar size were also major components of the protein present in oil bodies from seeds of mustard (Bergfeld *et al.*, 1978) and castor bean (Moreau *et al.*, 1980). At present there is no definitive proof that these polypeptides are components of either structural proteins or enzymes located in an oil body membrane. However, freeze-fracture electron microscopy has revealed that safflower oil bodies are bounded by a particle-free, protein-containing membrane (Fig. 11).

It is difficult to relate the various hypotheses about oil body development that have resulted from ultrastructural studies to present ideas about triacylglycerol biosynthesis derived from metabolic studies. It has been variously proposed that oil bodies develop in the cytoplasm from structures generated in the endoplasmic reticulum (Frey-Wyssling *et al.*, 1963), that triacylglycerol which is synthesized in the endoplasmic reticulum accumulates in the lipophilic region of this membrane from which mature oil bodies are released (Wanner and Theimer, 1978; Wanner *et al.*, 1981) and that oil bodies originate, then develop in the cytoplasm independently of the endoplasmic reticulum (Rest and Vaughan, 1972; Smith, 1974; Bergfeld *et al.*, 1978). Since oil bodies of rape and crambe retain considerable synthetic capability after isolation (Section V.B), whereas linseed and safflower cotyledons do not (C. R. Slack and J. A. Browse, unpublished data), it is tempting to speculate that the former could develop in the cytoplasm, whereas the latter might evolve in the endoplasmic reticulum. Unfortunately, this distinction between oil bodies that do and do not contain high proportions of polyunsaturated fatty acids is not supported by ultrastructural studies.

VII. ENVIRONMENTAL EFFECTS ON THE FATTY ACID COMPOSITION OF OIL SEED LIPIDS

The fatty acid compositions of a number of seed oils have been shown to vary with the temperature at which the seeds have developed, with lower temperatures favouring a more unsaturated oil (Hilditch and Williams,

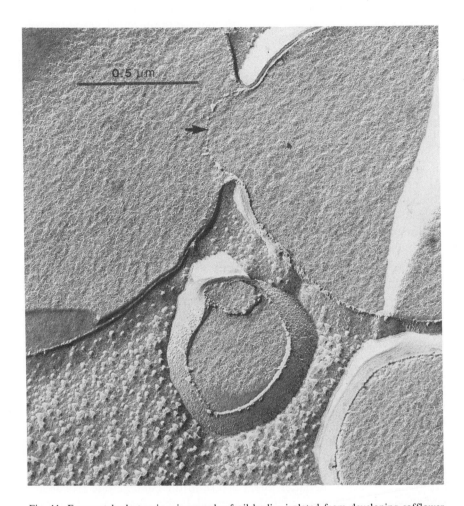

Fig. 11. Freeze-etch electronic micrograph of oil bodies isolated from developing safflower (*Carthamus tinctorius*) cotyledons. A raised rim at the junction of the fracture face and outer surface was a feature common to all oil bodies. Where partially fused oil bodies were observed, a line of raised particles occurred across the fusion zone (arrow) derived presumably from the surface of the oil body. The raised rim and particulate material at the zones of fusion are indicative of non-lipid material at the oil body surface (from Slack *et al.*, 1980).

1964; Canvin, 1965b). Temperature changes which in the long term result in alterations to the fatty acid composition of linseed and soybean oil also induce similar modifications to the fatty acid composition of PC and diacylglycerol in the developing cotyledons of these species, but much more rapidly (Slack and Roughan, 1978). This result is consistent with evidence which suggests that PC and diacylglycerol are intermediates of triacylglycerol synthesis and that PC is the substrate for oleate and linoleate desaturation (Section V,E).

Harris and James (1969a, b) have presented evidence that the extent of stearate conversion to unsaturated acids changes with temperature, suggesting that this effect is mediated by the solubility of oxygen in water, which is temperature dependent. However, the major change in seed oils occurs between oleate and polyunsaturated fatty acids (linoleate or linolenate depending on species), indicating that the oleate desaturase rather than the stearate desaturase is important in determining the temperature response in oil seeds. Dompert and Beringer (1970) showed that sunflower seeds allowed to develop under an atmosphere of oxygen contained a higher proportion of 18:2 than seeds developing under air. However, the change in dissolved oxygen concentration in that study would have been considerably greater than the change which could be expected from a drop in ambient temperature. It should be noted that the effect of temperature on the equilibrium solubility of oxygen in water is not the only way in which temperature might affect the oxygen concentration in the embryo. Increasing temperature will increase the rate of oxygen-consuming reactions in the seed tissue and if oxygen diffusion is restricted by the seed coat the concentration in the internal tissues will be lower than that predicted from temperature and the external oxygen partial pressure (Beevers, 1961). The rates of lipid synthesis and desaturation reactions will also both increase with rising temperature, so that it is the *relative* rates of these processes which must be considered. Recent evidence (Browse and Slack, 1983) suggests that the effect of temperature on fatty acid composition is the result of a difference between the temperature response of fatty acid and triacylglycerol synthesis and the temperature response of the oleate desaturase.

VIII. CONCLUSIONS

In this chapter we have given a brief account of what is known about the synthesis of storage lipids during seed development and emphasized where information is lacking. At present there are more gaps in our knowledge than firm facts and for many of the steps we have only vague ideas about what may occur. This is particularly true for the latter part of the pathway leading to the synthesis of triacylglycerols and their insertion into oil bodies. The hypotheses that have been proposed are based largely on the results of *in vivo*-labelling studies. What is now required is detailed information about individual enzymes. Progress in this area has been slow, probably because of the difficulties in studying membrane-bound enzymes that metabolize lipophilic substrates. However, techniques for the solubilization of membrane-bound enzymes and their insertion into artificial membranes are now available. Hopefully, by applying these techniques to study triacylglycerol synthesis, it might one day be possible to produce triacylglycerols in a reconstituted system.

ACKNOWLEDGEMENTS

We are grateful to Drs S. E. Gardiner and W. A. Laing for helpful discussions, Miss J. Bambery for typing and proofreading the manuscript, and to Miss M. Soulsby and Mrs V. Pech for preparation of illustrations.

REFERENCES

Agrawal, P. K., and Canvin, D. T. (1971a). *Plant Physiol.* **47**, 672–675.
Agrawal, P. K., and Canvin, D. T. (1971b). *Can. J. Bot.* **49**, 267–272.
Appelqvist, L.-A. (1975). *In* 'Recent Advances in the Chemistry and Biochemistry of Plant Lipids' (T. Galliard and E. I. Mercer, eds), pp. 247–286. Academic Press, London.
Appleby, R. S., Gurr, M. I., and Nichols, B. W. (1974). *Eur. J. Biochem.* **48**, 209–216.
Barron, E. J., and Stumpf, P. K. (1962). *Biochim. Biophys. Acta* **60**, 329–337.
Beevers, H. (1961). 'Respiratory Metabolism in Plants.' Harper and Row, New York.
Bergfeld, R., Hong, Y.-N., Kühnl, T., and Schopfer, P. (1978). *Planta* **143**, 297–307.
Brock, K., and Kannangara, C. G. (1976). *Carlsberg Res. Comm.* **41**, 121–129.
Brockeroff, H., and Yurkowski, M. (1966). *J. Lipid Res.* **7**, 62–64.
Browse, J. A., and Slack, C. R. (1981). *FEBS Letts.* **131**, 111–114.
Browse, J. A. and Slack, C. R. (1983). *Biochim. Biophys. Acta* **753**, 145–152.
Canvin, D. T. (1963). *Can. J. Biochem. Physiol.* **41**, 1879–1885.
Canvin, D. T. (1965a). *Can. J. Bot.* **43**, 49–62.
Canvin, D. T. (1965b). *Can. J. Bot.* **43**, 63–69.
de la Roche, I. A., Weber, E. J., and Alexander, D. E. (1971). *Lipids* **6**, 531–536.
De Luca, V., and Dennis, D. T. (1978). *Plant Physiol.* **61**, 1037–1039.
Dennis, D. T., and Miernyck, J. A. (1982). *Annu. Rev. Plant Physiol.* **33**, 27–50.
De Wit, C. T. (1978). 'Simulation of Assimilation, Respiration and Transpiration of Crops.' Halstead Press, Wiley, New York.
Dompert, W., and Beringer, H. (1970). *Naturwissenschaften* **57**, 40.
Dorne, A.-J., Carde, J.-P., Joyard, J., Borner, T., and Douce, J. (1982). *Plant Physiol.* **69**, 1467–1470.
Downey, R. K., and Craig, R. M. (1964). *J. Amer. Chem. Soc.* **41**, 475–478.
Downey, R. K., and McGregor, D. I. (1975). *Current Adv. Plant Sci.* **6**, 151–167.
Drennan, C. H., and Canvin, D. T. (1969). *Biochim. Biophys. Acta* **187**, 193–200.
Dybing, C. D., and Craig, B. M. (1970). *Lipids* **5**, 422–429.
Eckey, E. W. (1954). 'Vegetable Fats and Oils.' Reinhold, New York.
Fatemi, S. H., and Hammond, E. G. (1977a). *Lipids* **12**, 1032–1036.
Fatemi, S. H., and Hammond, E. F. (1977b). *Lipids* **12**, 1037–1041.

Frentzen, M., Heinz, E., McKeon, T. A., and Stumpf, P. K. (1983). *Eur. J. Biochem.* **129**, 629-636.
Frey-Wyssling, A., Grieshaber, E., and Mühlethaler, K. (1963). *J. Ultrastructure Res.* **8**, 506-516.
Galliard, T., and Stumpf, P. K. (1966). *J. Biol. Chem.* **241**, 5806-5812.
Gardiner, S. E., Roughan, P. G., and Slack, C. R. (1982). *Plant Physiol.* **70**, 1316-1320.
Gurr, M. I. (1980). In 'The Biochemistry of Plants' (P. K. Stumpf and E. E. Conn, eds), Vol. 4, pp. 205-248. Academic Press, New York.
Gurr, M. I., Blades, J., and Appleby, R. S. (1972). *Eur. J. Biochem.* **29**, 362-368.
Gurr, M. I., Blades, J., Appleby, R. S., Smith, C. G., Robinson, M. P., and Nichols, B. W. (1974). *Eur. J. Biochem.* **43**, 281-290.
Harris, P., and James, A. T. (1969a). *Biochem. J.* **112**, 325-329.
Harris, P., and James, A. T. (1969b). *Biochim. Biophys. Acta* **187**, 13-18.
Harvey, B. L., and Downey, R. K. (1964). *Can. J. Plant Sci.* **44**, 107-111.
Harwood, J. L. (1979). *Prog. Lipid Res.* **18**, 55-86.
Harwood, J. L., and Stumpf, P. K. (1972). *Arch. Biochem. Biophys.* **148**, 282-290.
Harwood, J. L., Sodja, A., Stumpf, P. K., and Spurr, A. R. (1971). *Lipids* **6**, 851-854.
Hatch, M. D., and Stumpf, P. K. (1961). *J Biol. Chem.* **236**, 2879-2885.
Hawker, J. S., and Buttrose, M. S. (1980). *Ann. Bot.* **46**, 313-321.
Heinstein, P. F., and Stumpf, P. K. (1969). *J. Biol. Chem.* **244**, 5374-5381.
Hilditch, T. P., and Williams, P. N. (1964). 'The Chemical Constitution of Natural Fats' (4rh edn). Chapman and Hall, London.
Hitchcock, C., and Nichols, B. W. (1971). 'Plant Lipid Biochemistry.' Academic Press, London.
Huang, K. P., and Stumpf, P. K. (1970). *Arch. Biochem. Biophys.* **140**, 158-173.
Ireland, R. J., and Dennis, D. T. (1980). *Planta* **149**, 476-478.
Ireland, R. J. and Dennis, D. T. (1981). *Can. J. Bot.* **59**, 1423-1425.
Jaworski, J. G., and Stumpf, P. K. (1974). *Arch. Biochem. Biophys.* **162**, 158-165.
Jaworski, J. G., Goldschmidt, E. E., and Stumpf, P. K. (1974). *Arch. Biochem. Biophys.* **163**, 769-776.
Jellum, M. D (1970). *J. Agric. Food Chem.* **18**, 365-370.
Joyard, J., and Douce, R. (1977). *Biochim. Biophys. Acta* **486**, 273-285.
Kannangara, C. G., and Jensen, C. J. (1975). *Eur. J. Biochem.* **54**, 25-30.
Kannangara, C. G, and Stumpf, P. K. (1972). *Arch. Biochem. Biophys.* **152**, 83-91.
Kannangara, C. G., and Stumpf, P. K. (1973). *Arch. Biochem. Biophys.* **155**, 391-399.
Kennedy, E. P. (1961). *Fed. Proc. Am. Soc. Exp. Biol.* **20**, 934-940.
Knowles, P. F. (1972). *J. Amer. Oil Chem. Soc.* **49**, 27-29.
Kolattukudy, P. E. (1980). In 'The Biochemistry of Plants' (P. K. Stumpf and E. E. Conn, eds), Vol. 4, pp. 571-645. Academic Press, New York.
Kuhn, D. N., Knauf, M, and Stumpf, P. K. (1981). *Arch. Biochem. Biophys.* **209**, 441-450.
Lane, M. D., Moss, J., and Polakis, S. E. (1974). *Curr. Top. Cell. Regul.* **8**, 139-195.
McDermitt, D. K., and Loomis, R. S. (1981). *Ann. Bot.* **48**, 275-290.
Macher, B. A., Brown, C. P., McManus, T. J., and Mudd, J. B. *Plant Physiol.* **55**, 130-136.
McMahon, V., and Stumpf, P. K. (1966). *Plant Physiol.* **41**, 148-156.
Mazliak, P., Oursel, A., Abdelkader, B., and Grosbois, M. (1972). *Eur. J. Biochem.* **28**, 399-411.
Mohan, S. B., and Kekwick, G. O. (1980). *Biochem. J.* **187**, 667-676.
Moreau, R. A., and Stumpf, P. K. (1981). *Plant Physiol.* **67**, 672-676.
Morris, L. J. (1967). *Biochem. Biophys. Res. Commun.* **29**, 311-315.

Mudd, B. J. (1980). *In* 'The Biochemistry of Plants' (P. K. Stumpf and E. E. Conn, eds), Vol. 4, pp. 249-282. Academic Press, New York.
Muller, B., Ziegler, I., and Ziegler, H. (1969). *Eur. J. Biochem.* **9**, 101-106.
Muller, L. L., Hensarling, T. P, and Jacks, T. J. (1975). *J. Amer. Oil Chem. Soc.* **52**, 164-165.
Murphy, D. J., and Stumpf, P. K. (1981). *Arch. Biochem. Biophys.* **212**, 730-739.
Nagai, J., and Bloch, K. (1965). *J. Biol. Chem.* **240**, 3702-3703.
Nagai, J., and Bloch, K. (1966). *J. Biol. Chem.* **241**, 1925-1927.
Nagai, J., and Bloch, K. (1968). *J. Biol. Chem.* **243**, 4626-4633.
Nakamura, Y., and Yamada, M. (1974). *Plant, Cell Physiol.* **15**, 37-48.
Nelson, D. R., and Rinne, R. W. (1977). *Plant, Cell Physiol.* **18**, 1021-1027.
Nikolau, B. S., Hawke, J. C., and Slack, C. R. (1981). *Arch. Biochem. Biophys.* **211**, 605-612.
Norton, G., and Harris, J. F. (1975). *Planta* **123**, 163-174.
Ohlrogge, J. B., Pollard, M. R., and Stumpf, P. K. (1978a). *Lipids* **13**, 203-210.
Ohlrogge, J. B., Shine, W. E., and Stumpf, P. K. (1978b). *Arch. Biochem. Biophys.* **189**, 382-391.
Ohlrogge, J. B., Kuhn, D. N., and Stumpf, P. K. (1979). *Proc. Natl Acad. Sci. U.S.A.* **76**, 1194-1198.
Oo, K. C., and Stumpf, P. K. (1979). *Lipids* **14**, 132-143.
Pollard, M. R., and Stumpf, P. K. (1980a). *Plant Physiol.* **66**, 641-548.
Pollard, M. R., and Stumpf, P. K. (1980b). *Plant Physiol.* **66**, 649-655.
Pollard, M. R., McKeon, T., Gupta, L. M., and Stumpf, P. K. (1979). *Lipids* **14**, 651-662.
Preiss, J., and Kosuge, T. (1977). *Annu. Rev. Plant Physiol.* **21**, 433-466.
Privett, O. S., Dougherty, K. A., Erdahl, W. L., and Stolyhwo, A. (1973). *J. Amer. Oil Chem. Soc.* **50**, 516-520.
Putt, E. D., Craig, B. M., and Carson, R. B. (1969). *J. Amer. Oil Chem. Soc.* **46**, 126-129.
Reid, E. E., Lyttle, C. R., Canvin, D. T., and Dennis, D. T. (1975). *Biochem. Biophys. Res. Commun.* **62**, 42-47.
Reid, E. E., Thompson, P., Lyttle, C. R., and Dennis, D. T. (1977). *Plant Physiol.* **59**, 842-848.
Reitzel, L., and Nielsen, N. C. (1976). *Eur. J. Biochem.* **65**, 131-138.
Rest, J. A., and Vaughan, J. G. (1972). *Planta* **105**, 245-262.
Röbbelen, G. (1975). *In* 'Crop Genetic Resources for Today and Tomorrow' (O. H. Frankel and J. G. Hawker, eds), pp. 231-247. Cambridge University Press, Cambridge.
Roughan, P. G., and Slack, C. R. (1977). *Biochem. J.* **162**, 457-459.
Roughan, P. G., and Slack, C. R. (1982). *Annu. Rev. Plant Physiol.* **33**, 97-132.
Roughan, P. G., Holland, R., and Slack, C. R. (1980). *Biochem. J.* **188**, 17-24.
Saito, K., Kawaguchi, A., Okuda, S., Seyama, Y., Yamakawa, T., Nakamura, Y., and Yamada, M. (1980). *Plant, Cell Physiol.* **21**, 9-19.
Shimakata, T., and Stumpf, P. K. (1982a). *Arch. Biochem. Biophys.* **217**, 144-154.
Shimakata, T., and Stumpf, P. K. (1982b). *Proc. Natl Acad. Sci. U.S.A.* **79**, 5808-5812.
Shine, W. E., Mancha, M., and Stumpf, P. K. (1976). *Arch. Biochem. Biophys.* **173**, 472-479.
Simcox, P. D., and Dennis, D. T. (1978a). *Plant Physiol.* **62**, 287-290.
Simcox, P. D., and Dennis, D. T. (1978b). *Plant Physiol.* **61**, 871-877.
Simcox, P. D., Reid, E. E., Canvin, D. T., and Dennis, D. T. (1977). *Plant Physiol.* **59**, 1128-1132.

Simcox, P. D., Garland, W., De Luca, V., Canvin, D. T., and Dennis, D. T. (1979). *Can. J. Bot.* **57**, 1008-1014.
Sims, R. P. A., McGregor, W. G., Plessers, A. G., and Mes, J. C. (1961). *J. Amer. Oil Chem. Soc.* **38**, 276-279.
Sinclair, T. R., and De Wit, C. T. (1975). *Science* **189**, 565-567.
Slabas, A. R., Roberts, P. A., Ormesher, J., and Hammond, E. W. (1982). *Biochim. Biophys. Acta* **711**, 411-420.
Slack, C. R., and Roughan, P. G. (1978). *Biochem. J.* **170**, 437-439.
Slack, C. R., and Roughan, P. G. (1980). *In* 'Biogenesis and Function of Plant Lipids' (P. Mazliak, P. Benveniste, C. Costes and R. Douce, eds), pp. 88-98. Elsevier, Amsterdam.
Slack, C. R., Roughan, P. G., and Terpstra, J. (1977). *Biochem. J.* **155**, 71-80.
Slack, C. R., Roughan, P. G., and Balasingham, N. (1978). *Biochem. J.* **170**, 421-433.
Slack, C. R., Roughan, P. G., and Browse, J. (1979). *Biochem. J.* **179**, 649-656.
Slack, C. R., Bertaud, W. S., Shaw, B. D., Holland, R., Browse, J., and Wright, H. (1980). *Biochem. J.* **190**, 551-561.
Slack, C. R., Campbell, L. C., Browse, J. A., and Roughan, P. G. (1983). *Biochim. Biophys. Acta* **754**, 10-20.
Smith, C. R. (1970). *In* 'Progress in the Chemistry of Fats and Other Lipids' (R. T. Holman, ed.), Vol. 11, Pt I, pp. 139-177. Pergamon, Oxford.
Smith, C. G. (1974). *Planta* **119**, 125-142.
Smith, S. (1980). *J. Dairy Sci.* **63**, 337-351.
Stumpf, P. K. (1980). *In* 'The Biochemistry of Plants' (P. K. Stumpf and E. E. Conn, eds), Vol. 4, pp. 177-204. Academic Press, New York.
Stymne, S. (1980). 'The Biosynthesis of Linoleic and Linolenic Acids and Plants.' Ph.D. Thesis, Swedish University of Agricultural Sciences, Uppsala, Sweden.
Stymne, S., and Appleqvist, L.-Å. (1978). *Eur. J. Biochem.* **90**, 223-229.
Stymne, S., and Appleqvist, L.-Å. (1980). *Plant Sci. Letts* **17**, 287-294.
Stymne, S., and Glad, G. (1981). *Lipids* **16**, 298-305.
Styme, S., Stobart, A. K., and Glad, G. (1983). *Biochim. Biophys. Acta* **752**, 198-208.
Vick, B., and Beevers, H. (1977). *Plant Physiol.* **59**, 459-463.
Vick, B., and Beevers, H. (1978). *Plant Physiol.* **62**, 173-178.
Wanner, G., and Theimer, R. R. (1978). *Planta* **140**, 163-169.
Wanner, G., Formanek, H., and Theimer, R. R. (1981). *Planta* **151**, 109-123.
Weaire, P. J., and Kekwick, R. G. B. (1975). *Biochem. J.* **146**, 425-437.
Weber, F. J. (1973). *In* 'Symposium Proceedings on Industrial Uses of Cereals' pp. 161-206. Amer. Assoc. Cereal Chem., St Paul, Minnesota.
Weber, E. J., de la Roche, I. A., and Alexander, D. E. (1971). *Lipids* **6**, 525-530.
Wilson, R. F. (1981). *Crop Sci.* **21**, 519-524.
Wilson, R. F., Weissinger, H. H., Buck, J. A., and Faulkner, G. D. (1980). *Plant Physiol.* **66**, 545-549.
Wright, H. C. (1980). 'Long Chain Fatty Acid Synthesis in Rape Seed Cotyledons.' M.Sc. Thesis, Massey University, New Zealand.
Wu, X.-Y., Moreau, R. A., and Stumpf, P. K. (1981). *Lipids* **16**, 897-902.
Yamada, M., and Usami, Q. (1975). *Plant, Cell Physiol.* **16**, 879-884.
Yamada, M., Usami, Q., and Nakajima, K. (1974). *Plant, Cell Physiol.* **15**, 49-58.
Yang, S. F., and Stumpf, P. K. (1965). *Biochim. Biophys. Acta* **98**, 19-26.
Yatsu, L. Y. (1965). *J. Cell Biol.* **25**, 193-199.
Yatsu, L. Y., and Jacks, T. J. (1972). *Plant Physiol.* **49**, 937-943.
Yatsu, L. Y., Jacks, T. J., and Hensarling, T. P. (1971). *Plant Physiol.* **48**, 675-682.
Yermanos, D. M. (1975). *J. Amer. Oil. Chem. Soc.* **52**, 115-117.

Yonuschot, G. R., Ortwerth, B. J., and Koeppe, O. J. (1970). *J. Biol. Chem.* **245**, 4193-4198.
Zilkey, B. F., and Canvin, D. T. (1969). *Biochem. Biophys. Res. Commun.* **34**, 646-653.
Zilkey, B. F., and Canvin, D. T. (1971). *Can. J. Bot.* **50**, 323-326.

CHAPTER 7

Toxic Compounds in Seeds

E. A. BELL

I.	Introduction	245
II.	The Alkaloids	246
III.	Cyanogenic Compounds	250
	A. Cyanogenic Glycosides	250
	B. Cyanogenic Cyanolipids	251
IV.	Non-Cyanogenic Nitriles	251
V.	Non-Protein Amino Acids	252
	A. Mimosine	252
	B. Oxalylamino Acids	253
	C. Arginine Analogues	253
	D. Other Amino Acids	254
VI.	Amines	254
VII.	Saponins	257
VIII.	Glucosinolates	258
IX.	Miscellaneous Toxins	258
	A. Cycasins	258
	B. Enzyme Inhibitors	259
	C. Phytohaemagglutinins (Lectins)	259
X.	Possible Roles of Toxic Compounds in Seeds	259
XI.	Seed Toxins and the Development of Food and Fodder Crops	261
	References	262

I. INTRODUCTION

Before discussing the various types of physiologically active compounds which have been found in seeds it is well worth reflecting on the meaning of the terms 'toxic' and 'poisonous'.

I suspect that we have all read detective stories in which cyanide, arsenic or a little-known arrow poison has been used by the vicar's wife, or someone else we didn't suspect, to eliminate the general, ballerina or visiting prime minister.

The victims, however, were all human beings sharing the same basic biochemistry and the cyanide, arsenic and little-known arrow poison were all chemicals capable of disrupting one or more of the biochemical processes that are essential for human life. Other forms of life do not necessarily share with humans the same metabolic pathways or even digestive processes and it should come as no surprise to us to discover that a compound that is highly toxic to one form of life may be harmless or even nutritious to another. When we describe a compound or a plant as 'poisonous' or 'toxic' we must therefore always be careful to identify the organism(s) in which toxicity has been observed. Too frequently the term 'poisonous' is used as it if were synonymous with 'poisonous to man or animals he has paid money for'. This narrow, if understandable, preoccupation with man and domestic animals is reflected in many papers on 'poisonous plants' and plant 'toxins' in the medical and veterinary literature. If, however, we seek a role for such toxins in the plants themselves we must conclude that it is unlikely that man or domestic animals exercised the environmental pressures that led to the natural selection of these toxin-accumulating species. It is far more likely that man and domestic animals are the victims of chemical defences that have evolved in response to environmental pressures such as those exercised by phytophagus insects. The presence in some plant species of compounds, such as the rotenones and pyrethrins which are highly toxic to cold-blooded animals, but relatively harmless to warm-blooded animals, strongly supports this supposition, as do the highly specialized biochemical adaptations shown by certain phytophagus insects that enable them to detoxify or metabolize compounds able to protect their host plants from other insect predators.

Seeds have frequently proved to be rich sources of physiologically active secondary compounds. As the science of chemical ecology develops it is becoming clearer that we may be seriously misled in trying to interpret the role of a secondary compound in a seed if we fail to realize that the compound may play many different biochemical roles in different organisms.

II. THE ALKALOIDS

The alkaloids are a heterogenous collection of nitrogen containing compounds that are largely of plant origin. Some are exceedingly poisonous to man; for example, physostigmine ex *Physostigma venenosum* (LD_{50} 3 mg/kg orally in mice) and cytisine ex *Laburnum anagyroides* (LD_{50} 4 mg/kg subcutaneously in dogs) (Cooper-Driver, 1983).

Alkaloids may occur in different organs of a plant but they are frequently found in highest concentration in the seeds. The seeds of *Datura* species, for example, contain major concentrations of the tropane alkaloids hyoscyamine **(1)** and hyoscine **(2)** and also atropine; their toxicity to man is well documented in both America and Asia. The practice of Malayan thieves of adding ground *D. fastuosa* seeds to the tea or coffee of their intended victims is described by Gimlette (1929), the quantity administered being sufficient to produce unconsciousness. Cases are on record, however, of the seeds being used with lethal effect, and hyoscine itself was used by the notorious British murderer, Dr Crippen (Gimlette, 1929).

The curare arrow poisons of the South American Indians are prepared from species of *Strychnos* of the family Loganiaceae and the seeds of *S. nux vomica* are the commercial source of the alkaloid strychnine **(3)** which is now largely used as a rat poison. The seeds of the tropical African legume *Physostigma venenosum*, better known as the ordeal bean of Calabar, are rich in the alkaloid physostigmine **(4)**. These seeds were used in West Africa to try suspected malefactors: failure to survive a dose was taken as proof of guilt (Dalziel, 1937). Physostigmine is now employed in ophthalmology to reduce the size of the pupil after treatment with atropine (Burger, 1970). Accidental poisoning of humans, particularly children, by alkaloid containing seeds is common. In Britain one of the most dangerous species in this respect is *Laburnum anagyroides* (common Laburnum, golden rain) which is grown as an ornamental (Forsyth, 1968). The seeds of this species contain cytisine **(5)** and numerous cases of human poisoning have been recorded.

The genus *Lupinus* is another legume genus in which the seeds of many species contain high concentrations of alkaloids. Lupin poisoning in sheep has usually been observed after the animals have fed on lupins on which the pods have been well formed. Over twenty alkaloids have been identified in lupin species, most are of the quinolizidine type like lupanine **(6)** which is a principal alkaloid of *L. angustifolius*, the narrow leafed lupin.

Several species of *Crotalaria* are toxic to animals and the leaves, stems and seeds are known to contain the pyrrolizidine alkaloid monocrotaline **(7)**. Among domestic fowl chronic poisoning has been caused by the ingestion of 80–160 seeds. Mixed with normal poultry food as little as 0.2 lb of *C. spectabilis* (rattle box) seed per ton of feed (0.01%) produced toxic effects in the birds and a concentration of 1% produced 100% mortality (Kingsbury, 1964).

The seeds of *Castanospermum australe*, the black bean of Queensland and New South Wales, are eaten by Aborigines after soaking in water for several days and then roasting (Francis and Southcott, 1967). In 1968, three

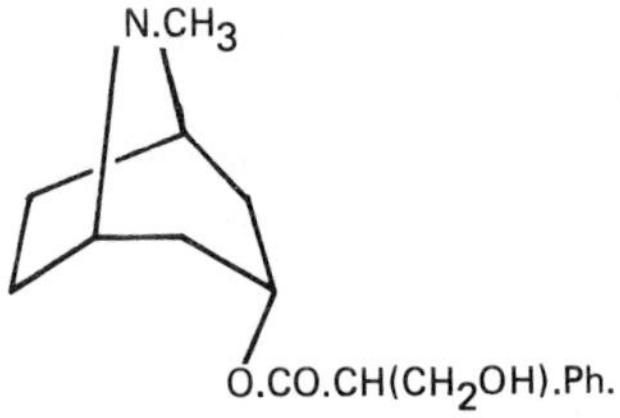

(1) Hyoscyamine

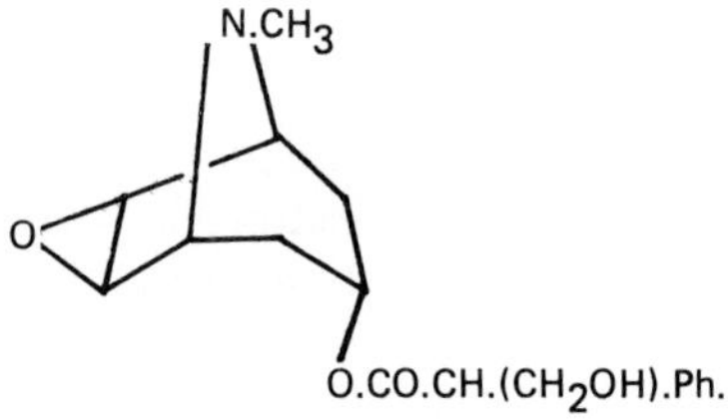

(2) Hyoscine

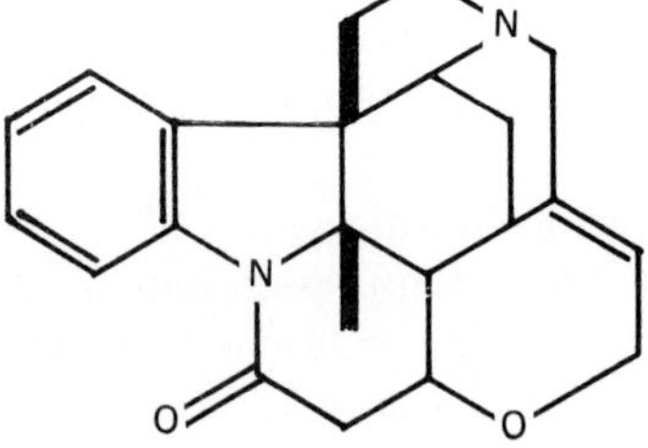

(3) Strychnine

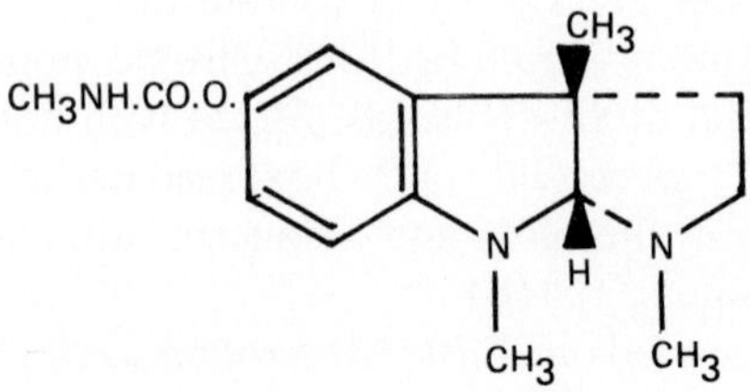

(4) Physostigmine

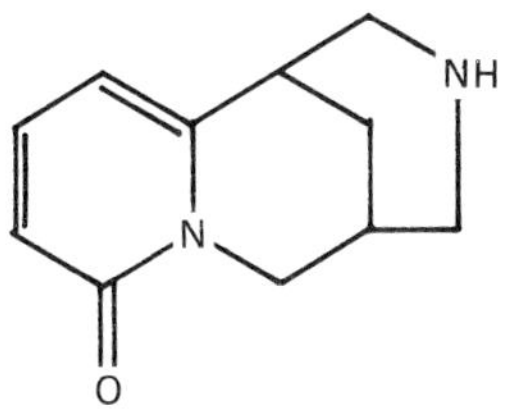

(5) Cytisine

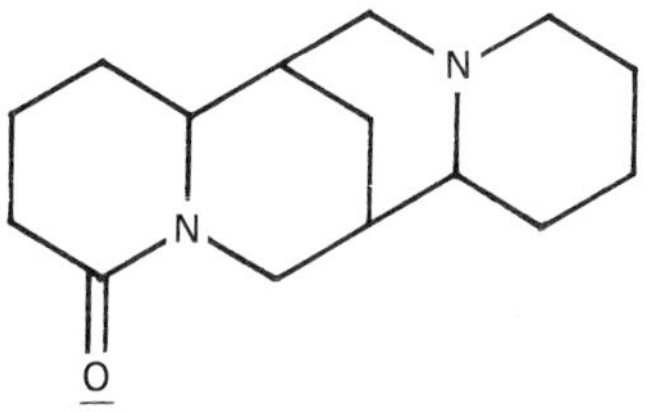

(6) Lupanine

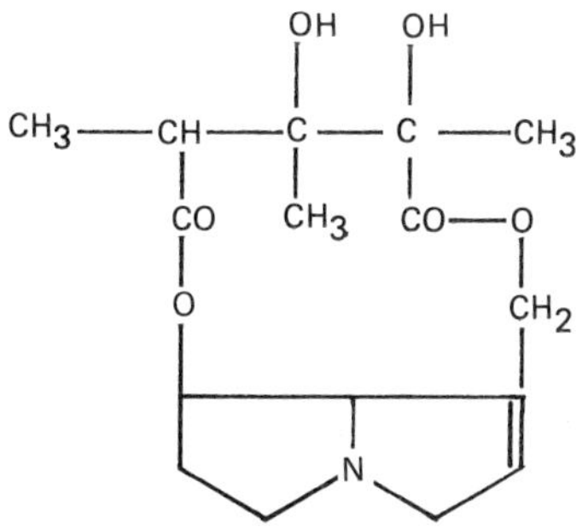

(7) Monocrotaline

servicemen in southeastern Queensland who ate small amounts of raw seeds became severely ill within two hours and were admitted to hospital with symptoms of vomiting, severe abdominal pain and dizziness (Everist, 1974). The raw seeds can, moreover, cause severe gastro-intestinal irritation and even death when eaten by cattle or horses (Everist, 1974). Although the seed extracts failed to give a positive alkaloid reaction with Dragendorff's reagent it has recently been shown that the seeds contain a novel type of higher plant alkaloid which has been named castanospermine **(8)** (Hohenschutz *et al.*, 1981).

The coffee bean (*Coffea arabica*) and the mescal bean (*Sophora secundiflora*) formerly used by the Indians of the American South West to induce hallucinations both owe their physiological effects to the presence of alkaloids, caffeine and sophorine, respectively. The Arapaho and Iowa tribes were reported using seeds as a medicine and narcotic in 1820. The beans have been excavated from archaeological sites in Texas with suggestions of ritualistic use as early as A.D. 1000 (Schultes, 1981). These and other seed alkaloids too numerous to be listed can be mildly intoxicating at one concentration and lethal at another.

III. CYANOGENIC COMPOUNDS

Cyanogenic compounds are those capable of liberating hydrogen cyanide on hydrolysis. Most cyanogenic compounds in plants are cyanogenic glycosides although a few are cyanolipids. However, all the cyanogenic compounds share a common structural feature: all are substituted α-hydroxynitriles. In the cyanogenic glycosides the α-hydroxy group is condensed through a glycosidic link to a sugar moiety; for example, the cyanogenic glycoside, amygdalin [Ph. CH(OGlc.OGlc). C:N] and in the cyanogenic cyanolipids the α-hydroxy group is esterified with a fatty acid; for example, $[CH_2{:}C(CH_3).CH(O.CO.(CH_2)_{18}CH_3).C{:}N]$.

A. Cyanogenic Glycosides

These compounds have been detected in species of over 100 plant families although the number of individual cyanogenic glycosides isolated and characterized is fewer than 30 (Conn, 1980). The hydrolysis of cyanogenic glycosides takes place when the plant tissue is damaged and β-glucosidase enzymes [EC 3.2.1.21] come into contact with the glycosides. The enzymes release α-hydroxynitriles which in turn break down to hydrogen cyanide and an aldehyde or ketone under the influence of an hydroxynitrile lyase, oxynitrilase [EC 4.1.2.11]. The breakdown of amygdalin, the cyanogenic glycoside found in the seeds of bitter almonds (*Prunus amygdalus*) is shown in Figure 1.

Other cyanogenic glycosides found in seeds of economic importance are linamarin $[(CH_3)_2.C(OGlc).C{:}N]$ and lotaustralin (**9**) which occur in *Phaseolus lunatus* (the Lima bean) and in *Linum usitatissimum* (linseed). Variable concentrations of cyanogenic glycosides have been reported from different varieties of *P. lunatus* and the significance of this variability is discussed later (Section XI).

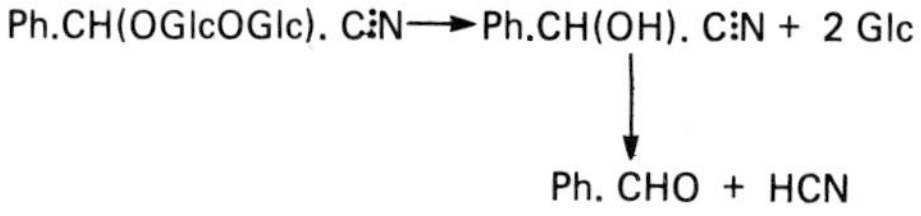

Ph.CH(OGlcOGlc).C:N ⟶ Ph.CH(OH).C:N + 2 Glc
⬇
Ph. CHO + HCN

Fig. 1. Release of hydrogen cyanide from the breakdown of amygdalin, a cyanogenic glucoside.

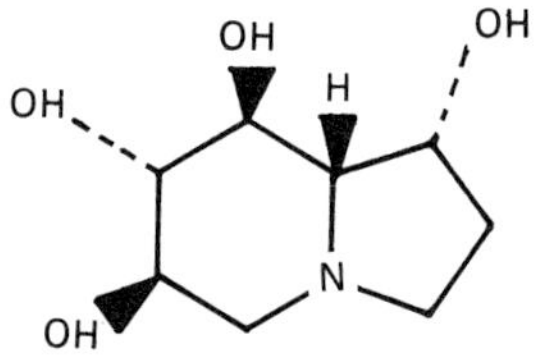

(8) Castanospermine

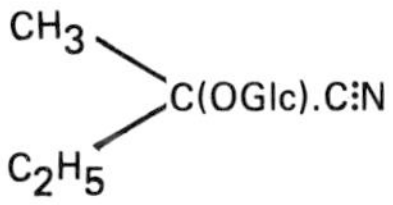

$$CH_3 \atop C_2H_5 {>} C(OGlc).C{:}N$$

(9) Lotaustralin

B. Cyanogenic Cyanolipids

Cyanolipids have only been found with certainty in one plant family, the Sapindaceae, where they are known to occur in the seed oils of six species. Two of the four cyanolipids that have been identified are esters of α-hydroxynitriles and give rise to hydrogen cyanide when hydrolysed.

The toxicity of hydrogen cyanide is related to its high affinity for cytochrome a which results in the blocking of respiration in aerobic organisms.

IV. NON-CYANOGENIC NITRILES

If a nitrile does not have an hydroxyl group in the α-position, that is, if it is not a cyanohydrin then it will not hydrolyse to give hydrogen cyanide. It has been found nevertheless that certain of these non-cyanogenic nitriles are toxins in their own right. Examples include β-cyanoalanine [NC. CH_2. $CH(NH_2)$. CO_2H] which was first isolated from seeds of vetch (*Vicia sativa*) and shown to be highly toxic to chicks and rats. The compound appears to act by inhibiting the pyridoxal-dependent conversion of methionine to cystine (Ressler et al., 1964). Another is β-aminopropionitrile [NC. CH_2. CH_2. NH_2] which occurs bound as the γ-glutamyl derivative in seeds of *Lathyrus odoratus* (the sweet pea). This compound has been found to produce aortic aneurisms in birds and skeletal deformation in rats (Schilling and Strong, 1955). Nitrile acts by inhibiting the enzymes responsible for the cross-linkages in elastin (O'Dell et al., 1966) and collagen (Levene, 1962).

V. NON-PROTEIN AMINO ACIDS

Plants synthesize over 200 non-protein amino acids (Bell, 1981a; Rosenthal, 1982) and these compounds are frequently found in major concentration in seeds. Little is known of the physiological activity of most of them as few have been isolated or synthesized in sufficient quantity for extensive biological assay. Some, however, have attracted attention by reason of their toxicity to man and others because of their agricultural or ecological importance.

A. Mimosine

Leucaena leucocephala is a mimosoid legume of wide distribution in the tropics. Cattle, goats and sheep that have grazed on this plant lose their hair (or wool) and suffer liver damage. Their thyroid glands also become enlarged. The compound responsible for hair loss and liver damage is the heterocyclic amino acid mimosine **(10)**. The development of goitres,

(10) Mimosine

however, is restricted to ruminants as the goitrogen is 3-hydroxy-4(1H)-pyridone (11) which is formed from mimosine by the action of micro-organisms present in the rumen (Hegarty *et al.*, 1979). Observations that some varieties of goat are able to consume *L. leucocephala* without ill effects have led to the discovery that the rumen of these animals contains micro-organisms capable of degrading mimosine beyond 3-hydroxy-4(1H)-pyridone. The possibility of introducing such micro-organisms into the rumen of ruminants which might otherwise be at risk to poisoning by *L. leucocephala* is being explored (R. J. Jones, in press).

B. Oxalylamino Acids

The seeds of four species of *Lathyrus* used as human food in the Indian sub-continent contain the neurotoxic amino acid α-amino-β-oxalylamino-propionic acid (ODAP [HO_2C. CO. NH. CH_2. CH(NH_2). CO_2H]). The same compound has also been identified in the seeds of other species of *Lathyrus*, *Crotalaria* and *Acacia* (Qureshi *et al.*, 1977). In man the effects of 'lathyrism' are an irreversible paralysis of the legs and in extreme cases, death.

The higher homologue of ODAP α-amino-γ-oxalylaminobutyric acid (ODAB [HO_2C. CO. NH. CH_2. CH_2. CH (NH_2). CO_2H]) also occurs in the seeds of certain *Lathyrus*, *Crotalaria* and *Acacia* species. This compound is not found in seeds used for human food and no information is available on its toxicity to mammals. It is neurotoxic in chicks, however, and acts as a feeding deterrent to locusts. The occurrence of oxalylamino acids and other physiologically active amino acids found in seeds of *Lathyrus* species, such as α,γdiaminobutyric acid [H_2N. CH_2. CH_2. CH (NH_2). CO_2H] which is toxic to mammals and homoarginine [N_2N. C(:NH). NH. CH_2. CH_2. CH_2. CH_2 CN(NH_2). CO_2 H] which is toxic to micro-organisms has been discussed recently (Bell, 1981b).

C. Arginine Analogues

Homoarginine, mentioned in the previous section, is the higher homologue of arginine [H_2N. C(:NH). NH. CH_2. CH_2. CH_2. CH(NH_2). CO_2H.] a protein amino acid of universal distribution. A close chemical analogue of any important biological intermediate clearly has high potential as a toxin as it may compete with the intermediate for enzyme sites or in other ways. As mentioned, homoarginine is toxic to certain micro-organisms, but it is not toxic to rats, being broken down by liver arginase [EC 3.5.3.1] to the essential amino acid, lysine (Stevens and Bush, 1950).

Canavanine [H_2N. C(:NH). NH. O. CH_2. CH(NH_2). CO_2H] which is found in the seeds of many species of the Papilionoideae (Bell et al., 1978) is another close analogue of arginine that is non-toxic to mammals at low concentration, but is highly toxic to a range of non-canavanine synthesizing plants, micro-organisms and insects (Rosenthal, 1977). In contrast indospicine [H_2N. C(:NH). CH_2. CH_2. CH_2. CH_2. CH(NH_2). CO_2H.] which was isolated from seed of *Indigofera spicata* (Hegarty and Pound, 1970) is a potent hepatotoxin and teratogen in mammals. Indospicine co-occurs with canavanine in the seeds of certain species of *Indigofera* (Charlwood and Bell, 1977) and it is possible that they complement one another as insecticidal or anti-fungal compounds.

D. Other Amino Acids

Other seed amino acids known to be toxic to mammals are α-amino-β-methylaminopropionic acid [CH_3.NH.CH_2CH(NH_2). CO_2H] from *Cycas circinalis* (Nunn et al., 1967) and selenocystathione [Se (CH_2. CH_2. CH(NH_2). CO_2H)$_2$] from *Lecythis ollaria* (coco de mono) (Aranow and Kerdel-Vegas, 1965). This seleno amino acid causes hair loss like mimosine.

VI. AMINES

In mammals the catecholamines, indoleamines and histamines are involved in important processes such as the transmission of nervous impulses and the regulation of blood pressure. The excessive build up of such amines in the mammalian system is normally prevented by the activity of a multiple enzyme system referred to as monamine oxidase (MAO) [EC 1.4.3.4] which oxidizes catecholamines, and a diamine oxidase [EC 1.4.3.6] which acts on histamine. These enzymes also act to destroy exogenous amines occurring in food. If the enzyme levels are for any reason unusually low or their

(11) 3-Hydroxy-4(1H)-pyridone

7. Toxins in Seeds   255

activities are suppressed by certain anti-depressant drugs then the amine levels may rise causing hypertension and other symptoms (Smith, 1981). Similar effects may result from eating foods which contain relatively high concentrations of these amines or related compounds. Tyramine (**12**), N-methyltyramine (**13**) and N-dimethyltyramine (hordenine, **14**) have been reported in sanwa millet (*Panicum miliaceum*) seeds (Sato *et al.*, 1970). The compound N-methyltyramine also occurs as a major component in seeds of *Acacia schweinfurthii* and has been found to induce migraine (Evans and Bell, 1980). The same compound occurs in the leaves of *A. berlandieri* (guajillo) and produces *ataxia* ('guajillo wobbles') in sheep and goats in Texas (Kingsbury, 1964).

The seeds of *Anadenanthera peregrina* (*Piptadenia peregrina*) from which hallucinogenic cohoba snuff is made have been shown to contain bufotenine (**15**) and a number of other tryptamine derivatives (Schultes, 1969).

In adult mammals the 'blood-brain barrier' protects the central nervous system against changing levels of physiologically active amines in the blood stream, but this barrier is not necessarily impervious to their amino acid precursors. It is interesting to note that L-DOPA (**16**) the precursor of dopamine (**17**) and 5-hydroxytryptophan (**18**) the precursor of serotonin (5-HT, **19**) constitute as much as 10% and 14% respectively of the dry seed weight of *Mucuna* and *Bandeiraea* (*Griffonia*) species (Bell and Janzen, 1971; Bell *et al.*, 1976). Both of these amino acids after crossing the blood-brain barrier will be decarboxylated by enzymes of the central nervous system to give their respective physiologically active amines.

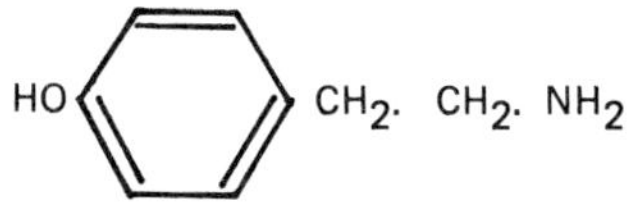

HO—⟨⟩—CH$_2$.CH$_2$.NH$_2$

(**12**)   Tyramine

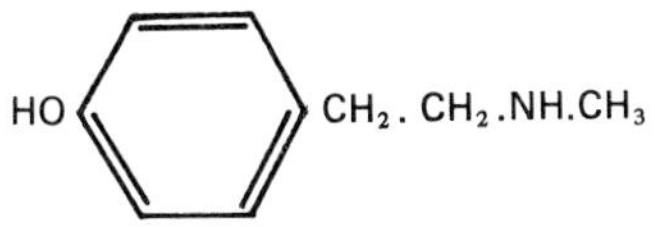

HO—⟨⟩—CH$_2$.CH$_2$.NH.CH$_3$

(**13**)   N-Methyltyramine

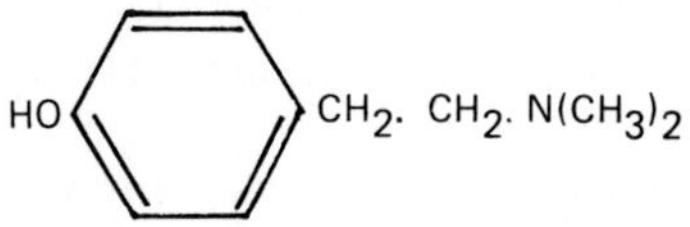

(14) Dimethyltyramine (hordenine)

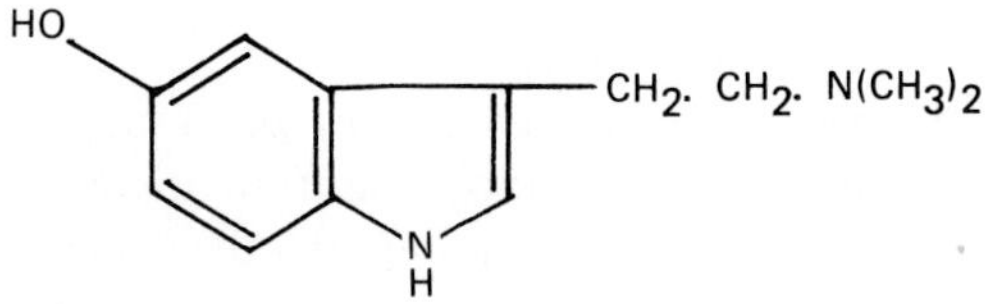

(15) Bufotenine

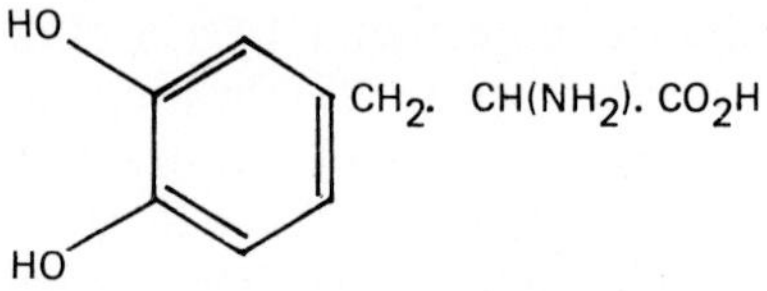

(16) 3,4-Dihydroxyphenylalanine (Dopa)

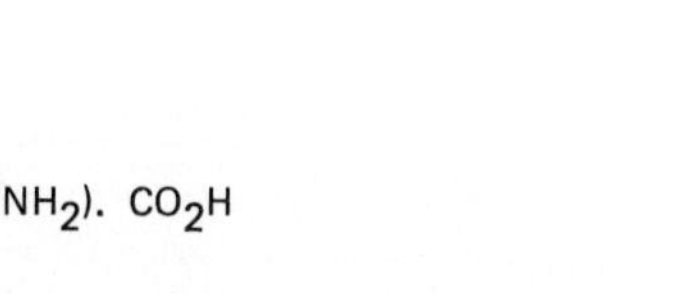

(17) Dopamine

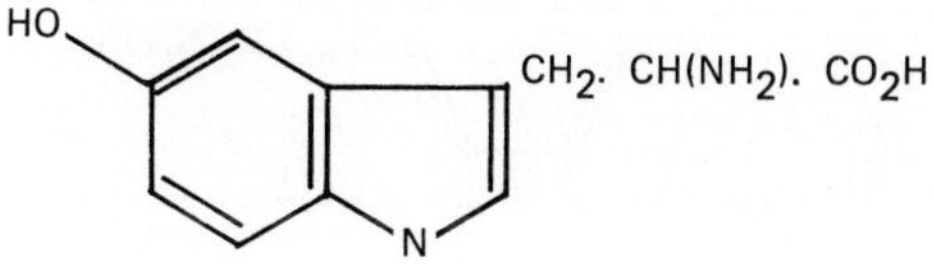

(18) 5-Hydroxytryptophan

(19) Serotonin

VII. SAPONINS

Saponins are bitter-tasting compounds which foam in aqueous solution. They haemolyse red blood cells and are particularly toxic to cold-blooded animals. The saponins are steroid or triterpenoid glycerides, are widely distributed and occur both in whole plants and in seeds. In a review of saponins in plant foodstuffs Birk and Peri (1980) gave an account of saponins found in the seeds of *Medicago sativa* (alfalfa) including medicagenic acid 3-β-0-triglucoside (20) and saponins from seeds of *Phaseolus vulgaris* (common bean), *Pisum sativum* var. *arvense* (field pea), *Lotus corniculatus* (birdsfoot trefoil) and *Glycine max* (soybean). The toxicity of saponins to fungi and other micro-organisms has been established and the subject has been reviewed by Birk and Peri (1980).

(20) Medicagenic acid 3-β-O-glucoside

VIII. GLUCOSINOLATES

Glucosinolates are the compounds responsible for the sharp taste of mustard and *Armoracea rusticana* (horseradish) and also contribute to the flavours of *Brassica oleracea* (cabbage) and other crucifers used as vegetables. The glucosinolates are found in highest concentration in the seed, sinigrin [R. C(GClc):N. OSO_3^- K^+ (R = CH_2: CH. CH_2.)] being the glucosinolate present in the seed of *Brassica nigra* (black mustard). Some 90 different glucosinolates are now known, these differ from each other in the nature of the grouping R. Associated naturally with the glucosinolates are the myrosinase enzymes [EC 3.2.3.1] which effect their hydrolysis to unstable aglycones and sugars. The aglycones can then undergo various changes often simultaneously. They can rearrange to give isothiocyanates (mustard oils), they can break down to give nitriles or less frequently they can rearrange to give thiocyanates. The factors that govern the formulation and proportions of the three types of end-product are not completely understood (Kjaer, 1981). These pathways are shown in Figure 2.

If there is an hydroxyl group at the C-2 position in the R group the glucosinolate can also give rise to an oxazolidine-2-thione (Tookey *et al.*, 1980). In this way 2-hydroxy-3-butenyl-glucosinolate found in rape (*Brassica napus*) seed can give rise to 5-vinyloxazolidine-2-thione (goitrin, **21**). Oxazolidinethiones and thiocyanates inhibit uptake of iodine by the thyroid gland but the toxicity of these goitrogens is relatively insignificant when compared with the toxicity of the nitriles. Enzyme-modified seed meal of crambe (*Crambe abyssinica*) which was lethal to rats at 28% of ration, supported growth at 88% of controls after nitriles had been removed from the meal by extraction with aqueous acetone (Tookey *et al.*, 1965).

IX. MISCELLANEOUS TOXINS

A. Cycasins

The seeds of a number of different species of *Cycas* have been found to contain glycosides of methylazoxymethanol. The first of these to be characterized was the glucoside cycasin (**22**) which was isolated from seeds of *Cycas circinalis* (Riggs, 1956). The glycoside itself is non-toxic in mammals but the aglycone that is liberated by the action of β-glucosidase enzymes present in the gut microflora is a powerful carcinogen. Carcinoma of the liver results from eating untreated seeds of *C. circinalis*. In Guam and neighbouring islands of the Pacific, the seeds are ground and extracted with water before use. Nevertheless, the ingestion of small amounts of the toxin over a prolonged period is thought to be responsible for the high incidence

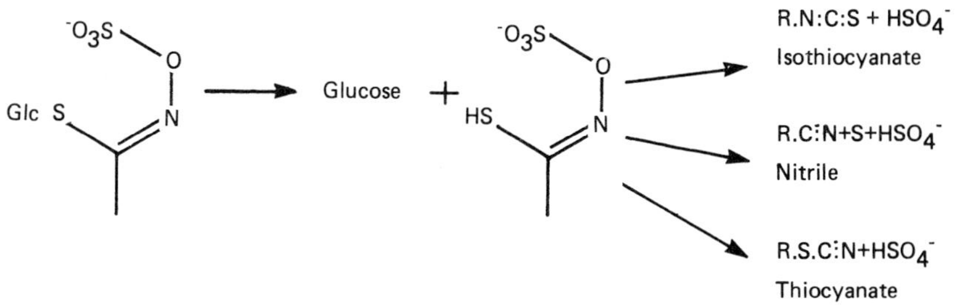

Fig. 2. Metabolism of glucosinolates: myrosinase catalysed hydrolysis of glucosinolate releases glucose plus aglycone, which may rearrange to form isothiocyanate or thiocyanate, or break down to nitrile.

of the neurological disease, amyotrophic lateral sclerosis, suffered by the islanders (Whiting, 1963).

B. Enzyme Inhibitors

Heat labile seed proteins that act as inhibitors of proteases (Section 3.III) and amylases (Section 5.III,A,7) are widely distributed in plants. Enzyme inhibitors of low molecular weight, however, are also found in seeds; for example, 2,5-dihydroxymethyl-3,4-dihydroxypyrrolidine from seeds of *Lonchocarpus sericeus*. This inhibits insect α-glucosidase [EC 3.2.1.20] (Evans *et al.*, 1984).

C. Phytohaemagglutinins (Lectins)

Ricin from *Ricinus communis* (castor bean) and abrin, from *Abrus precatorius* (rosary bean) are examples of highly toxic seed proteins which are related to phytohaemagglutinins (Section 3.II,C). Because they are proteins, the phytohaemagglutinins are destroyed by heating. This group of compounds is treated in Section 3.II.

X. POSSIBLE ROLES OF TOXIC COMPOUNDS IN SEEDS

It is apparent that seeds very frequently contain compounds that are toxic to one form of life or to another. Most results relate to mammals, but data on the toxicity of these compounds to other organisms are accumulating.
Experiments using the larvae of *Callosobruchus maculatus* (the cowpea

(21) 5-Vinyloxazolidine-2-thione

(22) Cycasin

weevil) which normally feed on the seeds of cowpea (*Vigna unguiculata*) have shown that these insects are poisoned by phytohaemagglutinins from *Phaseolus vulgaris* (Janzen *et al.*, 1976). Other experiments (Janzen *et al.*, 1977) showed that 0.1% of the alkaloids strychnine and atropine and the amino acid mimosine and β-cyanoalanine were lethal to the larvae of this insect. Clearly then the seeds of species accumulating these particular compounds are not at risk from *C. maculatus*. Other compounds used in the experiments proved toxic but only when supplied in higher concentration. Canavanine, for example, was significantly toxic to these larvae at 1% and lethal at 5%. This observation again leads us to the conclusion that seeds of *Canavalia* or *Dioclea* species with concentrations of canavanine in excess of 5% are as well protected as seeds of other species that contain lower concentrations of more toxic compounds. A study of the distribution of canavanine and quinolizidine alkaloids in the genera and tribes of the Papilionoideae has led to the suggestion (Bell, 1981c) that the alkaloids with a single role (that of protection) have been replaced during the course of evolution by compounds (including canavanine) that can function both in protecting the plant and in providing a readily available source of nitrogen to the germinating seedlings.

No compound can provide absolute protection to a seed. There are insects that eat bean seeds despite phytohaemagglutinins and some insects eat the seeds of *Mucuna* species despite their L-DOPA content. A seed beetle, *Caryedes brasiliensis*, lives on seeds of *Dioclea megacarpa* which may contain 13% of canavanine (Rosenthal, 1983). Rather than demolish the argument that canavanine protects seeds from potential predators, this observation, paradoxically, enhances it, for Rosenthal *et al.* (1976) have shown that this particular beetle has a specialist arginyl-tRNA synthetase [EC 6.1.1.19] which enables it to discriminate against canavanine and exclude this arginine analogue from its protein. If canavanine were not a barrier to most seed predators then there would be no reason to expect such a biochemical adaptation in the species which successfully feeds on a canavanine-containing species. That canavanine may have roles other than protection and nitrogen storage is suggested by recent findings (Wilson and Bell, 1978) that germinating seedlings of *Neonotonia wightii* (*Glycine wightii*) liberate canavanine through their roots in quantities that inhibit growth in lettuce (*Lactuca sativa*) seedlings. Canavanine may, therefore, play a role in discouraging competition from other plant species.

Although it would be groundless speculation to suggest that all seed toxins have an ecological role, a search for the biochemical responses which such toxins may have induced in predators or competitors may provide us with the best chances of identifying such roles, if they exist.

XI. SEED TOXINS AND THE DEVELOPMENT OF FOOD AND FODDER CROPS

Of the 250 000 flowering plants that have been described, fewer than 30 are used by man on a world-wide basis as food or fodder crops. One of the factors responsible for this restriction is the frequent occurrence in plants of compounds that are toxic to mammals. Plants that contain such compounds may, nevertheless, be valuable sources of food in other respects. Where the toxin is soluble in water it can often be partially removed by soaking the ground or pounded plant material as a preliminary preparation of food. This procedure is used in Africa to remove cyanogenic glycosides from cassava (*Manihot esculenta*) tubers, in Guam to remove cycasin from seeds of *Cycas circinalis* and in South America to remove alkaloids from lupin seeds (N. W. Simmonds, pers. comm.). Such procedures are unlikely to remove all the toxins present and chronic poisoning associated with the prolonged ingestion of sub-critical levels of toxins such as these are well known. During the course of agricultural history man has also attempted to increase the value of potential food crops by selecting the less-toxic individuals and varieties. In this way it has been possible over a long time to produce domesticated varieties that contain lower levels of toxins than their wild ancestors. The domesticated forms of *Phaseolus lunatus* contain levels of cyanogenic glycosides that are much lower than in the wild forms; for example, Puerto Rico, black variety (0.3% w/w) and American White (0.01% w/w). This is a great improvement, but it is one which has been achieved slowly using trial and error methods in which man himself has served as the experimental animal. If, however, we can identify a toxin by chemical means and develop methods for its quantitative estimation then there is no reason why the process of 'improving' a crop plant by reducing its toxic content can not be achieved with relative rapidity. Such methods also make it possible to identify toxin-free mutants whether of natural occurrence or produced by treatment with mutagenic agents.

As emphasized at the beginning of this chapter, it is also necessary to have a clear understanding of the relative toxicity of secondary compounds to different organisms. Experiments with artificially induced alkaloid-free strains of *Lupinus angustifolius* in Western Australia were not entirely successful because the alkaloid-free plants were much more vulnerable to insect attack than were the alkaloid-containing plants (J. S. Gladstones, pers. comm. 1977). In this example the toxin was a broad-spectrum toxin that affected both insects and mammals. Some toxins are more specific, however, or are toxic at different concentration levels to different organisms. Knowing then that one component of a seed is non-toxic to man but toxic to most seed beetles and that another component is toxic to both, we can recognise the desirability of producing seeds with high concentrations of the first but with little or none of the second. Natural genetic variability and analytical techniques permitting the rapid screening of populations can provide the basis for breeding programmes needed to achieve that aim.

Probably the most useful species on which to concentrate now are those that have not achieved world-wide status as crop plants but are used on a marginal basis in various parts of the world. Some of these crop plants are in effect partially domesticated and have reached a half-way stage of transformation into more nearly ideal food or fodder plants. Where toxicity is still a problem, application of first world techniques of analytical chemistry, toxicology, genetics and plant breeding to solve what are essentially third world problems could achieve major increases in world food supplies even in the short term.

Examples of the successful application of this approach to the production of less toxic seeds are seen in the development of strains of *Brassica napus* (rape) in which the toxic erucic acid [$CH_3.(CH_2)_7.CH:CH.(CH_2)_{11}.CO_2H$] of the seed oil has been replaced by oleic acid (Krzymanski and Downey, 1969; Section 6.V,B,*1*) and glucosinolate levels reduced. In *Lathyrus sativus* varieties that contain low levels of the neurotoxin α-amino-β-oxalylaminopropionic acid have also been identified for agricultural development.

REFERENCES

Aranow, L., and Kerdel-Vegas, F. (1965). *Nature* **205**, 1185–1186.
Bell, E. A. (1981a). *In* 'Progress in Phytochemistry' Vol. 7 (L. Reinhold, J. B. Harborne, and T. Swain, eds), pp. 171–196. Pergamon Press, Oxford.
Bell, E. A. (1981b). *Food Chemistry* **6**, 213–222.

Bell, E. A. (1981c). *In* 'Advances in Legume Systematics', Pt 2 (R. M. Polhill and P. H. Raven, eds), pp. 489–499. Royal Botanic Gardens, Kew.
Bell, E. A., and Janzen, D. H. (1971). *Nature* **229**, 136.
Bell, E. A., Fellows, L. E., and Qureshi, M. Y. (1976). *Phytochemistry* **15**, 823.
Bell, E. A., Lackey, J. A., and Polhill, R. M. (1978). *Biochem. System. Ecol.* **6**, 201–212.
Birk, Y., and Peri, I. (1980). *In* 'Toxic Constituents of Plant Foodstuffs' (I. E. Liener, ed.), pp. 161–182. Academic Press, New York.
Burger, A. (1970). 'Medicinal Chemistry', 3rd Edition. Wiley Interscience, New York.
Charlwood, B. V., and Bell, E. A. (1977). *J. Chromatog.* **135**, 377–384.
Conn, E. E. (1980). *In* 'Secondary Plant Products' (E. A. Bell and B. V. Charlwood, eds), pp. 461–492. Encyclopedia of Plant Physiology, New Series Vol. 8. Springer, Berlin.
Cooper-Driver, G. A. (1983). *In* 'Handbook of Naturally Occurring Food Toxicants' (M. Recheigl, ed.), p. 215. CRC Press, Boca Raton, Florida.
Dalziel, J. M. (1937). *In* 'Useful Plants of West Tropical Agriculture'. Crown Agents for the Colonies, London.
Evans, C. S., and Bell, E. A. (1980). *Trends in Neurosciences* **3**, 70–72.
Evans, S. V., Gatehouse, A. M. R., and Fellows, L. E. (in preparation).
Everist, S. L. (1974). 'Poisonous Plants of Australia.' Angus and Robertson, Sydney.
Forsyth, A. A. (1968). 'British Poisonous Plants' 2nd Edition. Ministry of Agriculture, Fisheries and Food, Bull. No. 161, HMSO, London.
Francis, D. F., and Southcott, R. V. (1967). 'Plants Harmful to Man in Australia.' Misc. Bull. No. 1 Botanic Garden, Adelaide, South Australia.
Gimlette, J. D. (1929). 'Malay Poisons and Charm Cures', 3rd Edition. Churchill, London.
Hegarty, M. P., and Pound, A. W. (1970). *Aust. J.Biol. Sci.* **23**, 831–842.
Hegarty, M. P., Lee, C. P., Christie, G. S., Court, R. D., and Haydock, K. P. (1979). *Aust. J. Biol. Sci.* **32**, 27–40.
Hohenschutz, L. D., Bell, E. A., Jewess, P. J., Leworthy, D. P., Pryce, R. J., Arnold, E., and Clardy, J. (1981). *Phytochemistry* **20**, 811–814.
Janzen, D. H., Juster, H. B., and Liener, I. E (1976). *Science* **192**, 795–796.
Janzen, D. H., Juster, H. B., and Bell, E. A. (1977). *Phytochemistry* **16**, 223–227.
Kingsbury, J. M. (1964). 'Poisonous Plants of the United States and Canada'. Prentice-Hall, Englewood Cliffs, New Jersey.
Kjaer, A. (1981). *Food Chemistry* **6**, 223–234.
Krzymanski, J., and Downey, R. K. (1969). *Can. J. Plant Sci.* **49**, 313–319.
Levene, C. I. (1962). *J. Exp. Med.* **116**, 119–130.
Montgomery, R. D. (1980). *In* 'Toxic Constituents of Plant Foodstuffs' 2nd edition (I. E. Liener, ed.), pp. 143–160. Academic Press, New York.
Nunn, P. B., Vega, A., and Bell, E. A. (1967). *Biochem. J.* **106**, 15.
O'Dell, B. L., Elsden, D. F., Thomas, J., Partridge, S. M., Smith, R. H., and Palmer, R. (1966). *Nature* **209**, 401–402.
Qureshi, Y. M., Pilbeam, D. J., Evans, C. S., and Bell, E. A. (1977). *Phytochemistry* **16**, 477–479.
Ressler, C., Nelson, J., and Pfeffer, M. (1964). *Nature* **203**, 1286–1287.
Riggs, N. V. (1956). *Chem. Ind.* p. 926.
Rosenthal, G. A. (1977). *Q. Rev. Biol.* **52**, 155–178.
Rosenthal, G. A. (1982). 'Plant Nonprotein Amino and Imino Acids.' Academic Press, New York.
Rosenthal, G. A. (1983). *Sci. Am.* **249**, 138–145.
Rosenthal, G. A., Dahlman, D. L., and Janzen, D. G. (1976). *Science* **192**, 256–258.
Sato, H., Sakamura, S., and Obata, Y. (1970). *Agric. Biol. Chem.* **34**, 1254–1255.

Schilling, E. D., and Strong, F. M. (1955). *J. Am. Chem. Soc.* **77**, 2843-2845.
Schultes, R. E. (1969). *Bull. Narcotics* **21** and **22**, 1-56.
Schultes, R. E. (1981). *In* 'Progress in Phytochemistry 7' (L. Rheinhold, J. B. Harborne and T. Swain, eds), pp. 301-331. Pergamon Press, Oxford.
Smith, T. A. (1981). *Food Chemistry* **6**, 169-200.
Stevens, C. M., and Bush, J. A. (1950). *J. Biol. Chem.* **183**, 139-149.
Tookey, H. L., VanEtten, C. H., Peters, J. E., and Wolff, I. A. (1965). *Cereal. Chem.* **42**, 507-514.
Tookey, H. L., VanEtten, C. H., and Daxenbichler, M. E. (1980). *In* 'Toxic Constituents of Plant Foodstuffs' 2nd Edition. (I. E. Liener, ed.), pp. 103-142. Academic Press, New York.
Whiting, M. G. (1963). *Econ. Bot.* **17**, 271-302.
Wilson, M. F., and Bell, E. A. (1978). *J. Exp. Bot.* **29**, 1243-1247.

Plant Species Index

A
Abelmoschus esculentus, 159
Abrus precatorius, 91, 94, 95, 97, 259
Acacia, 253
   berlandieri, 255
   elata, 98, 102
   schweinfurthii, 255
Acetabularia mediterranea, 197
Aegilops, 24, 32
   speltoides, 24
   squarrosa, see *Triticum tauschii*
Aegopodium podagraria, 172, 173, 174
Aesculus hippocastanum, 200
Albizzia julibrissin, 98, 102
alfalfa, see *Medicago sativa*
almond, see *Prunus dulcis*
Amaryllidaceae, 193
Amorphophallus konjac, 195, 196
Anadenanthera peregrina, see *Piptadenia peregrina*
Anemone coronaria, 14, 15
Annonaceae, 191, 193
Apiaceae, 169
Apium graveolens, 8, 159
apple, see *Malus sylvestris*
Araceae, 193
Arachis hypogaea, 91, 99, 104, 106, 140, 141, 142, 212, 236
Araliaceae, 172
Archaeanthus, 4, 5
Archontophoenix cunninghamiana, 192
Armoracia rusticana, 258
Arnica montana, 200
artichoke, Jerusalem, see *Helianthus tuberosus*

ash, European, see *Fraxinus excelsior*
Asparagus officinalis, 194, 201, 202
Aster tripolium, 200
Astragalus bisulcatus, 158
Avena sativa, 108, 109, 118, 123, 141, 202
avocado, see *Persea americana*

B
Bandeiraea, 255
   simplicifolia, 89, 91
Banksia
   coccinea, 149
   hookerana, 149
barley, see *Hordeum vulgare*
Bauhinia
   galpinii, 195
   purpurea, 195
   tomentosa, 195
bean, common, see *Phaseolus vulgaris*
beech, European, see *Fagus sylvatica*
Bertholletia excelsa, 141, 156, 157, 160, 161
birdsfoot trefoil, see *Lotus corniculatus*
bitter almond, see *Prunus amygdalus*
black bean, see *Castanospermum australe*
bluebell, see *Scylla non-scripta*
Botrychium lunaria, 169
Brassica, 18
   napus, 142, 191, 211, 212, 217, 222, 223, 225, 226, 228, 236, 237, 258
   nigra, 258
   oleracea, 169, 258
Brazil nut, see *Bertholletia excelsa*
broad bean, see *Vicia faba*

C

cabbage, *see Brassica oleracea*
Caesalpiniaceae, 101, 193
Canavalia, 260
    ensiformis, 88–91, 94, 170
    gladiata, 94
    maritima, 91, 94
Capsella bursa-pastoris, 8, 156, 158
caraway, *see Carum carvi*
carob, *see Ceratonia siliqua*
carrot, *see Daucus carota*
Carthamus tinctorius, 212, 213, 215, 218, 219, 221, 222, 231, 237, 238
Carum carvi, 146, 172, 175, 193
Caspiocarpus paniculiger, 4
cassava, *see Manihot esculenta*
Cassia, 194
    coluteoides, 181, 196
    corymbosa, 170, 196
Castanospermum australe, 247
castor bean, *see Ricinus communis*
Cecropia peltata, 178
Celastraceae, 188
celery, *see Apium graveolens*
Ceratonia siliqua, 8, 193, 194
Cercis siliquastrum, 195
chick pea, *see Cicer arietinum*
Cicer, 14
    arietinum, 16, 17, 19–24, 30, 33, 35, 56, 85, 87, 105, 106, 173
    bijugum, 16, 20
    cuneatum, 16, 20
    echinospermum, 16, 17, 20–24, 30
    judaicum, 16, 20
    pinnatifidum, 16, 20
    reticulatum, 16, 17, 20–24, 30, 33, 35
Clivia miniata, 193
clover, red, *see Trifolium pratense*
cocksfoot grass, *see Dactylis glomerata*
coco de mono, *see Lecythis ollaria*
coconut, *see Cocos nucifera*
Cocos nucifera, 12, 229
Coffea arabica, 36, 193, 250
coffee bean, *see Coffea arabica*
Convolvulaceae, 193
Cordeauxia edulis, 101
cotton, *see Gossypium hirsutum*
cowpea, *see Vigna unguiculata*
crambe, *see Crambe abyssinica*
Crambe abyssinica, 158, 213, 214, 217, 222, 225, 226, 236, 237, 258
cress, *see Lepidium sativum*
Crotalaria, 247
    spectabilis, 247

Cruciferae, 191, 211, 229
cucumber, *see Cucumis sativus*
Cucumis sativus, 170, 188
Cucurbita
    andreana, 151, 153
    digitata, 142
    foetidissima, 142, 151
    maxima, 149, 152, 153, 156
    mixta, 151
    moschata, 151
    palmata, 142
    pepo, 102, 105, 151, 174, 175
Cuphea pubescens, 229
Cyamopsis tetragonolobus, 194, 196, 199
Cycas circinalis, 254, 258, 261
Cymbidium, 197
Cytisus sessilifolius, 91

D

Dactylis glomerata, 202
darnel, *see Lolium temulentum*
date palm, *see Phoenix dactylifera*
Datura, 247
    fastuosa, 247
    stramonium, 91, 92, 95
Daucus carota, 190
Dioclea megacarpa, 260
Dolichos biflorus, 91
doum palm, *see Hyphaene thebaica*
Dracaena draco, 193

E

Ebenaceae, 193
Edwardsia microphylla, 87
Elephantorrhiza burkei, 101
Eriobotrya japonica, 188
Erythea edulis, 192
Eucalyptus, 159

F

Fabaceae, *see* Papilionaceae
Fagus sylvatica, 169
fennel, *see Foeniculum officinale*
fenugreek, *see Trigonella foenum-graecum*
fescue
    meadow, *see Festuca pratensis*
    red, *see Festuca rubra*
    sheep's, *see Festuca ovina*
    tall, *see Festuca arundinacea*
Festuca
    arundinacea, 172
    ovina, 172, 174, 175

Plant Species Index 267

pratensis, 172
rubra, 172, 174, 175
field pea, *see Pisum sativum* var. *arvense*
flax, *see Linum usitatissimum*
Foeniculum officinale, 172, 175
Fraxinus excelsior, 175

G
Ginkgo biloba, 2
Gleditsia
    ferox, 194
    macracantha, 197
    triacanthos, 175, 196
Glycine
    max, 56, 75, 92, 93, 95, 97, 98, 100, 101, 102, 104, 106, 110, 118, 120–122, 125, 141, 142, 144, 151, 181, 191, 192, 193, 196, 198, 212, 213, 215, 217, 222, 231, 232, 238, 257
    wightii, *see Neonotonia wightii*
golden rain, *see Laburnum anagyroides*
Gossypium hirsutum, 8, 36, 107, 112, 118, 122, 140–143, 236
Grevillea annulifera, 149
Griffonia, *see Bandeiraea*
guajillo, *see Acacia berlandieri*
guar, *see Cyamopsis tetragonolobus*

H
Haemanthus katherinae, 8
Hakea
    platysperma, 149
    victoriae, 149
Helianthus
    annuus, 8, 29–31, 84, 123, 181, 212, 215, 231, 239
    tuberosus, 170, 200, 201, 202
honey locust, *see Gleditsia triacanthos*
Hordeum vulgare, 36, 101, 102, 109, 112, 141, 142, 145, 168, 176, 178, 179, 202, 218
horse chestnut, *see Aesculus hippocastanum*
horseradish, *see Armoracia rusticana*
Hyacinthoides non-scripta, *see Scylla non-scripta*
Hyphaene thebaica, 192

I
Indigofera spicata, 254
Ipomoea batatas, 187

Iridaceae, 193, 194
Iris
    ochroleuca, 193
    sibirica, 193
Italian ryegrass, *see Lolium italicum*
ivory nut, *see Phytelephas macrocarpa*

J
jojoba bean, *see Simmondsia chinensis*

K
Kennedia prostrata, 145
konjac, *see Amorphophallus konjac*

L
Labiatae, 172
Lablab purpureus, 101
Laburnum anagyroides, 247
Lactuca sativa, 103, 193, 261
Lathyrus, 253
    odoratus, 91, 252
    sativus, 91, 101, 262
Lecythis ollaria, 254
Leguminosae, 191, 194
Lens, 14
    culinaris, 15, 16, 19, 29, 56, 91, 93, 95, 98, 125, 173
    ervoides, 19
    nigricans, 15, 16, 19
    orientalis, 15, 16, 19
lentil, *see Lens culinaris*
Lepidium sativum, 168
lettuce, *see Lactuca sativa*
Leucaena leucocephala, 194, 252, 253
Liliaceae, 193, 194
Lilium longiflorum, 169, 170, 192
Lima bean, *see Phaseolus lunatus*
Limnanthes alba, 217, 229
linseed, *see Linum usitatissimum*
Linum usitatissimum, 142, 168, 212, 221, 222, 231, 236, 237, 250
Livistona australis, 192
Loganiaceae, 193, 247
Lolium
    italicum, 172
    perenne, 172, 200
    temulentum, 172, 202
loquat, *see Eriobotrya japonica*
Lotus
    corniculatus, 257
    tetragonolobus, 91

Plant Species Index

lucerne, *see Medicago sativa*
lucerne, Townsville, *see Stylosanthes humilis*
lupin
   narrow leafed, *see Lupinus angustifolius*
   white, *see Lupinus albus*
   yellow, *see Lupinus luteus*
Lupinus
   albus, 42–47, 54–56, 60–62, 66, 67, 70–80, 85, 173
   angustifolius, 47, 48, 52, 54, 55, 60, 85, 111, 247, 262
   arboreus, 92
   cosentinii, 84
   luteus, 168, 172, 176, 178
Lycopersicon esculentum, 154, 155, 156, 158, 160, 190
Lyriophyllum, 4, 5

M

Machaerium sp., 101
maize, *see Zea mays*
Malus sylvestris, 36, 188, 189
Manihot esculenta, 182, 261
meadowfoam, *see Limnanthes alba*
Medicago sativa, 181, 196, 198, 257
mescal bean, *see Sophora secundiflora*
millet, *see Panicum miliaceum*
Mimosaceae, 98, 101, 193
Mucuna
   flagellipes, 94
   utilis, 106
mung bean, *see Vigna radiata*
mustard, *see Sinapis alba*
mustard, black, *see Brassica nigra*

N

nasturtium, *see Tropaeolum majus*
Neonotonia wightii, 261

O

oats, *see Avena sativa*
okra, *see Abelmoschus esculentus*
Oleaceae, 172, 188
orchid, *see Cymbidium; Orchis morio*
Orchidaceae, 193
Orchis morio, 197
ordeal bean, *see Physostigma venenosum*
Oryza sativa, 36, 109–111, 123, 124, 140, 141, 142, 143, 178, 182, 183, 187
Oxytropis lambertii, 158

P

Palmae, 193
Panicum miliaceum, 255
Papilionaceae, 101, 193, 254, 260
Papilionoideae, *see* Papilionaceae
pea, *see Pisum sativum*
peanut, *see Arachis hypogaea*
Pedaliaceae, 172
perennial ryegrass, *see Lolium perenne*
Persea americana, 225
Phaseolus
   lunatus, 92, 99, 100, 250, 261
   vulgaris, 34, 56, 89, 91, 92, 94–97, 99, 101, 102, 104, 108, 111, 125, 141, 142, 145, 174, 175, 183, 191, 192, 257, 260
Phoenix dactylifera, 192
Physostigma venenosum, 246, 247
Phytelephas macrocarpa, 192, 193
Pinus sylvestris, 103
Piptadenia peregrina, 255
Pisum
   abyssinicum, 18
   cinereum, 18
   elatius, 14, 15, 18, 19, 30, 33, 34, 97
   fulvum, 15, 18, 19
   humile, 14, 15, 18, 19, 30
   sativum, 9, 11, 12, 14, 15, 18, 19, 29, 30, 33, 34, 57–60, 64, 67, 74, 79, 80, 84–87, 102, 104, 106, 107, 108, 109, 110–112, 116–121, 124, 125, 126, 141, 142, 145, 169, 170, 171, 176, 178, 179, 180, 181, 182, 183, 184, 185, 186, 192, 197
   sativum, var. *arvense*, 44, 46, 47, 51–56, 61–63, 87, 104, 106, 257
Pittosporaceae, 172
Plantaginaceae, 172
potato, *see Solanum tuberosum*
Prunus
   amygdalus, 250
   dulcis, 8–10, 106, 141, 222
Psophocarpus tetragonolobus, 94, 98, 101
Pueraria thunbergiana, 92
pumpkin, *see Cucurbita pepo*

R

radish, *see Raphanus sativus*
Ranunculaecarpus quinquiecarpellatus, 4
rape, *see Brassica napus*
Raphanus sativus, 144
rattle box, *see Crotalaria spectabilis*

Plant Species Index 269

rice, see *Oryza sativa*
Ricinus communis, 8, 84, 91, 94, 95, 97, 105, 106, 110, 111, 112, 118, 122, 125-127, 140, 142, 143, 145, 148, 150, 155, 156, 162, 214, 215, 217, 218, 222, 230, 237, 259
Robinia pseudoacacia, 101
Rosaceae, 188
rosary bean, see *Abrus precatorius*
rye, see *Secale cereale*

S

safflower, see *Carthamus tinctorius*
Sapindaceae, 251
Scylla non-scripta, 193
Secale cereale, 142, 200, 202
Selaginella, 169, 172
  kraussiana, 169
serido bean, see *Vigna unguiculata*
sesame, see *Sesamum indicum*
Sesamum indicum, 173, 174
shepherd's purse, see *Capsella bursa-pastoris*
Simmondsia chinensis, 102, 210, 211, 217, 225, 226, 228, 229
Sinapis alba, 18, 191, 236, 237, 258
Solanaceae, 172
Solanum tuberosum, 36, 170, 182, 185, 189, 190, 226, 227
Sophora
  japonica, see *Styphnolobium japonica*
  microphylla, see *Edwardsia microphylla*
  secundiflora, 250
Sorghum bicolor, 170
soybean, see *Glycine max*
spinach, see *Spinacia oleracea*
Spinacia oleracea, 168, 169, 181, 186, 189, 218
squash, see *Cucurbita maxima*
Stellaria media, 8
Strychnos nux vomica, 247
Stylosanthes humilis, 195
Styphnolobium japonica, 91, 94
sunflower, see *Helianthus annuus*
sweet corn, see *Zea mays*
sweet pea, see *Lathyrus odoratus*
sweet potato, see *Ipomoea batatas*

T

Tamarindus indica, 101, 191
teosinte, annual, see *Zea luxurians*; *Zea mays*

teosinte, perennial, see *Zea diploperennis*
Thermopsis caroliniana, 92
tomato, see *Lycopersicon esculentum*
Trifolium pratense, 195
Trigonella foenum-graecum, 171, 176, 188, 194, 196, 197, 199
Triticum
  aestivum, 8, 9, 17, 24, 25, 29, 30, 32, 36, 91, 95, 106, 107, 108, 109, 112-118, 141, 142, 144, 147, 151, 153, 157, 158, 160, 170, 171, 173, 175, 176, 181, 182, 183, 190, 200, 201, 202, 228
  boeoticum, 16, 17
  dicoccoides, 17, 32, 112
  dicoccum, 16, 24, 25, 29, 32
  durum, 16, 32
  monococcum, 16, 29
  tauschii, 17, 24
  turgidum, see *T. dicoccum*
  urartu, 17
Tropaeolum majus, 217, 225, 226

U

Ulex europaeus, 91
Umbelliferae, 156, 159, 172

V

vetch, see *Vicia sativa*
Vicia
  cracca, 91
  faba, 25, 26, 91, 92, 93, 94, 102, 104, 105, 112, 118, 119, 120, 141, 157, 171, 173, 174
  galilaea, 26
  hyaeniscyamus, 26
  narbonensis, 26
  sativa, 14, 252
Vigna
  radiata, 44, 92, 102, 103, 111, 112, 170, 171, 181, 187, 188, 190, 191, 192, 197
  unguiculata
    subsp. *cylindrica*, 30
    subsp. *dekindtiana*, 30, 32
    subsp. *mensensis*, 30, 32
    subsp. *sesquipedalis*, 30
    subsp. *unguiculata*, 30, 32, 42-47, 49-51, 54-58, 60-66, 68-70, 74-80, 85, 87, 97, 102, 103, 105, 106, 260

W

wheat, *see Triticum aestivum*
Wistaria floribunda, 91

X

Xylomelum angustifolium, 149

Z

Zanthoxylum rhabdospermum, 4, 6, 7

Zea
  diploperennis, 28
  luxurians, 28
  mays, 8, 26, 27, 28, 36, 102, 108, 109, 118, 127, 141, 149, 176, 178, 180, 181, 182, 183, 184, 185, 187, 188, 190, 212, 213
  mays, subsp. *mexicana*, 26–28
  mays, subsp. *parviglumis*, var. *huehuetenangensis*, 26, 28
  mays, subsp. *parviglumis*, var. *parviglumis*, 26, 28

Subject Index

A
abrin, 97, 259
[^{14}C]acetate, labelling seed fatty acids, 215, 217, 223, 225, 226, 227, 229, 231, 233
acetyl-CoA
 as fatty acid precursor, 216, 217-221, 226
 synthesis from sucrose, 214-216
acetyl-CoA carboxylase, 216, 217, 218, 226
acetyl-CoA synthetase, 215, 226
N-acetyl galactosamine, 91, 92
N-acetyl glucosamine, 91, 125
N-acetyl glucosaminidase, 112
acid phosphatase
 of seed coats, 145
 of seed storage organs, 112
acyl carrier protein (ACP), 216, 218, 225
acyl-CoA alcohol transacylase, 228, 229
acyl-CoA phospholipid acyltransferase, 223, 232, 233-235
ADP, 180, 182, 186, 187, 188
ADP-glucose, 180, 182, 183, 185, 189, 191
ADP-glucose pyrophosphorylase, 181-183
ajugose, 174
alanine
 in embryo sac liquid, 11-13, 67
 in proteins, 9
  albumins, 22, 85
  lectins, 90, 119, 120
  proteinase inhibitors, 99, 100
 translocation, 63, 64
Albian epoch, 4, 5
albumins
 electrophoretic profiles, 18-21, 24
 as reserve proteins, 22, 84-87, 101, 105, 106

aldolase, 215, 216
aleurone cell layers
 differentiation, 7, 8, 144, 175, 176, 199
 protein bodies, 109, 141, 144-147, 151, 153
 protein composition, 9
alkaloids, in seeds, 246-250, 260
allantoic acid, translocation, 61, 62, 65
allantoin, translocation, 61, 62, 65, 68, 69
allantoinase, 68
alleles
 seed dispersal, 28, 29
 seed protein composition, 18, 86, 97, 101
 starch synthesis in seeds, 178, 180, 183, 184, 187
 toxins in seeds, 97, 101
 triacylglycerol composition, 212, 213
amides
 role in fruit growth, 62, 63, 68, 69, 80, 104
 translocation, 61-69, 80
 see also asparagine; glutamine
amino acids
 accumulation in fruits, 12, 62, 63, 104
 accumulation in seeds, 11-13, 67, 104
 essential in human diet, 84, 85
 metabolism, 67
 non protein, 12, 13, 252-254
 translocation, 61-63
 see also individual amino acids
γ-aminobutyrate, translocation, 63
α-amino-β-methylaminopropionic acid, 254
α-amino-γ-oxalylaminobutyric acid (ODAB), 253
α-amino-β-oxalylaminopropionic acid (ODAP), 253, 262

β-aminopropionitrile, 252
ammonium ions, in embryo sac liquid, 11, 67
amygdalin, 250, 251
α-amylase, 24
β-amylase, 112
amylase inhibitors, 190
amylopectin
  properties, 177-180
  synthesis, 180-191
amyloplasts, see starch granules
amylose
  properties, 177-180
  synthesis, 180-191
Anacystis nidulans, 178
Angiosperm evolution, 3-5, 13
  fossil seeds, 3-7
Aptian epoch, 4, 5
arginase, 253
arginine
  in proteins, 9
    albumins, 22, 85
    lectins, 90
    proteinase inhibitors, 98, 99, 100, 101
  translocation, 63, 66
arginine analogues, 253, 254, 260
  in insect proteins, 260
arginyl-tRNA synthetase, 260
asparaginase
  in the embryo, 67
  in seed coats, 11, 67
asparagine
  in proteins
    albumins, 22, 85
    glycosylation, 93, 125
    lectins, 90, 93, 119, 120
    proteinase inhibitors, 100
  metabolism in seeds, 11-13, 67
Aspergillus oryzae, 98
ataxia, tyramine-induced, 225
ATP, 180, 182, 183, 186, 187, 188, 189, 215, 216, 221, 227
atropine, 247, 260

B
barium, in phytin, 156, 157, 158, 160
biotin, 217, 218
branching enzymes, 181, 184, 185, 190
bufotenine, 255, 256

C
cadmium, in phytin, 158

caffeine, 250
calcium
  carbonate, 159
  oxalate, 146, 159, 160
  in phytin, 140, 143, 149, 151, 152, 153, 154, 155, 156, 157, 158
  translocation, 61
Callosobruchus maculatus, 103, 259, 260
canavalin, 126
canavanine, 254, 260, 261
carbon-14, use as tracer, 46-60, 64-69, 72, 169, 173, 176, 179, 190, 191, 198, 214, 217, 221, see also [^{14}C]acetate
carboxypeptidase, 112
  inhibited by phytate, 147
Caryedes brasiliensis, 260
castanospermine, 249, 251
catechol oxidase, 33
cation exchange, during phytin synthesis, 143
cell division
  in embryo, 104, 175
  in endosperm, 8
cell expansion, 104
cell walls, as reserve polysaccharide, 55, 187, 192, 193, 199
Cenomanian epoch, 4, 5
chloroplasts
  in fruit walls, 54, 56, 176
  in seeds, 54, 176
  starch synthesis, 54, 176, 180, 181, 182, 191
choline phosphotransferase, 233-255
chromium, as artifact in EDX analysis, 152, 155
chromosome number, 14, 16, 26
chymotrypsin, 98, 103
  inhibitors, 98-104
ciceritol, 173
citrate, 181, 183
cobalt, as cofactor, 171, 188, 196, 197
CO_2 concentration, within fruits, 58
CO_2 fixation
  by fruit walls, 54-60
  by leaves, 46-54
  by seeds, 60
codon, 99, 100, 120
Colletotrichum lindemuthianum, 98
concanavalin A
  binding specificity, 91, 123, 177
  structure, 88-94
β-conglycinin, 104
copper
  as artifact in EDX analysis, 152
  as cofactor, 196
  in phytin, 158

Subject Index   273

cotyledon, as storage organ, 10, 11, 95, 102, 104, 111, 168, 181, 211, 231
Cretaceous period, 3–5
β-cyanoalanine, 252, 260
cyanogenic glycosides, 250, 251, 261, see also amygdalin
cyanolipids, 250
cycasin, 258, 260
cycloheximide, 116, 117
cysteine
    in proteins
        albumins, 22, 84, 85, 86
        enzymes, 95, 103
        globulins, 22, 23, 120, 122, 123
        lectins, 90, 91, 95, 97
        proteinase inhibitors, 98, 99, 100
    see also disulphide bonds
cytisine, 247, 249
cytoplasm, as intracellular compartment, 95, 102, 169, 170, 181, 186, 187, 215–217, 221, 223, 226, 227

D

debranching enzymes, 177, 190
D-enzyme, 190
desaturase, 219, 220, 221, 229, 231, 232, 234, 235, 239
desiccation
    of fruits, 74
    of seeds, 74, 104, 106, 107, 223
Devonian period, 2, 3
diacylglycerol acyltransferase, 224, 225
diacylglycerols, 223–225, 228, 230, 231, 232, 233, 234, 235, 239
diamine oxidase, 254
α,γ-diaminobutyric acid, 253
dictyosome, 112, 114–117, 124, 126
2,5-dihydroxymethyl-3,4-dihydroxypyrrolidone, 259
3,4-dihydroxyphenylalanine (DOPA), 255, 256, 260
N-dimethyltyramine, 255, 256
disaccharides, 168–171
disulphide bonds, of reserve proteins, 22, 23, 95, 97, 98, 99, 120, 122, 123
DNA
    chloroplast, 28
    mitochondrial, 28
    nuclear, 7, 93, 104
    synthesis, see endoreduplication
domestication, 28, 32
    anemone, 14, 15
    broad bean, 25
    chick pea, 30, 33
    cowpea, 30, 44–46
    lentil, 19
    Lima bean, 261
    lupin, 44–46, 262
    maize, 26–28
    pea, 30, 33
    rape, 212, 227, 228, 262
    sunflower, 30, 31
    wheat, 30, 32
dopamine, 255, 256

E

eicosenoate, 211, 212, 225, 229
eicosenoyl-CoA, 227
electrophoresis
    enzyme detection, 24
    seed proteins, 18–26
embryo
    development, 10, 11, 75, 79, 106
    oil storage, 55, 79, 80, 211, 222, 223, 225, 229, 231
    oligosaccharide storage, 172–176
    phytin storage, 141
    polysaccharide storage, 54, 55, 80, 178–202
    protein storage, 55, 80, 104, 116, 117
embryo sac, 8, 9, 11, 67
endoplasmic reticulum
    oil body origin, 237
    polypeptide processing, 108, 112, 115, 124, 125, 126
    protein deposition, 108
    site for acyl-PC desaturases, 231, 232, 233
    site of phytate synthesis, 143, 148–151
endoreduplication
    in embryo cells, 104
    in endosperm cells, 7, 11
endosperm
    cell walls, 8, 168, 192, 193, 199
    definition, 7
    development, 8–13, 30, 32, 115, 143
    oil reserves, 8, 214, 222, 223, 229, 230
    oligosaccharide reserves, 175
    protein reserves, 8, 9, 32, 94, 101, 105, 108–117
    starch reserves, 8, 176–178, 181–183
    see also aleurone cell layers
enzyme activity
    of canavalin, 126
    of lectins, 92
erucic acid, 211, 212, 225, 226, 229, 262
erucyl-CoA, 227, 228
Escherichia coli, 217, 218

Subject Index

F
Famennian epoch, 2
fatty acids
    chain elongation, 216–221
    desaturation, 219, 220, 221, 231–235
    hydroxylation, 230
    in phospholipids, 222, 230
    synthesis, 214–222, 229, 230, 231, 239
    in triacylglycerols, 211–214, 230
    see also individual fatty acids
fatty alcohols
    synthesis, 228, 229
    in wax esters, 211
ferredoxin, 220, 221
flowering, position of buds, 29, 30, 45, 47
fructofuranan fructosyl transferase, 201
fructofuranans, 167, 168, 200, 202
fructokinase, 186
fructose bisphosphatase, 187
fructose-6-phosphate, 168, 169, 180, 181, 182, 185, 186, 187, 195, 215, 216
fructosylsucroses, see kestoses
1^F-fructosyl transferase, 201, 202
6^G-fructosyl transferase, 201
fucose, 91, 92

G
galactinol raffinose galactosyl transferase, 174, 175
galactinol sucrose galactosyl transferase, 173
galactinols, 170, 173, 174
galactoglucomannans, 194, 195
galactokinase, 187, 188
galactomannans, 167
    structure, 192, 193
    synthesis, 195–199
galactopinitol, 171
galactose, 91, 92, 97
galactose-1-phosphate uridylyl transferase, see UDP-galactose pyrophosphorylase
α-galactosidase, 92, 126, 171, 173, 187, 188
galactosylsucroses, 172–176, 187
α-6-galactosyl transferase, 198
gametophyte
    female (mega), 2, 6, 7, 103
    male, 2, 6
gas exchange, between fruits and atmosphere, 57–60, 75–77
GDP-glucose, 169, 195
GDP-mannose pyrophosphorylase, 195, 197
gliadin, 9, 24, 25, 114, 118
globoid crystals, 140–163
globulin, 18, 21, 90, 92, 105, 106, 109, 110, 118, 123
4-α-D-glucanotransferase, see D-enzyme
α-glucans, 176–191
β-glucans, 191, 192
glucomannans, 193, 198
glucosamine, 125
glucose, 91, 92
α-glucosidase, 112, 190, 259
β-glucosidase, 112, 250, 258
glucose-1-phosphate, 170, 180, 182, 183, 185, 187, 188, 189, 214, 216
glucose-6-phosphate, 169, 170, 171, 180, 181, 185, 214, 215, 216
glucose-1-phosphate adenylyl transferase, see ADP-glucose pyrophosphorylase
glucose-6-phosphate dehydrogenase, 215, 220, 222
glucose-1-phosphate uridylyl transferase, see UDP-glucose pyrophosphorylase
glucose phosphate isomerase, see phosphoglucoisomerase
glucosinolates, 258, 259, see also sinigrin
glucosyltrehalose, see selaginose
glutamate
    in proteins, 9
        albumins, 22, 85
        lectins, 90, 93, 94, 119, 120, 125
        proteinase inhibitors, 100
    translocation, 66
glutamine
    in embryo sac liquid, 11–13, 67
    in fruits, 62, 63
    in proteins, 9, 67
        albumins, 22, 85
        lectins, 90, 93
        proteinase inhibitors, 100
    translocation, 63, 66
glutelin, 9, 109, 114, 123
glutenin, 9, 24, 114
glyceraldehyde phosphate dehydrogenase, 216, 220
[^{3}H]glycerol, labelling seed lipids, 225, 232, 233
glycerol-3-phosphate
    acylation, 228
    pathway, 223–225, 227, 228, 232, 234, 235
glycine
    in proteins, 9
        albumins, 22, 85
        lectins, 90, 93
        proteinase inhibitors, 100
    translocation, 64
glycinin, 118, 120, 121, 127

Subject Index 275

glycolysis
    in developing oil seed cotyledons, 215, 216, 221
    in developing oil seed endosperm, 214, 215, 216, 221, 223
glycosylation, of seed proteins, 93, 122–127
goitrin, see 5-vinyloxazolidine-2-thione
goitrogens, 253, see also 3-hydroxy-4-(1H)-pyridone
Golgi body, see dictyosome
grain beetles, see Tenebrio molitor; Tribolium

H

haemagglutinins, see lectins
harvest index, 45
    definition, 44
helianthin, 123
Helminthosporium maydis, 36
hexitol, translocation, 188
hexokinase, 187, 216, 217
hexose phosphate isomerase, 223, see also phosphoglucoisomerase
histidine
    in proteins, 9
        albumins, 22, 85
        lectins, 90
        proteinase inhibitors, 100
    translocation, 63
homoarginine, 253
homoserine, in embryo sac liquid, 12, 13
hordenine, see N-dimethyltyramine
hybridization, 15, 16, 17, 28
hydroxynitrile lyase, 250
3-hydroxy-4(1H)-pyridone, 253, 254
5-hydroxytryptophan, 255, 256
hyoscine, 247, 248
hyoscyamine, 247, 248

I

immunocytochemistry, 95, 96, 109–116, 117
indospicine, 254
chiro-inositol, 171
myo-inositol, 142, 170, 171, 173, 174
myo-inositol hexaphosphate, see phytic acid; phytin
myo-inositol-1-phosphatase, 171
myo-inositol-1-phosphate synthase, 171
scyllo-inositol, 171
invertase, 187, 217
iron
    as cofactor, 171, 196
    in phytin, 145, 151, 154, 155, 156, 157, 158

isoelectric focusing, 25
isoenzymes
    enzymes of starch synthesis, 183, 184
    fructokinase, 186
    glyceraldehyde-3-phosphate dehydrogenase, 220
    in identification, 24
    phosphoglucoisomerase, 181
    phosphoglucomutase, 170, 181
    of plastids, 215, 218, 223
    starch phosphorylase, 186
    UDP-glucuronate decarboxylase, 192
isoleucine, in proteins, 9
    albumins, 22, 85
    lectins, 90, 119
    proteinase inhibitors, 100
isothiocyanate, 258, 259

K

kestoses, 200–202

L

lead, excluded from phytin, 158
leaf flap feeding technique, 65, 68, 69
lectins
    definition, 88, 96
    specificity, 91, 92
    subunit structure, 88–91, 92–95, 118–121, 125
    toxicity, 259
legumin, 19, 22, 23, 24, 86, 106, 110, 120, 124, 127
leucine, in proteins, 9
    albumins, 22, 85
    lectins, 90, 119, 120
    proteinase inhibitors, 99, 100, 101
linamarin, 250
link peptide, 23, 120, 122
linoleate
    desaturation, 231–235, 239
    in lipids, 211, 214
    synthesis, 231
linolenate
    in lipids, 211–214
    synthesis, 231–235
lipid storage in seeds
    effects of temperature, 237–239
    energy value, 210
    see also oil bodies; triacylglycerols; wax esters
lotaustralin, 250, 251
lupanine, 247, 249
lychnose, 174

L

lysine, in proteins, 9
  albumins, 22, 85
  lectins, 90, 93, 94, 119
  proteinase inhibitors, 99, 100, 101

M

magnesium
  as cofactor, 170, 171, 182, 188, 196, 197
  in phytin, 140, 143, 149, 151, 152, 153, 154, 155, 156, 157
  translocation, 61
malate dehydrogenase, 112
malonyl-CoA, 216, 217, 218, 219, 226, 228
manganese
  as cofactor, 171, 188, 196
  in phytin, 145, 151, 153, 154, 157, 158
  in proteins, 88, 91, 140
β-mannanase, 194, 195
β-mannans, 167, 192-199
mannan synthase, 195
mannose, 91, 125
mannose phosphate isomerase, 195, 196
mannose phosphate mutase, 195, 196
α-mannosidase, 92, 111, 112, 126
β-4-mannosyl transferase, *see* mannan synthase
mealworm, *see* Tenebrio molitor
medicagenic acid 3-β-O-glucoside, 257
methionine, in proteins, 9
  albumins, 22, 84, 85
  globulins, 120, 121
  lectins, 90, 91, 119
  proteinase inhibitors, 100
N-methyltyramine, 255
mimosine, 252, 260
mineral analysis, methods for seed tissues, 159-162
mitochondrion, 217
modelling budgets
  for carbon, 49-54, 58-60, 70-80
  for nitrogen, 70-80
  for water, *see* water economy
monensin, 116, 117
monoamine oxidase, 254
monocrotaline, 247, 249
mucilages, 168, 196
myrosinase, 258

N

NAD, 171, 189, 192, 220
NADH, 171, 218-221
NADP, 187, 188, 215, 216, 220, 227

NADPH, 187, 215, 216, 218-221, 226, 227
Neocomian epoch, 4, 5
neurotoxins, from seeds, 253, 259
nigericin, 116, 117
nitrate
  assimilation, 52, 54, 61, 62, 63, 65, 67
  reductase, of fruits, 78
nitriles, 251-258, 259
nitrogen-15, use as tracer, 52, 54, 64-67
nitrogen fixation, in nodulated roots, 43, 62, 63, 65, 71

O

oil bodies
  composition, 236, 237
  membrane proteins, 236-238
  origin, 236
  synthetic capacity, 226, 228, 237
oleate
  chain elongation, 225-228, 229
  desaturation, 219, 231-235, 239
  hydroxylation, 230
  in lipids, 211-214
  synthesis, 216
Oligocene, 4
oligosaccharides
  accumulation in seeds
    fructosylsucroses, *see* kestoses
    galactosylsucroses, 172-176
  see also individual compounds
organic acids, 12, 61, 64
osmium tetroxide, fixation artifacts, 156, 161
oxygen, 221, 230, 232, 239

P

palmitate
  chain elongation, 216, 229
  in lipids, 211-214
  synthesis, 216
pentose phosphate pathway, 196
  in developing oil seeds, 215, 222
PEP, 216
PEP carboxylase
  in fruit walls, 54, 58
  in seeds, 58
pepstatin A, 103
perisperm, 8, 9, 10
phenylalanine
  in proteins, 9
    albumins, 22, 85
    lectins, 90, 119
    proteinase inhibitors, 99, 100, 101

translocation, 63
phloem, 60
phloem sap
　composition, 58, 61, 62, 63, 64, 68, 69, 70, 76-80
　sampling methods, 60, 61, 63, 64, 65
phosphate, inorganic
　as inhibitor, 169, 182, 189
　as source of seed phosphorus, 145
phosphatidate, 223, 224, 228
phosphatidate phosphatase, 223, 224
phosphatidylcholine (PC), 224, 225, 230-236
　in polyunsaturated fatty acid synthesis, 231, 239
phosphoglucoisomerase, 181, 196
phosphoglucomutase, 169, 181, 185, 196
6-phosphogluconate, 215, 216
6-phosphogluconate dehydrogenase, 216, 220, 222, 223
3-phosphoglycerate, 182, 216
phospholipids
　composition in developing seeds, 222, 231
　content in developing seeds, 222, 223, 236
　synthesis in endoplasmic reticulum, 231
　see also phosphatidylcholine
phosphorus, storage in seeds, 139-165
photosynthate
　composition, 64
　partitioning, 42-54, 70, 71
physostigmine, 246, 247, 248
phytase, 112
phytate, see phytic acid; phytin
phytic acid
　determination, 161, 162
　synthesis, 141, 142
phytin, 140-163
　accumulation, 141-143
　cation composition, 157
　　environmental effects, 157-159
　　tissue and cell specificity, 147, 149-157
　complex with reserve proteins, 141, 142, 162
　nutrition, animal and human, 147
　proportion of seed dry matter, 140
phytoglycogen
　properties, 177, 178, 180
　synthesis, 184
phytohaemagglutinins, see lectins
planteose, accumulation, 172, 173, 175, 176
plastids, fatty acid synthesis, 214-222, 232
pod structure, 29, 54-60, 79

polypeptide
　processing, 23, 98, 102, 108-127
　synthesis, 118
　see also protein
polyphenols, of seed coats, 33, 34
potassium
　as cofactor, 183
　in phytin, 140, 143, 149, 151, 152, 153, 154, 155, 156, 157
　translocation, 61
primer, in starch synthesis, 183, 189
prolamin, 9, 108, 109, 114, 118, 126, 127
proline, in proteins, 9
　albumins, 22, 85
　lectins, 90
　proteinase inhibitors, 100
protein
　accumulation in seeds, 10, 55, 79, 80, 104-127
　composition, 9, 22, 85, 119, 121
　mobilization, 84, 95, 102, 103
　rate of synthesis, 9, 86, 104
proteinases
　of insect larvae, 32, 103
　in vacuoles, 112
proteinase inhibitors, 32, 98-104, 259
　Bowman-Birk type, 85, 86, 99-102
　Kunitz type, 86, 98, 101, 102
protein body, 107-127
　hydrolases, 112
　inclusions
　　crystalloid proteins, 105, 109, 121, 142, 143
　　globoid crystals, 105, 140-163
　　matrix proteins, 95, 105, 126, 144, 146, 148
　　niacin, 142, 144, 145
　　see also calcium
　membrane, 95, 110-114
provascular cells, mineral storage, 149, 152-156
pullulanase, 190
pyrophosphate, 169, 170, 182, 183, 187, 188, 195
pyruvate, 214, 215, 216
pyruvate dehydrogenase, 214, 215, 216, 220
pyruvate kinase, 223

R

rachis, 27-29
radicle, mineral storage, 149, 151, 153-155
raffinose
　accumulation in seeds, 173-175
　translocation, 172

Subject Index

respiration rate, of developing seeds, 104
Rhizobium, association with legume roots, 97
ribonuclease (RNase), 112, 118
ribosomes, 98, 104, 107, 126, 218
ricin, 97, 259
ricinoleate
    synthesis, 230
    in triacylglycerols, 230
RNA, 104
    mRNA, 86, 104, 107, 108, 118, 120, 127
    rRNA, 104, 106
    tRNA, 260
roots
    function during fruit development, 42, 43, 51
    as sinks for photosynthate, 42–46
RuBP carboxylase, of fruit walls and seeds, 54

S

saponins, 257
seed beetles, see *Callosobruchus maculatus*; *Caryedes brasiliensis*; *Tenebrio molitor*; *Tribolium*
seed coats
    anatomy, 4, 7, 11, 34
    impermeability, 4, 33, 239
    secretion, 11–13, 147
    surface, 6, 7, 33–35
seeds
    definition, 2
    food production, 34, 36, 37
seed size, 4, 29–32
    inverse relation to calcium content of globoids, 151
selaginose, 172
selenium, excluded from phytin, 158
selenocystathionine, 254
serine
    in embryo sac liquid, 12, 13
    in proteins, 9
        albumins, 22, 85
        glycosylation, 125
        lectins, 90, 94, 119, 120
        proteinase inhibitors, 99, 100
    translocation, 63, 64, 66
serotonin, 255, 257
sesamose, 174
signal peptides, 118, 119, 126
sinigrin, 258
sodium, in phytin, 157, 158, 159
sophorine, 250

sorbitol dehydrogenase, 189
sorbitol oxidase, 189
sorbitol-6-phosphate dehydrogenase, 189
soyin (soybean agglutinin), 92, 93, 97, 125
spherosomes, see oil bodies
stachyose
    accumulation, 173, 174
    translocation, 174
starch, 167
    granules, 54, 177–180, 189–191
    in oil seeds, 216, 217, 226
    synthesis, 176–191
    see also amylose; amylopectin; phytoglycogen
starch phosphorylase, 185, 186
starch-sugar cycle of leaves, 64
starch synthase, 181, 183, 184, 185, 189
stearate, in lipids, 211, 212, 213
stem, vascular organization, 47, 72
strychnine, 247, 248, 260
sucrose
    accumulation in seeds, 167, 168
    in embryo sac liquid, 11, 12
    metabolism in seeds, 8, 173, 186, 187, 200, 201, 216, 217
    translocation, 8, 58, 61, 64, 168, 186, 214
sucrose phosphatase, 168
sucrose phosphate, 168, 169
sucrose phosphate synthase, 168, 169
sucrose sucrose 1^F-fructosyl transferase, 201
sucrose synthase, 186, 187, 217
sulphur amino acids, see cysteine; methionine

T

Tenebrio molitor, 103
thioesterase, 230
threonine
    in embryo sac liquid, 12, 13
    in proteins, 9
        albumins, 22, 85
        glycosylation, 125
        lectins, 90, 93, 94, 119
        proteinase inhibitors, 100
    translocation, 63, 66
tonoplast, 110–114
toxic compounds of seeds
    detoxification procedures, 247, 249, 258, 261
    elimination by breeding, 228, 261, 262
    evolutionary significance, 32, 102–104, 259, 260
tracheal, sap, see xylem sap

Subject Index 279

transpiration
  of fruits, see water economy, of fruits
  ratio, see water economy, of whole plants
trehalose, 169
trehalose phosphatase, 169
trehalose phosphate synthase, 169
triacylglycerols
  acyl composition, 210–214
    effects of temperature, 237
  content in seeds, 212, 213
  as energy reserve, 210
  molecular species, 213, 214
  synthesis, 223–240
Tribolium
  castaneum, 32, 103
  confusum, 103
triose phosphate isomerase, 215, 216
trypsin, 98, 103
  inhibitors, 34, 86, 98–104
tryptophan, in proteins
  albumins, 85
  lectins, 90
  proteinase inhibitors, 99
tunicamycin, 125
tyramine, 255
tyrosine, in proteins, 9
  albumins, 22, 85
  lectins, 90, 119
  proteinase inhibitors, 100, 101

U

UDP, 169, 170, 186, 187
UDP-galactose, 170, 171, 173, 174, 187
UDP-galactose myo-inositol-1-galactosyl transferase, 170, 171
UDP-galactose pyrophosphorylase, 187, 188
UDP-glucose, 168, 169, 170, 171, 186, 187, 188, 189, 191, 192, 198, 216
UDP-glucose dehydrogenase, 192
UDP-glucose 4′-epimerase, 170, 188
UDP-glucose pyrophosphorylase, 169, 170, 187
UDP-glucuronate decarboxylase, 192
UTP, 169, 170, 188
umbelliferose, accumulation, 172, 175
unusual fatty acids
  synthesis, 219, 229
  in triacylglycerols, 230, 231
urea, 46, 64
ureolytic activity, 66, 68

V

vacuole, 8, 109–113, 126, 169
  enzymes, 112, 126
  inclusions, 109–113, 116–124, 159
  see also protein body; tonoplast
valine
  in embryo sac liquid, 12, 13
  in proteins, 9
    albumins, 22, 85
    lectins, 90, 94, 119
    proteinase inhibitors, 99, 100
  translocation, 63, 66
vascular function, 47, 74–79
  xylem to phloem transfer, 66–72
  xylem to xylem transfer, 70–73
verbascose, 174
vicilin, 86, 106, 110, 124
5-vinyloxazolidine-2-thione, 258

W

water content
  of fruits, 74
  of seeds, 74, 104, 106, 107
water economy
  of fruits, 74–79
  of leaves, 74
  of whole plants, 70, 74
wax esters, of jojoba bean
  composition, 211
  synthesis, 226, 228–229, 236

X

X-ray, energy dispersive (EDX) analysis, 149, 152–158, 160, 161
xylem
  organization, 3
  role in water removal from fruits and seeds, 78, 79
  see also vascular function
xylem sap
  composition, 61, 62, 63, 70, 76–79
  methods of sampling, 60
xyloglucan, 191, 192

Z

zinc
  as cofactor, 195
  in phytin, 143, 145, 157, 158